Thermofluids

Thermofluids
An Integrated Approach to Thermodynamics and Fluid Mechanics Principles

C. Marquand
University of Westminster, London, UK

D. Croft
Sheffield Hallam University, UK

JOHN WILEY & SONS
Chichester • New York • Brisbane • Toronto • Singapore

Copyright © 1994 by John Wiley & Sons Ltd,
Baffins Lane, Chichester,
West Sussex PO19 IUD, England

Reprinted September 1994

All rights reserved.

No part of this book may be reproduced by any means,
or transmitted, or translated into a machine language
without the written permission of the publisher.

Other Wiley Editorial Offices

John Wiley & Sons, Inc., 605 Third Avenue,
New York, NY 10158-0012, USA

Jacaranda Wiley Ltd, G.P.O. Box 859, Brisbane,
Queensland 4001, Australia

John Wiley & Sons (Canada) Ltd, 22 Worcester Road,
Rexdale, Ontario M9W 1L1, Canada

John Wiley & Sons (SEA) Pte Ltd, 37 Jalan Pemimpin #05-04,
Block B, Union Industrial Building, Singapore 2057

British Library Cataloguing in Publication Data

A catalogue record for this book is available from the British Library

ISBN 0 471 94357 6; 0 471 94184 0 (pbk)

Typeset in 10/12pt Times by Laser Words, Madras
Printed and bound in Great Britain by Bookcraft (Bath) Ltd

Contents

Acknowledgement		xi
Preface		xiii

1 Introduction to Thermal Systems — 1
  1.1 The idea of a system — 4
    1.1.1 Two types of system — 5
    1.1.2 Mass and mass flow rate — 6
  1.2 The Language of Thermofluids — 6
    1.2.1 Thermal energy — 8
  1.3 Energy and Power — 11
  1.4 Energy and the Environment — 12

2 Some Thermofluid Properties — 15
  2.1 The Property Mass — 16
  2.2 The Property Volume — 17
  2.3 The Property Pressure — 20
    2.3.1 Pressure variation with height/depth — 23
    2.3.2 Static and total pressures — 26
    2.3.3 Pressure manometry — 28
    2.3.4 Determination of atmospheric pressure — 32
    2.3.5 Inclined pressure manometers — 34
    2.3.6 Other pressure manometers — 36
    2.3.7 Pressure gauges and pressure transducers — 39
  2.4 The Property Temperature — 39
  2.5 The Property Viscosity — 42
    2.5.1 Streamline flow — 42
    2.5.2 The dynamic and kinematic viscosity — 43
    2.5.3 Molecular explanation of viscosity — 45
    2.5.4 Laminar and turbulent flow — 45
    2.5.5 The Reynolds number — 46
    2.5.6 Shear stresses in laminar and turbulent flow — 49
    2.5.7 The measurement of viscosity — 50

3 Work and Heat Transfer in a Thermal System — 57

3.1 Work Transfer in a Non-flow Process — 60
3.2 Heat Transfer in a Non-flow Process — 65
3.3 Special Characteristics of Work and Heat Transfer in a Non-flow Process — 68
    3.3.1 Work and heat transfer must cross a boundary — 68
    3.3.2 The formulae for work and heat transfer only apply to reversible non-flow processes — 69
    3.3.3 Work and heat transfer are not properties of the fluid — 72
    3.3.4 Sign convention for work and heat transfer — 74
3.4 Work and Heat Transfer in a Flow Process — 78
    3.4.1 Time-dependent values — 78

4 The First Law of Thermodynamics — 83

4.1 The First Law of Thermodynamics in Action — 84
    4.1.1 Methods for determining work transfer — 85
    4.1.2 Methods for determining heat transfer — 88
4.2 Efficiency of a Thermal System — 90
    4.2.1 Overall thermal efficiency — 90
    4.2.2 Energy transfer efficiency — 91
    4.2.3 Efficiencies and costs — 95
4.3 Examples of the First Law of Thermodynamics — 97
    4.3.1 Energy balance on a reciprocating engine — 97
    4.3.2 Energy balance on a steam plant — 101
    4.3.3 Energy balance on a refrigerator/heat pump — 104
    4.3.4 Energy balance on a closed cycle gas turbine — 107
4.4 Combined Heat and Power Plants — 110

5 The Non-flow Energy Equation — 113

5.1 Five Basic Non-flow Processes for Changing the State of a Fluid — 116
    5.1.1 Constant pressure non-flow process — 117
    5.1.2 Constant volume non-flow process — 119
    5.1.3 Constant temperature non-flow process — 121
    5.1.4 Constant entropy non-flow process — 122
    5.1.5 Polytropic non-flow process — 124
5.2 Fluid Specific Heats — 127
5.3 Cycles Consisting of Non-flow Processes — 128
5.4 Effect of the Type of Fluid — 132

6 The Fluid as a Perfect Gas — 133

6.1 Formulae which Apply to a Perfect Gas — 133
    6.1.1 The equation of state — 133
    6.1.2 Joule's law — 135
    6.1.3 Relationship between specific heats — 138
    6.1.4 Entropy of a perfect gas — 139

6.2	Determination of Property Values		140
6.3	Non-flow Processes with Perfect Gases		143
	6.3.1	Constant pressure non-flow process with a perfect gas	143
	6.3.2	Constant volume non-flow process with a perfect gas	143
	6.3.3	Constant temperature non-flow process with a perfect gas	144
	6.3.4	Constant entropy non-flow process with a perfect gas	145
	6.3.5	Polytropic non-flow process with a perfect gas	145
6.4	Flow Processes with Perfect Gases		161

7 The Fluid as Water/Steam — 163

7.1	Temperature/Volume Relationship for Water/Steam		163
7.2	Pressure/Volume Relationship for Water/Steam		165
7.3	Other Property Relationships for Water/Steam		166
	7.3.1	Specific heat	167
7.4	Temperature/Entropy Relationship for Water/Steam		168
7.5	Diagrams for Perfect Gases		168
7.6	Dryness Fraction		169
7.7	Steam Tables		170
	7.7.1	Superheated steam	170
	7.7.2	Saturated water and steam	173
	7.7.3	Unsaturated water	176
7.8	How to Determine the Condition of Water/Steam		177
7.9	Water/Steam in Non-flow Processes		178
7.10	Water/Steam in Flow Processes		182

8 Steady Flow Processes — 183

8.1	The Continuity Equation		185
8.2	The Conservation of Energy		189
	8.2.1	Heat transfer term	190
	8.2.2	Work transfer term	190
	8.2.3	Flow work	190
	8.2.4	Kinetic energy	191
	8.2.5	Potential energy	191
	8.2.6	Internal energy	191
	8.2.7	Other energies	191
8.3	The Steady Flow Energy Equation (SFEE)		191
8.4	The SFEE Applied to a Boiler/Heat Exchanger		195
	8.4.1	When the fluid is a perfect gas	197
8.5	The SFEE Applied to a Nozzle/Diffuser		201
	8.5.1	When the fluid is a perfect gas	205
8.6	The SFEE Applied to a Rotary Turbine/Compressor		210
	8.6.1	When the fluid is a perfect gas	213
8.7	The SFEE Applied to an Expansion Valve		217
	8.7.1	When the fluid is a perfect gas	218
8.8	The SFEE Applied to a Mixing Process		219
	8.8.1	When the fluid is a perfect gas	220

8.9	The SFEE Applied to Pipe/Ductwork	222
8.10	Some Common Thermal Systems	222
8.11	Simplified Forms of the SFEE	226

9 Effects of a Fluid in Motion — 229

9.1	Fluid Flow in a Pipe or Duct	229
	9.1.1 Equivalent diameter of a duct	231
9.2	Fluid Flow over a Flat Plate	232
9.3	Separation of the Boundary Layer on a Flat Plate	239
9.4	Drag on a Bluff Body	239
	9.4.1 Ideal fluid theory	239
	9.4.2 Real fluid	240
9.5	Streamline Bodies	242
9.6	Lift and Drag Coefficients	242
	9.6.1 Bluff bodies	242
	9.6.2 Streamline bodies	244
9.7	Flow Measurement	246
9.8	Flow-measuring Devices Based upon the Continuity and Bernoulli Equations	247
	9.8.1 Venturi meter	250
	9.8.2 Nozzle meter	253
	9.8.3 Orifice plate meter	256
9.9	Other Methods of Flow Measurement Based Upon the Bernoulli Equation	258
	9.9.1 Pitot static tube	258
	9.9.2 Entry nozzle	261
9.10	Other Methods of Flow Measurement	264
	9.10.1 Turbine meter	264
	9.10.2 Rotameter	264
	9.10.3 Hot wire anemometer	265
9.11	Application in Thermal Systems	266

10 The Steady Flow Momentum Equation — 267

10.1	Forces Due to the Flow of a Fluid	267
	10.1.1 Momentum forces	268
	10.1.2 Pressure forces	270
	10.1.3 Body forces	271
10.2	The Kinetic Energy and Momentum Correction Factor	271
10.3	Applications of the SFME	272
	10.3.1 Jet striking a perpendicular flat plate	272
	10.3.2 Jet striking a plate at an angle	276
	10.3.3 Reaction of a jet	280
	10.3.4 Force when fluid flows over a surface	284
	10.3.5 Force on a solid body in a flowing fluid	284
	10.3.6 Fluid flowing through a straight pipe or duct	285
	10.3.7 Fluid flowing in a pipe or duct in which the cross-section changes	289
	10.3.8 Fluid flowing around a bend in a pipe or duct	292
10.4	Application in Thermal Systems	299

11	**The Steady Flow Energy Equation Applied to Pipe Flow**		**301**
	11.1	Steady Laminar Flow of Fluid in a Straight Pipe	303
	11.2	Steady Turbulent Flow of Fluid in a Straight Pipe	307
	11.3	Conditions at Entry to a Pipe	307
	11.4	Variation of Friction Factor	307
	11.5	Empirical Formula for Friction Factor	308
	11.6	Compressible Flow Through a Straight Pipe	309
	11.7	Other 'Head Losses' in Pipes	318
	11.8	Equivalent Length	321
	11.9	Combinations of Pipes	323
		11.9.1 Pipes in series	323
		11.9.2 Pipes in parallel	326
		11.9.3 Branched pipes	328
	11.10	Computer Programs	334
	11.11	Application in Thermal Systems	334
12	**The Second Law of Thermodynamics**		**335**
	12.1	Implications of the Second Law of Thermodynamics	335
		12.1.1 The efficiency of an engine	335
		12.1.2 The heat pump	337
		12.1.3 The efficiency of a reversible engine	339
		12.1.4 Absolute scale of temperature	343
		12.1.5 The fluid property entropy	343
		12.1.6 Reversible and irreversible processes	346
13	**Problems**		**349**
	Appendices		373
	1 Formulae		373
	2 Nomenclature		387
	Further Reading		395
	Index		397

Acknowledgement

Our gratitude goes to David Lim for his help in preparing the drawings for this book. His mastery of the computer package and attention to detail have been invaluable.

Preface

Thermofluids is an intriguing subject. It is very conceptual, being based upon two laws which are really statements of observation. In addition, the theoretical development requires the manipulation of some fluid properties which have no physical presence. But at the same time, it is a very practical subject. It enables the energy transfer to be determined in many useful thermal systems with some degree of confidence.

This book has been written to help readers through the hurdles of thermofluids. The difficult and confusing aspects are introduced in a logical and progressive manner which is simple to follow. A systematic approach has been adopted, with each chapter providing a little more knowledge to fit into the jigsaw.

In presenting a book that combines the basic principles of thermodynamics and fluid mechanics, much care has been taken to ensure that the end result is an integrated work, not a text in which half the chapters are about thermodynamics and the other half about fluid mechanics. The examination of a thermal system enables this to be achieved. Methods are proposed for predicting how much energy in the form of work and heat is available in the components that make up common thermal systems, how much can be transferred or converted, and the efficiency and cost when this occurs. Consideration is given to the act of combining these components into a useful system through the use of pipework and ductwork, and the consequences that this has upon the fluid.

No attempt is made to disguise the fluid property entropy. It is introduced early on in the book and treated exactly like any other property, which helps to remove some of its mystique. It can then be used to its full advantage.

As with any subject, there are a number of words of jargon which crop up regularly and may take some getting used to. Terms such as 'property', 'boundary', and 'state' all refer to a thermal system and need to be understood in this context. Familiarisation with the subject matter is the only way to feel at ease with the terminology employed.

In addition, a large number of formulae need to be addressed which can be an overwhelming prospect. Even deciding which one to use in a particular situation can be confusing. However, a list of all the formulae quoted is included in Appendix 1, under the appropriate headings, which should help both to keep them all in perspective and to avoid the mistakes that are commonly made, such as applying perfect gas equations to steam plant!

Over the years, certain nomenclature has become fairly standardised for the study of thermodynamics and certain nomenclature for the study of fluid mechanics. Unfortunately, when the two subjects are combined, various clashes are inevitable, for example, v for velocity in fluids and v for volume in thermodynamics. A complete list of all the nomenclature chosen for this book, which avoids any overlap and, therefore, confusion, is given in

Appendix 2, along with the appropriate units (the SI system of units is employed throughout the text).

At all stages, relevant and meaningful worked examples are integrated into the text in order to illustrate the theoretical implications of the subject. This is by far the best way to understand thermofluids. Wherever possible, the examples relate principle to practice.

The authors hope that readers will find this book both informative and interesting. It is assumed that what is needed are explanations of the difficult concepts in a clear and unpretentious way. The subject matter covered is suitable as an introductory text in thermofluids for all branches of engineering. There is sufficient material in the book, however, to be able to understand and predict with confidence the efficiency and cost of work and heat transfer in many typical thermal systems.

1

Introduction to Thermal Systems

This book is about energy.

You will already have many ideas about energy and the way it is used in daily life; this book will explain the scientific basis of some of those ideas.

Before going into any detailed analysis of the subject, it is helpful to set the scene by having a look at the way energy is used in two well-known examples. Imagine that you pull into a garage and fill up your car with petrol. The energy contained in the fuel is converted by the engine into propulsive power which drives the car. It may surprise you to know that less than one-third of all the energy in the fuel is converted into useful power; the rest is lost as heat — mainly in the exhaust gas and the engine cooling water. This enormous loss is not caused by a fault in the design of the engine — it is an inherent loss in this type of system. Why?

Before answering that question, consider another type of energy: electricity. Most electricity is produced by thermal power stations using coal. Again, it may surprise you to know that not more than 40% of the energy in the coal is converted into electricity; the rest is lost as heat, which is delivered to the atmosphere via the large cooling towers which surround power stations. Again, this enormous loss is not caused by a fault in the design of the units inside the power station — it is an inherent loss in this type of system. Again, why?

The simple answer to either question is that both of the above are examples of energy being converted from one form to another, not directly, but via an intermediate process which produces heat. This 'indirect' method of converting energy always requires heat to be thrown away for the process to continue.

In the case of the car engine, the fuel burns very rapidly, virtually explodes, inside the engine, producing high-pressure and high-temperature gas, which pushes the engine pistons down the cylinder and, via the transmission system, drives the car wheels. The engine is arranged in such a way that, once the pistons have reached the end of their travel, the gas is released so that the pistons can return to their initial positions in order to be ready for another push from more exploding gas. The important idea here is that the released gas still contains energy because of its temperature and pressure and is simply exhausted to atmosphere. No matter how well the engine is designed, a large amount of energy will be thrown away in the exhaust gas. In addition, the engine is cooled, usually with water, and the water is itself cooled in the car radiator. This energy, too, is simply thrown away to the atmosphere. Thus, a large part of the fuel energy is dissipated as heat.

A thermal power station operates in a slightly different way, but suffers the same limitation on how much of the fuel energy appears as electrical energy leaving the station. In simple terms, the coal is burned in a boiler and the heat is transferred to high-pressure water, converting it into high-pressure and high-temperature steam. Not all of the energy in the coal is collected by the water, some leaves as hot exhaust gas up the chimney of the boiler. The high-energy steam enters the turbine and, via the turbine blades, causes the turbine to turn. The steam flows along the turbine, continuously forcing all parts of the turbine to rotate, falling in temperature and pressure until it reaches the turbine end. At this point, the steam temperature and pressure are both low and it would seem that the turbine had extracted most of the energy content of the steam. So it has, in one sense, but to return the steam to the starting-point of the cycle of events, that is, as high-pressure water, it must first be condensed back to water. The condensing operation involves getting rid of a large amount of energy — the latent heat of the steam — and it is this energy which is eventually dissipated in the cooling towers so commonly seen at power station sites. Yet again, the process inherently requires a large percentage of the energy in the fuel to be wasted as heat exhausted to the atmosphere.

A good starting-point, therefore, is the general idea that the conversion of energy from one form to another, which includes the production of heat, is probably going to involve significant wastage of energy. This idea provokes a question: what happens in a conversion process which does not use heat? For example, in the thermal power station, the final production of electricity is achieved by converting the rotational energy of the turbine (which is called mechanical energy) to electrical energy using a device called a generator. In this case, only about 5% of the input mechanical energy is not converted to electricity. Such a tiny wastage is caused by friction and electrical losses inside the generator. The conversion process does not involve the intermediate production of heat and, thereby, is much more effective.

The same is true when electricity is produced by a means not involving heat. This is the case in a hydroelectric plant where high-pressure water drives a turbine which in turn powers a generator. In effect, the potential energy of the water (due to its height above the axis of the turbine) is converted to mechanical energy of the turbine and then into electrical energy by the generator. There is no intermediate production of heat in the process and, therefore, there is little wastage of energy, only that due to friction.

All the previous comments are summarised in the energy triangles of Figures 1.1, 1.2, and 1.3. Figure 1.1 shows an open triangle with the three corners labelled with three basic sorts of energy commonly involved in energy-conversion processes: stored, mechanical and electrical energy. For clarity, stored energy here is defined as the energy contained in a substance like water, air, or a fuel which is waiting to be used; it could take the form of the height of a water reservoir above a turbine, or the chemical energy within a hydrocarbon fuel like coal, oil, or natural gas, or the air at high-pressure at the exit from an air compressor. Mechanical energy can be pictured as a rotating turbine or a reciprocating engine. Figure 1.1 will be used as a base to represent energy-conversion processes. Figure 1.2 is labelled with the names of devices which change one form of energy directly into another. All these devices have one thing in common; they are good converters. Most of the initial energy is transferred into the final form. Figure 1.3 shows the most common conversion process which involves thermal energy in the form of heat, usually from the combustion of a fuel. This is then changed to mechanical energy by a number of well-known devices. All these processes have one thing in common: they are poor at converting thermal to mechanical energy and the bulk of the initial thermal energy is wasted.

There are some basic rules of science which describe the ideas of energy conversion between stored and mechanical energy. The science itself has two names, Thermodynamics

INTRODUCTION TO THERMAL SYSTEMS 3

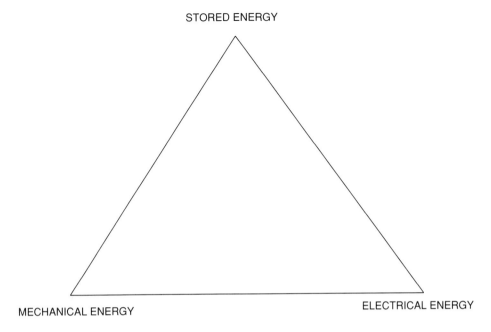

Fig. 1.1 *Energy conversion triangle.*

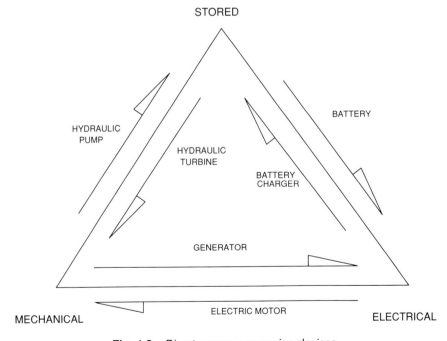

Fig. 1.2 *Direct energy conversion devices.*

4 INTRODUCTION TO THERMAL SYSTEMS

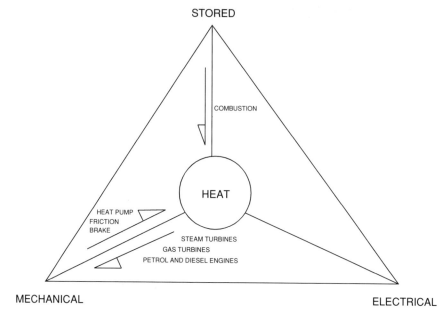

Fig. 1.3 *Energy conversion triangle (indirect).*

and Fluid Mechanics. Fluid mechanics is the area of study relating to the conversion process not involving the production of heat; thermodynamics is the study of conversion from thermal to mechanical energy. This book covers both subjects since they are closely related and a composite title, Thermofluids, is used to describe what is really a single discipline of study.

1.1 THE IDEA OF A SYSTEM

In the previous section, a coal-fired thermal power station was discussed in terms of how much of the energy from the coal was converted to electrical energy.

Imagine that a line is drawn around the perimeter of the power station and quantities of energy, which either enter or leave the station, are examined; whatever happens inside the station is not considered. A diagram which shows the perimeter and energy terms would appear as in Figure 1.4. This figure allows the introduction of some of the basic language of thermofluids. For example, the name given to the conversion activities within the perimeter is a System, and the perimeter is called the System Boundary. Outside the boundary is the Surroundings. The boundary can be drawn to suit the application.

The idea of representing a sequence of energy-conversion events in this way is a useful one because it shows how to assess the system in terms of what it is required to produce (electrical energy in this case), and what has to be put into the system to make it happen (stored energy in the coal). Again, thermofluids has a specific word for the 'assessment' of a system; it is called efficiency. For the power station, the efficiency could be defined by

$$\text{Power station efficiency} = \frac{\text{net electrical output}}{\text{energy input}}$$

THE IDEA OF A SYSTEM   5

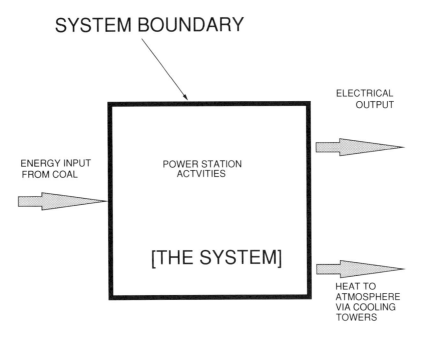

Fig. 1.4 *Power station system.*

The efficiency formula appears to exclude the heat loss to the ambient. In one sense this is true because there is no specific mention of the losses, but the 'efficiency' is a ratio of the desired output from the system to the energy input to the system; the heat loss to the ambient is not a desired output.

1.1.1 Two types of system

In Section 1.1, the thermal power station was said to be a system which could be pictured as one with coal flowing into the station at one end and electricity flowing out at the other. Such a system is termed an Open System; its characteristic is that material can cross the system boundary.

Another type of system often occurs in thermofluids — a Closed System. In this case, there is no flow of material across the boundary and all the energy exchanges are subjected to a fixed mass of material within the system. An everyday example of a closed system is the behaviour of a special type of air compressor in which a piston slides inside a cylinder and, in so doing, compresses the air. The system is sketched in Figure 1.5. Note that it is the behaviour of the system when both the inlet and exhaust valves are closed and the air is trapped inside the cylinder that is of interest (of course, the inlet valve must open at some point to let atmospheric air into the system and the exhaust valve must open to allow the system to deliver compressed air; at these times the compressor is operating as an open system — only the period in between when the air is trapped is under examination).

The sequence of events is that the piston is forced down the cylinder. At the end of its travel, the piston has squashed the air, raising its pressure and temperature. The exact way in which the pressure and temperature have increased is the subject of later analysis,

6 INTRODUCTION TO THERMAL SYSTEMS

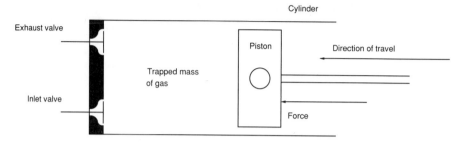

Fig. 1.5 *A closed system.*

suffice it to say at this stage that there are one or two available formulae that enable the exact calculation of the pressure and temperature to be made. As far as energy exchange is concerned, ideally all the mechanical energy from the movement of the piston would be contained in the air in the form of energy expressed in terms of pressure and temperature. In fact, some of this energy is used to overcome friction and some heat is transferred from the gas to the surrounding cylinder.

All the systems that will be examined and which occur within the energy triangles as a consequence of the conversion between stored and mechanical energy, whether open or closed, may be grouped under the general heading of thermal systems. The term thermal implies a heat transfer. This need not involve the intermediate production of heat as in Figure 1.3. In a hydroelectric plant, some of the energy contained in the water as a result of its height above the turbine is used to overcome friction, and friction effects result in heat transfer.

1.1.2 Mass and mass flow rate

The two types of system, open and closed, involve another idea which is basic. In the air compressor example of Section 1.1.1, the energy exchanges occurred to a fixed mass of air. In contrast, when analysing the energy exchanges inside a steam turbine, it should be remembered that the steam is flowing continuously through the turbine. It can be said that the mass is passing through the turbine at a certain rate, and the phrase mass flow rate is used to define this concept.

1.2 THE LANGUAGE OF THERMOFLUIDS

Up to now, everyday language has been employed to describe thermofluid events. Like every other scientific discipline, thermofluids has lots of words which have a specific meaning. The difficulty for a person new to the subject is that many of these words are used in daily conversation in a very ordinary way. Energy is such a word; 'you seem full of energy today', someone may say to you. In general, energy can be described as an ability to achieve some sort of activity. In thermofluids, energy has a specific meaning and also has specific units. Consider the following quantities, their SI units and symbols, as listed in Table 1.1, which will help to give a clear idea about the units of energy. The unit of force must be related to the other units in Table 1.1 by Newton's second law of motion which, simply expressed,

Table 1.1 SI units of common quantities.

Quantity	SI Unit	SI Symbol
Mass	Kilogram	kg
Mass flow rate	Kilogram per second	kg/s
Distance	Metre	m
Velocity	Metre per second	m/s
Acceleration	Metre per second^2	m/s^2
Force	Newton	N

states that
$$\text{Force} = \text{mass} \times \text{acceleration} \qquad (1.1)$$

In terms of units:
$$\text{Newton} = \text{kg} \times \text{m/s}^2 = \text{kg m/s}^2 = \text{N}$$

The newton is one of a limited number of combinations of primary units which are given their own names. Others that are used in this book are the joule, the watt and the pascal.

In thermofluids, the interest lies with forces moving objects for, in this way, energy is transferred from one object to another. If a force in newtons is applied to an object and the object moves through a distance in metres, the force is doing work on the object and this work transfer or work done is given by the formula

$$\text{Work transfer (work done)} = \text{force} \times \text{distance moved} \qquad (1.2)$$

In terms of units:
$$\text{Work transfer (work done)} = \text{N} \times \text{m} = \text{N m}$$

The group unit (N m) is also given a separate name — a joule — which is the unit of energy. This means that all types of energy have the units of joules (denoted by the symbol J). The joule is a group unit of N m; but the newton is itself a group unit of kg m/s^2 so the basic units of a joule will be

$$\text{Joule} = \text{N m} = \text{kg m/s}^2 \times \text{m} = \text{kg m}^2/\text{s}^2 \qquad (1.3)$$

Pursuing the idea that all forms of energy have the same units, by examining some well-known energies, another building block in the study of thermofluids can be introduced. Consider the terms potential energy and kinetic energy to make the point. Take the case of a mass at a particular height travelling with a given velocity. Its potential energy can be defined as being equal to the product of the mass in kilograms, the acceleration due to gravity in metres per second squared, and the vertical height of the mass above a datum level in metres.

In terms of units:

$$\text{Potential energy} = \text{kg} \times \text{m/s}^2 \times \text{m} = \text{kg} \times \text{m}^2/\text{s}^2 = \text{kg m}^2/\text{s}^2 = \text{N m} = \text{J}$$

From Equation (1.3), the units of potential energy can be seen to be joules. Its kinetic energy can be defined as a function of the product of the mass in kilograms and its velocity squared. In fact the function has a value of 0.5.

8 INTRODUCTION TO THERMAL SYSTEMS

In terms of units:

$$\text{Kinetic energy} = \text{kg} \times (\text{m/s})^2 = \text{kg m}^2/\text{s}^2 = \text{N m} = \text{J}$$

Again, from Equation (1.3), the units of kinetic energy can be seen to be joules.

EXAMPLE 1.1
An object of mass 15 kg is raised through a height of 2 m and then released. Determine its velocity after it has fallen back to its original position.

Solution
This is an energy exchange process which involves work transfer, the work done required to raise the mass initially through a height of 2 m, and the subsequent transfer of potential energy, due to the mass being 2 m above a datum level, into kinetic energy when the mass is released and allowed to fall back down through the 2 m distance.

The gravitational force on the object is given by Equation (1.1) as

$$\text{Gravitational force on object} = \text{mass of object} \times \text{acceleration due to gravity}$$

$$\therefore \text{Gravitational force on object} = 15 \times 9.81 = 147.15 \text{ N}$$

The work done in raising the object is given by Equation (1.2) as

$$\text{Work done in raising object} = \text{gravitational force on object} \times \text{distance raised}$$

$$\therefore \text{Work done in raising object} = 147.15 \times 2 = 294.3 \text{ N m} = 294.3 \text{ J}$$

This amount of energy is now effectively stored as potential energy until the object is released and allowed to fall 2 m. The kinetic energy at the height of 2 m is zero. When the object falls, it gains kinetic energy and loses potential energy. During the fall there is a continuous exchange of potential to kinetic energy until the object hits the ground after a travel of 2 m. At this point the kinetic energy of the object will be 294.3 J. But, from the definition of kinetic energy:

$$\text{Kinetic energy of object} = 0.5 \times \text{mass of object} \times (\text{velocity of object})^2$$

$$\therefore 294.3 = 0.5 \times 15 \times (\text{velocity of object})^2$$

$$\therefore \text{Velocity of object} = 6.26 \text{ m/s}$$

This is an example of an energy-exchange process, one in which all the energy is converted from one form to another. Note that there is no conversion to heat (thermal energy); otherwise the final kinetic energy would not have been equal to 294.3 J. Even so, the analysis is slightly idealised, in that friction losses due to drag forces on the object as it falls have been ignored.

1.2.1 Thermal energy

After the last example, it is the logical next step to consider energy exchange processes in which heat is involved.

Firstly, in a closed system, when a piston is forced back along a cylinder by some high-pressure and high-temperature gas. The arrangement is as shown in Figure 1.6. The piston initially is at one extreme of its travel towards the top end of the cylinder. The high-energy gas, therefore, will push the piston back along the cylinder and will do work on the piston.

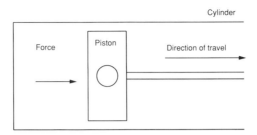

Fig. 1.6 *A reciprocating piston cylinder mechanism.*

The gas is effectively exchanging its thermal energy (evaluated in terms of its pressure and temperature) for the mechanical energy of the piston. This type of process occurs many millions of times in the life of a car engine; it is the operation in which a fuel explodes inside the cylinder, thereby creating a small pocket of high-pressure and high-temperature gas. Basically the piston is connected by a set of links and gears to the driving wheels of the car and motion is obtained. The point was made at the beginning of this introduction that such an operation involving heat being converted to mechanical energy is inherently one of low efficiency; most of the thermal energy of the gas is not converted to mechanical energy. The reason is simple. When the piston has reached the other extreme of its travel towards the bottom end of the cylinder, the gas still contains a lot of thermal energy. This has to be exhausted so that the piston can return to the top end of the cylinder and again receive another push in order to keep the car in motion. There are other reasons for inefficient conversion. The hot gas will, of course, heat up the surrounding material of the cylinder and the piston assembly. If these items are not cooled in some way, they reach too high a temperature for the lubrication system to work properly and the piston seals will rub directly on the cylinder walls, creating high friction effects which will cause damage to the seals. The cooling system also takes some of the thermal energy of the gas away from the conversion process. To summarise, in such a system, no more than a third of the input fuel energy is converted to useful desired output, namely work output. In practical terms, remember that the next time you put 30 litres of petrol in your car, no more than 10 litres will propel the car; the rest is largely lost to the atmosphere via the exhaust pipe and the engine cooling system.

Now an open system, such as an open cycle gas turbine plant, as in Figure 1.7. In an open cycle gas turbine plant, the air is first compressed, heat is added in a combustion chamber, the high energy gas is expanded in a turbine, and finally the gas is exhausted to atmosphere. Figure 1.7 also shows values of energy transfers for the various components; for example, the compressor requires 20 units of energy input (this is supplied by the turbine through a shaft connecting the two components), while the turbine produces 40 units of energy output. It is not important at this stage to know exactly how to calculate these amounts; this will be done later in the book. There are some straightforward thermofluids formulae which ease such calculations. Here it is important to notice that the exhaust gas energy is a massive part (80 units) of the input energy from the fuel (100 units) and that the overall energy-conversion process has a low efficiency. The reason is again fairly simple but not avoidable. The gas is expanded in the turbine from a high-pressure and temperature to a lower pressure and temperature. You might be thinking that less energy would be rejected to atmosphere if the final temperature was as low as possible. You would be absolutely right to think in such a way, but remember that the gas exhausts to atmosphere and so the

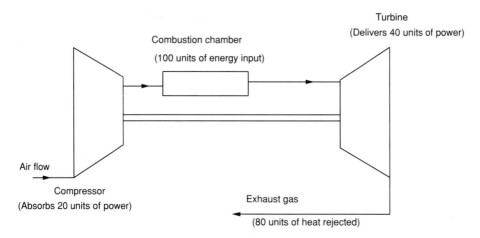

Fig. 1.7 *An open cycle gas turbine plant.*

final pressure can be no lower than this value, otherwise the gas could not escape from the turbine. The final pressure is, therefore, limited and, by the laws of thermofluids, so is the temperature. Later in the book you will learn how to calculate the change in temperature for a given change in pressure; suffice it to say here that, if the pressure changes by a factor of 5, the temperature will only change by a factor of about 1.5. In practical terms, the exhaust gas may be as hot as 900°C degrees Centigrade as it leaves the turbine. Now it can be seen why so much of the inlet energy leaves in the exhaust gas instead of making a bigger contribution to the desired energy-conversion process.

From the consideration of the closed and open system, it can be deduced that thermal energy is associated with the temperature of the fluid. In fact, heat transfer is an energy that arises due to a temperature difference.

EXAMPLE 1.2

Referring to Figure 1.7, and using the data in the figure, calculate the overall conversion efficiency of the open cycle gas turbine plant.

Solution

From the basic definition in Section 1.1, the overall conversion efficiency of the plant will be

$$\text{Overall conversion efficiency of gas turbine plant} = \frac{\text{net mechanical output}}{\text{energy input to system}}$$

The net mechanical output from the plant is the net work transfer or net work done, which is equal to the output from the turbine less the input required to drive the compressor. Hence:

$$\text{Net mechanical output} = \text{net work transfer} = 40 \text{ units} - 20 \text{ units} = 20 \text{ units}$$

The energy input to the system is the heat addition in the combustion chamber. Hence:

$$\text{Energy input to system} = 100 \text{ units}$$

$$\therefore \quad \text{Overall conversion efficiency of gas turbine plant} = \frac{20}{100} = 20\%$$

1.3 ENERGY AND POWER

Another small building block is needed before the study of thermofluids can begin in earnest. In Section 1.2.1, the closed system inside the cylinder involved several events: work done on the piston by the gas, reduction of the thermal energy of the gas during the expansion, and heat loss through the cylinder wall. All these activities are expressed in the units of energy, namely joules. By contrast, in an open system, the activities are related to time; mass flow rate has the units of kilogram per second, whereas in a closed system, the mass was considered without any reference to time. It follows that all the other activities in an open system, such as work and heat transfer, also must be related to time. The units of these items will no longer be joules, but joules per unit time — joules per second. The standard SI name for 1 joule per second is 1 watt. It would be perfectly correct to describe a watt as the unit of energy flow rate, but it is most usual to use the word Power, or in the case of heat transfer, Rate of Heat Flow or Heat Flow Rate.

To illustrate this idea, consider the events which take place in a power station which has a very high overall efficiency — a hydroelectric power station. Figure 1.8 shows the major pieces of plant which comprise such a station.

The basic idea is that water flows from a high-level reservoir to a lower one. In so doing, the initial potential energy of the water is converted to work output by the turbine, but at a certain rate because the water is flowing with a specified mass flow rate. In other words, the final output of the power station is expressed in watts not joules. Note that the process

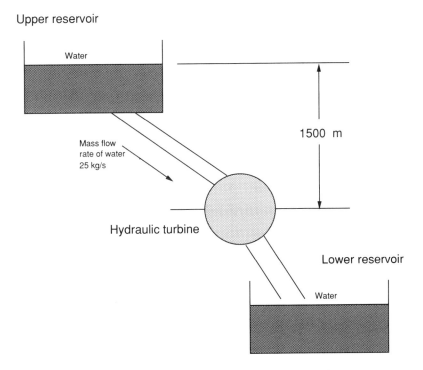

Fig. 1.8 *A hydroelectric power station.*

does not involve the generation of heat and, therefore, is likely to have a high value of conversion efficiency. In fact, the main reason for the loss in efficiency is friction — friction between the water and the pipe which links the reservoir and turbine, and friction inside the turbine between the water and the turbine parts. The following example illustrates the energy-conversion processes involved in a hydroelectric plant and how the units of quantities involved are those of power (watts).

EXAMPLE 1.3

Referring to Figure 1.8, calculate the power generated by the turbine and the overall efficiency of the station if the following data apply:

Mass flow rate of water	= 25 kg/s
Height of water in reservoir above turbine axis	= 1 500 m
Total power loss due to friction	= 45 000 W = 45 kW

Solution

From the definition of potential energy, the rate at which the potential energy available in the water in the reservoir is converted into power output by the turbine is given by the product of the mass flow rate of the water supplied from the upper reservoir, the acceleration due to gravity, and the reservoir height above the turbine axis. Hence:

$$\text{Rate of potential energy conversion} = 25 \times 9.81 \times 1500$$

$$\therefore \quad \text{Rate of potential energy conversion} = 367\,875 \text{ watts} = 367\,875 \text{ W}$$

This would be the power output of the turbine but for the fact that 45 kW of the available power is lost to friction. Therefore, the net power output from the system is

$$\text{Net power output from system} = 367\,875 - 45\,000 = 322\,875 \text{ W} = 322.9 \text{ kW}$$

The overall efficiency of the station is the ratio at which net power is produced from the available potential energy. Hence:

$$\text{Overall efficiency of station} = \frac{\text{net power output}}{\text{potential energy input}}$$

$$\therefore \quad \text{Overall efficiency of station} = \frac{322\,875}{367\,875} = 87.8\%$$

1.4 ENERGY AND THE ENVIRONMENT

Thermofluids is the study of the conversion between stored and mechanical energy, and the aim of this book is to provide the theoretical background to enable that study to take place. It is clear that the efficiency of any energy-conversion plant can only be evaluated from this theoretical base. Although it is not a matter for great discussion in this book, the authors are keen to point out that the way in which energy is used has a significant effect upon the environment.

There are two main problems, both associated with the use of fossil fuels for generating heat. At the lowest level, it is of great concern that many of the world's forests are being removed to provide heat and raw material for paper-making. At a higher level, the depletion of oil and gas reserves and the subsequent use of these fuels to produce heat, thereby generating major pollution problems, is a matter which the study of thermofluids might provide answers to. Certainly there will be less energy usage and hence less pollution if the conversion efficiency of the process is increased. The alternative is to use the logic of the energy-conversion triangles of Figures 1.1, 1.2 and 1.3, and move towards energy-conversion processes which do not involve heat but do use renewable energy (wind, wave, tidal, solar) and so retard the rate of depletion of fossil fuels.

2

Some Thermofluid Properties

All the thermal systems discussed in Chapter 1, the car engine, the thermal power station, the hydroelectric scheme, the gas turbine plant, require a fluid for the transfer of energy to take place. In the hydroelectric scheme, the fluid is water in the liquid phase; in the gas turbine it is air in the gaseous phase; and in the thermal power station, it is both water in the liquid phase and steam in the vapour phase. In other words, the fluid can be a gas, liquid or vapour, or indeed change phase within the system, but it is because the fluid is present and because something happens to it that energy is realised.

It is important, therefore, to be able to say what the condition of the fluid is at any point, and this is achieved from a knowledge of the values of the fluid properties. The main ones are those that can be sensed, namely the mass, pressure, temperature, and volume, but there are also others which are useful and relevant in certain applications. Any point at which the fluid properties are determined is called a State, and when the properties at a given state are constant with time, equilibrium is said to have been established. If the fluid is identified as being at state 1, the properties there are given the subscript 1, and at state 2, the subscript 2, etc. The act whereby a fluid changes the value of one or a number of its properties is referred to as a change of state or Process. Sometimes in a thermal system, the fluid passes through a number of states, usually more than two at least, and ends up in the same condition in which it began. It has completed a cycle.

It is one thing to describe a fluid as a liquid or a vapour, and to ascribe properties to it, but what exactly is a fluid? The essential idea is that it is composed of millions of molecules. Each molecule is vibrating randomly about a point and is able to rotate and move from point to point, giving rise to a certain amount of energy which is inherent in the fluid (Figure 2.1). The principle of the continuum is that, because there are so many molecules which make up a fluid, the space between the molecules does not matter and need not be considered. The fluid can exist in two distinct phases, liquid and vapour, or as a mixture of the two phases. In the liquid phase, the molecules are more closely packed together than in the vapour phase (a fluid is called a Gas when it exists in the vapour phase at normal temperatures and pressures (NTP), that is, under normally prevailing atmospheric conditions). The properties describe what happens to the molecules under different circumstances, such as whether they are made to compress together more tightly or perhaps made to vibrate more energetically.

Some fluid properties are mainly relevant when the fluid is flowing from one place to another, such as the dynamic viscosity. Others can only be shown to exist theoretically, and yet are extremely important to the understanding of thermofluids. This is what makes the study of the subject somewhat enigmatic and yet, at the same time, so fascinating. Generally, only two independent properties need be identified to adequately describe the condition of a

16 SOME THERMOFLUID PROPERTIES

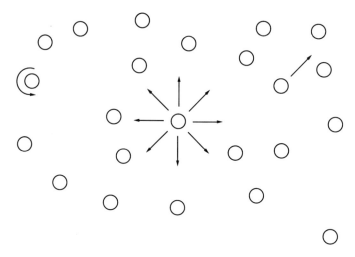

Fig. 2.1 *Fluids consist of molecules which vibrate, rotate and translate.*

fluid in any state, and this is known as the Two-property Rule. This chapter considers some of the more readily accessible properties of a fluid which must be explained before further analysis of a thermal system is possible.

A number of the formulae developed are applicable to the fluid only when it is in a particular phase, gas, liquid, vapour or indeed changing phase. In an attempt to remind readers to use the correct formulae in given situations, whenever a particular fluid is used in the solution to a problem, the principal fluid properties, such as the mass, volume, pressure and temperature, are identified by a subscript and the brackets { }. In order not to make it too confusing, only eight fluids have been utilised in the text. These are as follows:

{ }$_{air}$ air as the fluid
{ }$_{ex}$ exhaust products of combustion as the fluid
{ }$_{fuel}$ fuel as the fluid
{ }$_{he}$ helium as the fluid
{ }$_{m}$ mercury as the fluid
{ }$_{oil}$ oil as the fluid
{ }$_{st}$ steam as the fluid
{ }$_{w}$ water as the fluid

2.1 THE PROPERTY MASS

The mass of a fluid is a measure of the amount of substance present, that is, the molecules and the gaps that separate them. This can be obtained from the weight which is a force because the mass must act in some direction, usually vertically downwards (Figure 2.2). The mass has the symbol m, the weight is the mass multiplied by the acceleration due to gravity g, giving it the units of force. Hence:

$$\text{Mass: } m \text{ (units kg)}$$

$$\text{Weight: } mg \text{ (units N)}$$

THE PROPERTY VOLUME   17

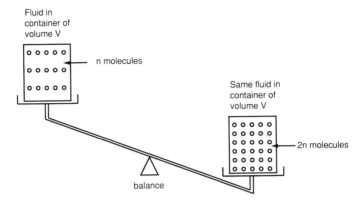

Fig. 2.2 *The property mass is a measure of the amount of substance present.*

Masses can be determined by comparison with standard measures. In fact, the kilogram is the mass of a lump of platinum preserved in the International Bureau of Weights and Measures. The mass flow rate was referred to in Chapter 1 for an open system, when the fluid is moving from place to place. It has the symbol $\dot{m}$. Thus:

$$\text{Mass flow rate: } \dot{m} \text{ (units kg/s)}$$

All the systems considered in this book will have either a constant mass of fluid or a constant mass flow rate. If the fluid is moving with a certain mass flow rate, it will obviously have a velocity which is given the symbol C. Thus:

$$\text{Fluid velocity: } C \text{ (units m/s)}$$

2.2 THE PROPERTY VOLUME

The volume of a fluid is a measure of the space occupied by a given mass of the molecules (Figure 2.3). It has the symbol V. Volume as a property needs to be treated with care as the same mass of fluid can occupy different volumes, particularly when the fluid is a gas or vapour. This is because the fluid molecules can be squeezed up together when they are compressed or can be allowed to have greater distances between them and thus fill up any space when they are expanded. For example, consider two containers, both one cubic metre in volume, one containing the gas oxygen and the other the gas nitrogen at NTP. Together, the total volume occupied by the two gases is two cubic metres and the total mass is the mass of oxygen and the mass of nitrogen. However, all the nitrogen could be pushed into the oxygen container. The mass of the two gases remains the same but the volume occupied by them is now only one cubic metre.

The property volume can be seen to be mass dependent and such properties are called extensive (Figure 2.4). They can be added together. This gives rise to a further property, the Specific Volume, which is the volume per unit mass, given the symbol v. Thus:

$$\text{Volume: } V \text{ (units m}^3\text{)}$$

$$\text{Specific Volume: } v \text{ (units m}^3\text{/kg)}$$

18 SOME THERMOFLUID PROPERTIES

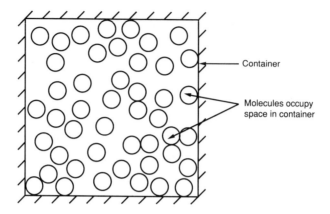

Fig. 2.3 *The property volume is a measure of the space occupied by the molecules of a fluid.*

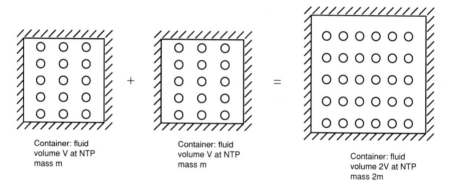

Fig. 2.4 *Volume is an extensive (mass dependent) property.*

where $V = mv$. Volumes of simple shapes can be easily determined through the measurement of lengths with standard rules.

A further property associated with volume is the Density. This is defined as 'the mass of fluid per unit volume'. Hence, the density is a measure of the mass of different fluids all occupying the same volume. It is given the symbol ρ and is the inverse of the specific volume. Thus:

$$\text{Density: } \rho \text{ (units kg/m}^3\text{)}$$

where

$$\rho = \frac{m}{V} = \frac{m}{mv} = \frac{1}{v}$$

For liquids, the density remains reasonably constant through large variations of pressure. For example, consider a 2 m³ cylindrical tank containing 1 m³ of a liquid, with a flat plate resting on the top surface of the liquid. It would not matter how many weights were added on to the plate, the volume of the liquid would remain approximately 1 m³. Therefore,

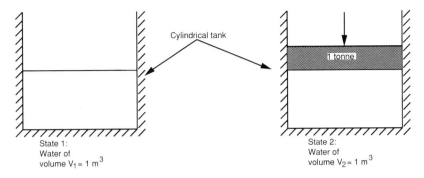

Fig. 2.5 *Liquids are incompressible, their volume varies little with pressure.*

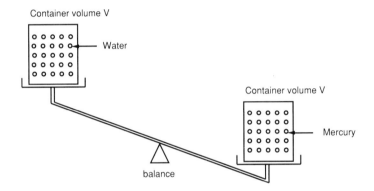

Fig. 2.6 *Mercury is denser than water.*

liquids are considered to be incompressible (Figure 2.5) and to have approximately constant densities. For example, the density of water is taken as 1000 kg/m³, and the density of mercury as 13 600 kg/m³. Hence a volume of mercury is 13.6 times the mass of an equal volume of water (Figure 2.6).

Gases and vapours, on the other hand, do change their volume as the pressure is varied. Therefore, gases and vapours are said to be Compressible. For example, if the density of air at NTP is 1.225 kg/m³, at a pressure 100 times greater and the same temperature it is 122.5 kg/m³ (Figure 2.7).

Sometimes the Relative Density of a fluid is used, which is the ratio of the density of the fluid to the density of water, assumed to be 1000 kg/m³. If the relative density of an oil is stated to be 0.85, its density is 850 kg/m³. It is still sometimes called the specific gravity.

When a fluid is moving from place to place, the volumetric flow rate, given the symbol $\dot{V}$, becomes relevant. Thus:

$$\text{Volumetric flow rate: } \dot{V} \text{ (units m}^3\text{/s)}$$

The relationship between the mass flow rate and volumetric flow rate of a fluid becomes $\dot{m} = \rho \dot{V}$.

20 SOME THERMOFLUID PROPERTIES

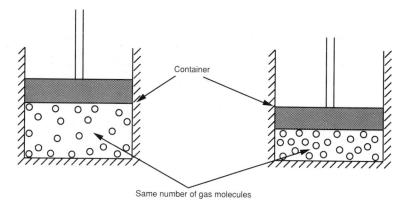

Fig. 2.7 *Gases are compressible, their volume depends upon pressure.*

As stated earlier, only constant mass flow rates will be considered in this book. The volume flow rates of incompressible fluids must also remain constant, but the volume flow rates of compressible fluids will depend upon the variation in the fluid density.

2.3 THE PROPERTY PRESSURE

Pressure is defined as 'the force exerted per unit area'.

In the case of a fluid, the force arises because, as the vibrating molecules of the fluid move and come into contact with a solid surface, they rebound from the surface and the resulting momentum change creates a force (Figure 2.8). The pressure p is the ratio of the force exerted F to the perpendicular area A upon which the force is applied. Thus:

$$\text{Pressure: } p \text{ (units N/m}^2\text{)}$$

where

$$p = \frac{\text{force}}{\text{unit area}} = \frac{F}{A}$$

The group unit N/m² is given its own name, the pascal, written as Pa.

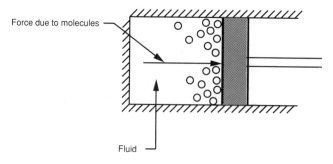

Fig. 2.8 *Fluid pressure is the ratio of the force exerted on a perpendicular area as the fluid molecules strike a surface.*

Unfortunately, there are a number of pressures which arise in thermofluids. The symbol p is used for the absolute static pressure. To explain this, firstly consider the meaning of the term absolute. A container which has been evacuated has no molecules in it and the pressure exerted is zero. This is the start of the absolute pressure scale.

The atmospheric pressure p_{atm} is the pressure exerted by the air molecules in the atmosphere. The air in a room contains millions of molecules and any solid object in the room will be bombarded by the molecules next to the solid surface, the force exerted per unit area of solid surface being the atmospheric pressure. The value of the atmospheric pressure at sea-level is 101 325 Pa. How this value comes about is dealt with in Section 2.3.4 but, because it is rather a large number, it is more convenient to use the scale of kPa, MPa or bar when dealing with pressures, where

$$1 \text{ bar} = 1.01325 \times 10^5 \text{ N/m}^2 = 101.325 \text{ kPa} = 0.101325 \text{ MPa}$$

However, it is Pa or N/m^2 which is in the SI system of units and should be used in all calculations.

Pressures measured on a gauge are relative to atmospheric. Therefore, the atmospheric pressure measured on a gauge is zero and pressures below atmospheric are negative (Figure 2.9). Gauge pressures p_{ga} are converted to absolute pressures by adding the atmospheric pressure:

$$\text{Absolute pressure} = p_{ga} + p_{atm} \tag{2.1}$$

Pressures used in calculations should be absolute pressures with one exception, considered in Chapter 10, and that will always be assumed so unless otherwise stated.

Under the kinetic theory, individual molecules of a fluid are considered to vibrate, rotate, and diffuse, in other words move from point to point even though the fluid itself can be considered to be stationary. But it is important to distinguish this movement, diffusion, from

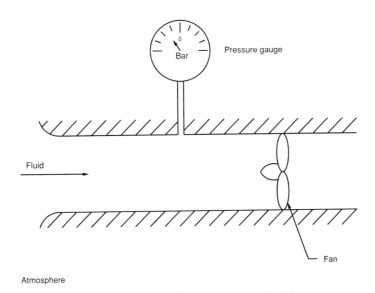

Fig. 2.9 *Fluid pressures less than atmospheric are negative on a gauge.*

the movement of the molecules when the whole fluid has a bulk velocity and is flowing from place to place in some manner. The pressure exerted by the molecules vibrating, rotating and diffusing is the static pressure p. As the molecules undergo this activity in a random manner, the static pressure is considered to be exerted equally in all directions about a point. But if the fluid, and so the molecules which comprise it, is also flowing with a velocity C, an additional force is exerted on any perpendicular surface that the molecules come into contact with. The total pressure or stagnation pressure p_0 is the sum of the static pressure and the dynamic pressure p_{dyn} which arises when the fluid motion is arrested by the surface (Figure 2.10):

$$\text{Total pressure} = \text{static pressure} + \text{dynamic pressure}$$

$$\therefore \quad p_0 = p + p_{dyn} \tag{2.2}$$

$$\therefore \quad p_0 = p + 0.5\rho C^2 \tag{2.3}$$

This last equation must be accepted for the time being, its validity will be demonstrated in Section 8.11. The dynamic pressure expressed as $0.5\rho C^2$ obviously has the same units as the static pressure p and total pressure p_0. Rearranging Equation (2.3) gives

$$\frac{p_0}{\rho} = \frac{p}{\rho} + 0.5 C^2$$

The last term in the above equation is the kinetic energy of the fluid per unit mass or mass flow rate. Equation (2.3), therefore, is a simplified expression for the conservation of energy and in this format the energy level of the fluid is accounted for by two terms, the kinetic energy and the energy due to the pressure of the fluid, usually called the Flow Work. Flow work is considered further in the derivation of the Steady Flow Energy Equation, Section 8.2.3.

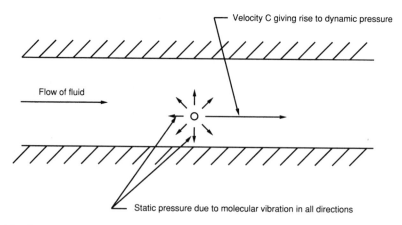

Fig. 2.10 *Fluid molecules exert a static pressure due to their vibration and a dynamic pressure due to their velocity.*

2.3.1 Pressure variation with height/depth

Before the measurement of pressure can be considered, its variation with height must be examined. It can be sensed by a mountain climber that the pressure exerted by the air on his or her ears decreases with height, or by a deep-sea diver that it increases with depth in water. The change in pressure is due to gravitational effects.

Consider a cylindrical column of fluid within a larger mass of the same fluid, as shown in Figure 2.11. The column is at an angle θ to the horizontal, it has a constant cross-sectional area A_{xs} and is of length L. One end is at a vertical height of z above some datum and the other end is at a vertical height of $(z + dz)$ above the same datum. There is a static pressure p exerted on the cross sectional area A_{xs} at the bottom end of the column, and a static pressure $(p + dp)$ on the cross-sectional area A_{xs} at the top end. By convention, pressures are drawn to hold the fluid in the shape under consideration. Therefore, at the bottom, the pressure force is drawn upwards along the axis, and at the top, downwards along the axis of the column.

There is a force due to the weight of the column acting vertically downwards from the centre of gravity, equal to the mass of the column m multiplied by the acceleration due to gravity g. In addition, there will be static pressures acting on the sides of the column, from the top and the underneath. However, the intention is to resolve forces along the axis of the column which will make forces due to these pressures zero and they have been omitted from the figure.

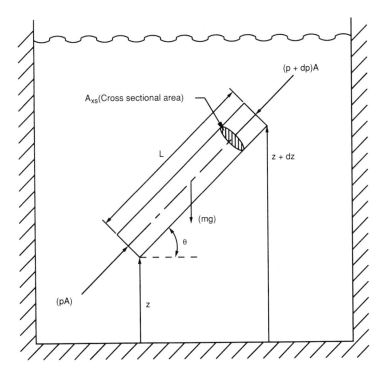

Fig. 2.11 *Column of fluid within a larger mass of the same fluid.*

SOME THERMOFLUID PROPERTIES

Given that the column is stationary, there will be no net force upon it. Resolving the forces along the axis of the column gives

$$pA_{xs} - (p + dp)A_{xs} - mg\cos\theta = 0 \tag{2.4}$$

For a fluid of density ρ, the mass of the column is

$$m = \rho A_{xs} L = \frac{\rho A_{xs}\, dz}{\cos\theta} \tag{2.5}$$

Substituting Equation (2.5) into Equation (2.4) gives

$$dp = -\rho g\, dz \tag{2.6}$$

This equation can be integrated in three ways. Firstly, assuming $dz = 0$, in other words that the column is horizontal and that the pressures under examination at either end of the column are on the same level. This gives $dp = 0$.

$$\therefore\quad p = \text{const}. \tag{2.7}$$

In other words, the pressures in a continuous fluid are equal at the same level, but may vary from level to level. Therefore, in Figure 2.12, the pressure of the fluid in the container is the same on level XX. This is also true for Figure 2.13, which, in fact, provides a device with which to measure pressures.

Equation (2.6) can also be integrated between state 1 and state 2, assuming that the density of the fluid is constant, as follows:

$$p_2 - p_1 = \rho g(z_1 - z_2) \tag{2.8}$$

This equation demonstrates the effects felt by the mountain climber or the deep-sea diver mentioned before, namely that the pressure of a fluid decreases with height and increases

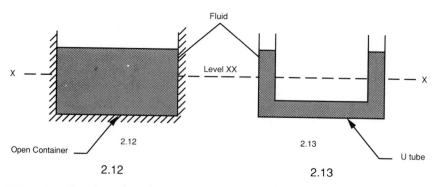

Fig. 2.12 and 2.13 The pressures in a continuous fluid are equal at the same level in a fluid container and in a U-tube.

with depth. It provides a mathematical statement of the relationship between the pressure change and the height for an incompressible fluid, and note that it is the vertical height that is important. The equation can be applied to an instrument called a manometer which will give a reading of pressure difference, as detailed in Section 2.3.3. Pressure is often expressed as a head, in other words as a height of liquid in metres. Equation (2.8) provides the conversion from head in metres to pressure in pascals or vice versa. The symbol Z is used for a head (or height) of fluid. Hence:

$$\text{Head/height: } Z \text{ (units m)}$$

Equation (2.8) also shows that the pressure of a fluid varies gradually with height, there are no sudden steps where the pressure leaps from one value to another. In fact, this is another important characteristic of the property pressure. The only time that the pressure can suddenly go through a step change in value is in a shock wave, and these only occur in special circumstances such as supersonic flow and in water-hammer situations. Neither need be of concern in this book. For most applications, fluid pressures try to equalise themselves, resulting in gradual changes. Pressure is, therefore, said to be an Intensive property (Figure 2.14) because of this equalising habit, as opposed to volume which was an extensive property, meaning that volumes could be added together and, as such, were mass dependent. Pressures are not added together and are not mass dependent.

Equation (2.6) can be integrated too, even when the density varies, if an expression relating the pressure and the density is known. The density can be eliminated leaving the pressure to be integrated with respect to the vertical height. It is found, in practice, that some fluids, especially gases, obey a law whereby

$$pv^n = \frac{p}{\rho^n} = \text{const}.$$

Use of this expression allows the integration to proceed, but it is not worth expounding further here.

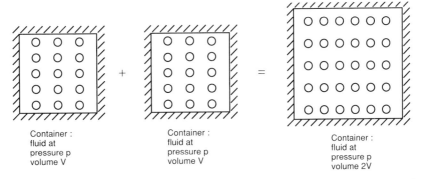

Fig. 2.14 *Fluid pressure is an intensive property, pressures are not added together.*

2.3.2 Static and total pressures

Equation (2.3) gives the relationship between the static and total pressures of the fluid, but it is important to realise their significance when it comes to their measurement. If a hole is drilled in the side of a duct through which a fluid is flowing, and fitted with a tapping perpendicular to the flow, with the tapping connected to a device which will measure pressure, the pressure recorded will be the static pressure of the fluid. Consider Figure 2.15.

A molecule flowing past the hole drilled in the duct will be exerting both a static pressure due to its vibration and a dynamic pressure due to its velocity. But the velocity is a vector and is in the direction of flow parallel to the axis of the duct. Therefore, the component of velocity which will act through the hole, at $90°$ to the flow direction, is $C \cos 90$, which is zero. The component of pressure due to the molecular velocity recorded on the pressure-measuring device is zero. On the other hand, the force exerted due to the molecular vibration above the hole acts equally in all directions. Thus, there will be a static component of pressure recorded by the instrument connected to the static pressure tapping. Hence, such a tapping perpendicular to the flow allows the measurement of static pressure to be made. If the hole is too large in diameter, some elements of the dynamic pressure will affect the reading. In general, there are national standards which specify the size of hole allowed and the order of accuracy with which the static pressure can be measured, but around 2 mm diameter is usually satisfactory.

To measure the static pressure in the above manner requires the assumption to be made that it is uniform and constant across the cross-section of the duct. This is not strictly true, but is a reasonable assumption for most applications. The actual pressure and velocity profiles of a fluid in a duct will be considered in Chapters 9 and 11.

To measure the total pressure, it is necessary to place a probe facing into the fluid stream, as shown in Figure 2.16. The pressure recorded at the probe entry will be due to both the molecular vibration and the molecular velocity. Where the fluid is brought to rest at the

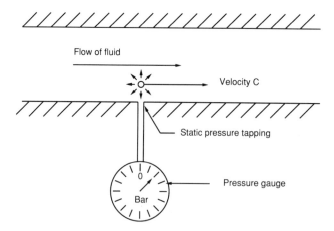

Fig. 2.15 *A static pressure tapping perpendicular to a fluid flow for the measurement of the fluid static pressure.*

THE PROPERTY PRESSURE

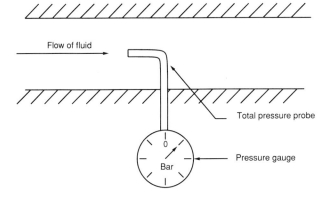

Fig. 2.16 *A total pressure probe facing into a fluid flow for the measurement of the fluid total pressure.*

entrance of the probe is known as a stagnation point. Once the total pressure and static pressure in a fluid stream are known, the velocity can be determined from Equation (2.3).

EXAMPLE 2.1

Air of constant density 1.2 kg/m³ is flowing through a horizontal circular pipe. At a given cross-section of the pipe, the static pressure is measured from a static pressure tapping to be 70 kPa gauge, and the total pressure is measured with a total pressure probe to be 90 kPa gauge (Figure 2.17). What is the average velocity of flow at that particular pipe cross-section if the atmospheric pressure is 100 kPa?

Some metres further down the pipe, the velocity of the air still has the same value, but the static pressure is now 60 kPa gauge. What is the decrease in the total pressure between the two measuring stations if the density of the air can be assumed to remain constant?

Repeat the calculation for water of density 1000 kg/m³.

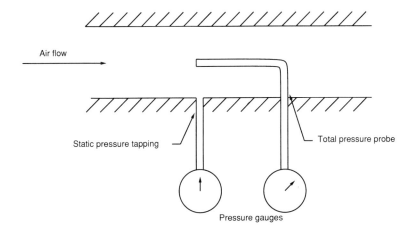

Fig. 2.17 *Example 2.1—a total pressure probe and a static pressure tapping are used to measure the velocity of air in a pipe.*

Solution

For air, first convert the gauge pressures into absolute pressures. At the initial measuring station, using Equation (2.1), the absolute static pressure of the air $\{p\}_{air}$ and the absolute total pressure $\{p_0\}_{air}$ are given by

$$\{p\}_{air} = 70 + 100 = 170 \text{ kPa} = 1.7 \times 10^5 \text{ Pa}$$

$$\{p_0\}_{air} = 90 + 100 = 190 \text{ kPa} = 1.9 \times 10^5 \text{ Pa}$$

Knowing the density of the air $\{\rho\}_{air}$, and given that it is constant, the air velocity $\{C\}_{air}$ can be determined from Equation (2.3):

$$\{p_0\}_{air} = \{p\}_{air} + 0.5\{\rho\}_{air}\{C\}_{air}^2$$

$$\therefore \quad 1.9 \times 10^5 = 1.7 \times 10^5 + 0.5 \times 1.2 \{C\}_{air}^2$$

$$\therefore \quad \{C\}_{air} = 182.6 \text{ m/s}$$

Repeating at the second measuring station, identified by superscript ′:

$$\{p'\}_{air} = 60 + 100 = 160 \text{ kPa} = 1.6 \times 10^5 \text{ Pa}$$

$$\{C'\}_{air} = \{C\}_{air} = 182.6 \text{ m/s}$$

$$\therefore \quad \{p'_0\}_{air} = 1.6 \times 10^5 + 0.5 \times 1.2 \times 182.6^2 = 1.8 \times 10^5 \text{ Pa}$$

$$\therefore \quad \text{Decrease in air total pressure} = \{p'_0\}_{air} - \{p_0\}_{air} = 10 \text{ kPa}$$

If the fluid is water but the measured pressures remain the same, at the first measuring station the velocity of the water $\{C\}_w$ is given by Equation (2.3) as

$$1.9 \times 10^5 = 1.7 \times 10^5 + 0.5 \times 1000\{C\}_w^2$$

$$\therefore \quad \{C\}_w = 6.3 \text{ m/s}$$

The velocity of the water is much lower than the velocity of the air, although it has the same dynamic pressure, because the density of water is much greater than the density of air.

At the second measuring station, the absolute total pressure of the water $\{p'_0\}_w$, again from Equation (2.3), is

$$\{p'_0\}_w = 1.6 \times 10^5 + 0.5 \times 1000 \times 6.3^2 = 1.8 \times 10^5 \text{ Pa}$$

The decrease in total pressure of the water is the same as the decrease in total pressure of the air because the static pressure decrease for both fluids is the same and the dynamic pressure remains constant.

2.3.3 Pressure manometry

Pressures can be determined very simply by means of a device called a U-tube manometer, as shown in Figure 2.18, where the static pressure of a fluid in a duct is being measured.

The U-tube is usually made of glass and is filled with a liquid of known density. As the fluid flows through the duct, its static pressure is registered through the static pressure tapping. If the fluid is anything other than air, the air must be evacuated and the fluid made to occupy the volume above the liquid in the left-hand limb of the manometer. The liquid in the manometer and the fluid from the duct must not react or mix together and there must be a distinct meniscus between the two so that the scale can be read clearly. Mercury is an

THE PROPERTY PRESSURE

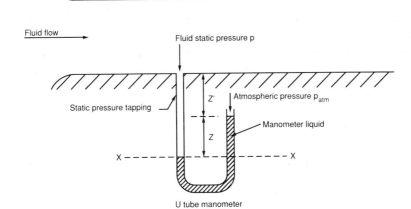

Fig. 2.18 *U-tube manometer for measuring the fluid static pressure.*

excellent liquid to use in a manometer for this reason. As it has a large density, it also does not give large differences in height either side of the U-tube at reasonable pressures.

Equation (2.6) led to the conclusion that the pressure of a continuous fluid at a given level is equal, but that the pressure varied from level to level. In all manometer applications, it is necessary to look for a level at which the pressure on the left-hand side (LHS) of the U-tube is equal to the pressure on the right-hand side (RHS). Clearly, in Figure 2.18, this is level XX because the manometer liquid is continuous through the U-bend. The idea is to equate the pressures acting on the LHS with those acting on the RHS at the chosen level.

As mentioned before, the assumption must be made that the static pressure of the fluid in the duct, which is being measured, is uniform all the way across the duct at any measuring position. Therefore, on the LHS of the U-tube, there is a static pressure p acting at the static pressure tapping point. In moving to the level XX, the pressure must increase according to Equation (2.8), whereby pressure increases with depth and decreases with height. The amount that it increases by can be determined utilising Equation (2.8) which relates the pressure to a vertical height multiplied by the density of the fluid and the acceleration due to gravity. If the fluid flowing through the duct has a density ρ, the pressure at level XX on the LHS of the U-tube is

$$p + \rho g(Z' + Z)$$

On the RHS of the U-tube, the liquid in the manometer is exposed to the atmosphere. Therefore, it must be subject to the atmospheric pressure at the surface. To get to level XX, the pressure must increase by an amount equivalent to the depth of manometer liquid expressed as a pressure. If the manometer liquid has a density of ρ', the pressure at level XX on the RHS of the U-tube is

$$p_{atm} + \rho' g Z$$

According to Equation (2.7), these two pressures are equal, giving

$$p - p_{atm} = gZ(\rho' - \rho) - \rho g Z' \tag{2.9}$$

SOME THERMOFLUID PROPERTIES

The LHS of Equation (2.9) is the static pressure of the fluid in the duct expressed as a gauge pressure. If the absolute value of the static pressure is needed, the atmospheric pressure must be found, as shown in Section 2.3.4. The other terms in the equation can be easily measured, allowing the pressure of the fluid to be determined.

In Figure 2.18, the static pressure of the fluid in the duct is greater than atmospheric. If the U-tube manometer is used to measure pressures which are less than atmospheric, the height of the liquid in the manometer in the right-hand limb will be below the height of the liquid in the left-hand limb.

Static pressures below atmospheric occur regularly in systems where air is drawn in from the atmosphere through a duct by a fan. Referring to Equation (2.2), if the total pressure at entry to the duct is the atmospheric pressure then, as the fluid moves with a given velocity, it achieves a certain dynamic pressure and the static pressure component must be less than atmospheric. In fact, this is a simplified approach to the situation because in practice the total pressure will not remain constant, as will be seen in Chapter 8, but the static pressure will still be below atmospheric.

EXAMPLE 2.2

Water of density 1000 kg/m³ is flowing through a horizontal pipe. The left-hand limb of a U-tube manometer is connected to a static pressure tapping at a certain cross-section of the pipe, and the right-hand limb is open to the atmosphere. The manometer contains mercury of density 13 600 kg/m³ (Figure 2.19). If the mercury in the left-hand limb is at a vertical height of 400 mm below the static pressure tapping and the mercury in the right-hand limb is at a vertical height 230 mm above the mercury in the left-hand limb, what is the gauge static pressure of the water in the pipe at that cross-section?

If the atmosphere in which the manometer is being used has a pressure of 100 kPa, what is the absolute static pressure of the water in the pipe?

If the pipe was moved to a position where the atmospheric pressure was 60 kPa and the heights of mercury in the manometer remained the same, what would now be the absolute static pressure of the water in the pipe?

If the water was replaced by an oil of relative density 0.8, and again the heights of mercury in the manometer did not change, what would be the absolute static pressure of the oil in the pipe?

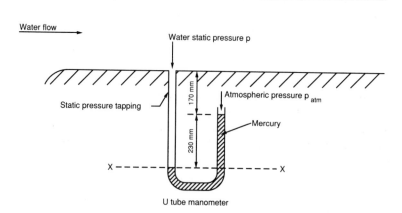

Fig. 2.19 *Example 2.2—a U-tube manometer is used to measure the static pressure of water in a pipe.*

Solution
Utilising Equation (2.9), the gauge static pressure of the water $\{p_{ga}\}_w$ is

$$\{p_{ga}\}_w = \{p\}_w - p_{atm} = gZ(\{\rho\}_m - \{\rho\}_w) - \{\rho\}_w gZ'$$
$$\therefore \{p_{ga}\}_w = 9.81 \times 0.23 \times (13\,600 - 1000) - 1000 \times 9.81 \times 0.17 = 26.8 \text{ kPa}$$

If the atmospheric pressure p_{atm} is 100 kPa, the absolute static pressure of the water $\{p\}_w$, from Equation (2.1), is

$$\{p\}_w = \{p_{ga}\}_w + p_{atm} = 26.8 + 100 = 126.8 \text{ kPa}$$

If the pipe is moved to an atmosphere where the pressure is 60 kPa, the absolute static pressure of the water, identified by superscript ', and again using Equation (2.1), becomes

$$\{p'\}_w = 26.8 + 60 = 86.8 \text{ kPa}$$

If the water is replaced by oil of relative density 0.8, the density of the oil $\{\rho\}_{oil}$ is

$$\{\rho\}_{oil} = 0.8 \times 1000 = 800 \text{ kg/m}^3$$

When the oil is flowing in the pipe and the atmospheric pressure is 60 kPa, the gauge static pressure of the oil $\{p_{ga}\}_{oil}$, again from Equation (2.9), is

$$\{p_{ga}\}_{oil} = \{p\}_{oil} - p_{atm} = gZ(\{\rho\}_m - \{\rho\}_{oil}) - \{\rho\}_{oil} gZ'$$
$$\therefore \{p_{ga}\}_{oil} = 9.81 \times 0.23 \times (13\,600 - 800) - 800 \times 9.81 \times 0.17 = 27.6 \text{ kPa}$$

Therefore, the absolute static pressure of the oil $\{p\}_{oil}$, from Equation (2.1), is

$$\{p\}_{oil} = 27.6 + 60 = 87.6 \text{ kPa}$$

Mercury is an excellent liquid to use in a manometer where high fluid pressures are being recorded because the height of mercury required is relatively small owing to its large density.

EXAMPLE 2.3
Air, assumed to be at a constant density of 1.2 kg/m³, is drawn by a fan into a duct. The left-hand limb of a U-tube manometer is connected to a static pressure tapping in the side of the duct at a point where it is required to measure the static pressure. The right-hand limb of the manometer is open to the atmosphere where the pressure is 100 kPa. The manometer liquid is coloured water of density 1000 kg/m³ (Figure 2.20). If the vertical height from the static pressure tapping to the water in the left-hand limb of the manometer is 150 mm, and the water in the right-hand limb of the manometer is 600 mm below the height in the left-hand limb, what is the static pressure of the air in the duct at that point?

If the manometer is physically raised so that the water in the left-hand limb of the manometer is at the same level as the static pressure tapping in the side of the duct, what is now the difference in height of the water either side of the manometer?

Solution
The pressure must be equal either side of the U-tube on level XX. Equating the pressures there and making use of Equation (2.8), the absolute static pressure of the air $\{p\}_{air}$ is given by

$$\{p\}_{air} + \{\rho\}_{air} gZ' + \{\rho\}_w gZ = p_{atm}$$

32 SOME THERMOFLUID PROPERTIES

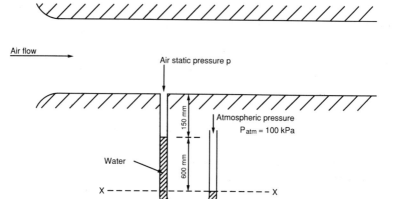

Fig. 2.20 *Example 2.3—a U-tube manometer is used to measure the static pressure of air in a pipe.*

$$\therefore \quad \{p\}_{air} + 1.2 \times 9.81 \times 0.15 + 1000 \times 9.81 \times 0.6 = 10^5$$

$$\therefore \quad \{p\}_{air} = 94.1 \text{ kPa}$$

The air static pressure is less than atmospheric as it has been drawn into the duct from the atmosphere by the fan. The manometer liquid in the left-hand limb of the U-tube is at a higher level than in the right-hand limb.

If the manometer is raised so that the water in the left-hand limb is at the same height as the static pressure tapping, the pressure of the air acting on the water in the left-hand limb must be the same as the static pressure of the air in the duct as measured at the static pressure tapping. Equating the pressures either side of the U-tube manometer at the level of the water in the right-hand limb, and making use of Equation (2.8), the difference in height of the water either side of the manometer Z is, therefore, given by

$$\{p\}_{air} + \{\rho\}_w g Z = p_{atm}$$

$$\therefore \quad 94.1 \times 10^3 + 1000 \times 9.81 Z = 10^5$$

$$\therefore \quad Z = 601 \text{ mm}$$

The answer is approximately the same because the pressure contribution of the air in the manometer is relatively insignificant compared to the water. This is a result of the density of air being so small compared to the density of water. A further conclusion must be that, although the air density has been assumed to be constant, it would not actually matter much if it was treated as compressible with a variable density.

Raising the manometer so that the level of liquid in the left-hand limb is at the same height as the static pressure tapping where the fluid pressure is being measured simplifies the manometer equation somewhat.

2.3.4 Determination of atmospheric pressure

The value of the atmospheric pressure can be found by means of a device called a Barometer, shown in Figure 2.21.

A glass tube is evacuated and placed upside down such that its open end is just below the surface level of a liquid in a container. This can be easily achieved if the tube is initially

THE PROPERTY PRESSURE

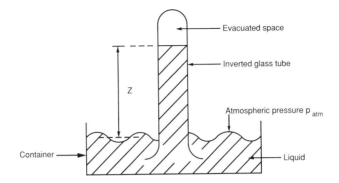

Fig. 2.21 A barometer measures the atmospheric pressure.

immersed in the liquid container, thereby filling with liquid, and withdrawn to the required position. The liquid will fall in the tube leaving an evacuated space above it. In practice, the atmospheric pressure is acting on the surface level of the liquid in the container and, as there is no pressure acting on the liquid in the tube, the liquid rises in the tube to a certain level.

If the liquid has a density ρ, Equation (2.8) gives

$$p_{atm} - 0 = \rho g Z \qquad (2.10)$$

Measuring the vertical height Z allows the atmospheric pressure to be determined. Practical barometers use mercury as the liquid because it has a large density, which means that the value of Z is reasonable.

EXAMPLE 2.4

What height does the liquid in a barometer rise to if the atmospheric pressure is 101.325 kPa and the liquid is as follows:
(a) Mercury of density 13 600 kg/m³?
(b) Water of density 1000 kg/m³?

What will be the percentage change in height recorded of either liquid if the barometer is used at the top of a mountain where the atmospheric pressure is 65 kPa?

Solution
From Equation (2.10):

$$Z = \frac{p_{atm}}{\rho g}$$

For mercury:

$$Z = \frac{1.013\,25 \times 10^5}{13\,600 \times 9.81} = 0.76 \text{ m of mercury}$$

For water:

$$Z = \frac{1.013\,25 \times 10^5}{1000 \times 9.81} = 10.33 \text{ m of water}$$

When the atmospheric pressure is 65 kPa, identified by superscript ′, for mercury:

$$Z' = \frac{0.65 \times 10^5}{13\,600 \times 9.81} = 0.49 \text{ m of mercury}$$

$$\therefore \text{ percentage change in height for mercury} = \frac{Z - Z'}{Z} = \frac{0.76 - 0.49}{0.76} = 35.5\%$$

Mercury is a practical liquid for a barometer as the atmospheric pressure is equivalent to a height of approximately 760 mm as opposed to over 10 m for water, but it is somewhat expensive.

2.3.5 Inclined pressure manometers

Smaller pressures can be measured if the manometer is inclined. The vertical height is required in Equation (2.8), but an inclined length can be measured and converted to a vertical height. Consider Figure 2.22.

Under a fluid pressure p, which is greater than atmospheric, the liquid level in the container drops slightly through a height Z, while in the tube it rises through some length L. If the liquid is of density ρ, using Equation (2.8):

$$p = p_{\text{atm}} + \rho g L \sin\theta + \rho g Z$$

Usually Z is insignificant and that term may be neglected. The equation becomes

$$p = p_{\text{atm}} + \rho g L \sin\theta \qquad (2.11)$$

EXAMPLE 2.5

Air flows through a horizontal duct. The left-hand limb of a U-tube manometer is connected to a static pressure tapping in the duct at a position where the static pressure of the air is 102 kPa. The right-hand limb of the manometer is open to the atmosphere where the pressure is 101 kPa. The position of the manometer is adjusted so that the liquid in the left-hand limb is at the same vertical height as the static pressure tapping (Figure 2.23). If the liquid

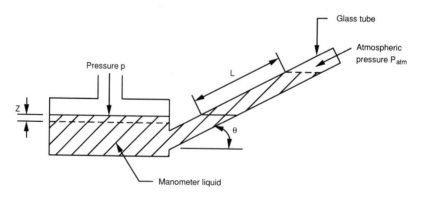

Fig. 2.22 An inclined pressure manometer enables small pressure differences to be measured.

THE PROPERTY PRESSURE   35

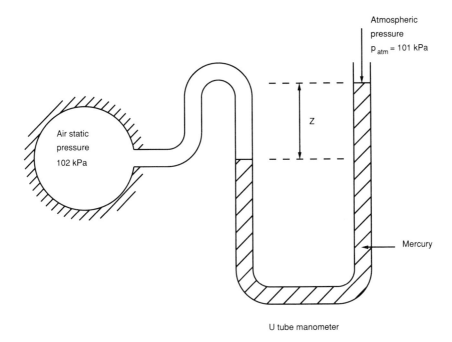

Fig. 2.23 *Example 2.5— a U-tube manometer is used to measure the static pressure of air in a duct.*

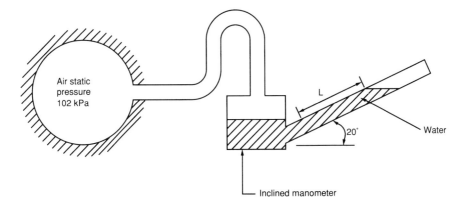

Fig. 2.24 *Example 2.5— an inclined manometer is used to measure the static pressure of air in a duct.*

in the manometer is mercury of density 13 600 kg/m³, what is the difference in height of the mercury either side of the U-tube?

If the mercury is replaced by water of density 1000 kg/m³, what is now the difference in height of the water either side of the U-tube?

If the U-tube manometer is replaced by an inclined manometer at an angle of 20 degrees to the horizontal, and the liquid in the manometer is again water, what will be the reading of length of water on the scale of the manometer (Figure 2.24)?

Solution

Making use of Equation (2.8) and equating the pressures either side of the U-tube manometer at the level of the liquid, the difference in height of the mercury Z either side of the U-tube is given by

$$\{p\}_{\text{air}} = p_{\text{atm}} + \{\rho\}_{\text{m}} g Z$$

$$\therefore \quad 1.02 \times 10^5 = 1.01 \times 10^5 + 13\,600 \times 9.81 Z$$

$$\therefore \quad Z = 7.5 \times 10^{-3} \text{ m}$$

As this is too small to measure accurately, water is used to replace the mercury in the manometer. The difference in the height of the water either side of the U-tube Z' is given by

$$1.02 \times 10^5 = 1.01 \times 10^5 + 1000 \times 9.81 Z'$$

$$\therefore \quad Z' = 0.102 \text{ m}$$

This height can be measured accurately. Alternatively, on an inclined manometer and using Equation (2.11), the length of water on the inclined manometer L is

$$\{p\}_{\text{air}} = p_{\text{atm}} + \{\rho\}_{\text{w}} g L \, \sin \theta$$

$$\therefore \quad 1.02 \times 10^5 = 1.01 \times 10^5 + 1000 \times 9.81 L \, \sin 20$$

$$\therefore \quad L = 298 \text{ mm}$$

Longer lengths of manometer liquid on an inclined manometer can be measured more accurately than shorter vertical heights and this allows reasonably small pressures to be determined with some confidence.

2.3.6 Other pressure manometers

Manometers can be used to measure the pressure difference between two fluids in separate pipes. All that is necessary is to connect the left-hand limb of the manometer to a static pressure tapping on one pipe, and the right-hand limb of the manometer to a static pressure tapping on the other pipe. The manometer liquid must not mix with either pipe fluid.

It is also possible to use a gas, typically air, as the fluid in the manometer. In this case, the manometer must be inverted and the air sit in the U-tube at the top. Although the air is compressible, any changes in its density with pressure are insignificant.

EXAMPLE 2.6

Water of density 1000 kg/m³ is flowing through a horizontal pipeline. Oil of density 800 kg/m³ flows through another horizontal pipeline near by. The left-hand limb of a U-tube manometer is connected to a static pressure tapping in the side of the water pipe, and the right-hand limb is connected to a static pressure tapping in the side of the oil pipe. The water pipe static pressure tapping is at a vertical height 1.2 m above the oil pipe static pressure tapping. The liquid in the manometer is mercury of density 13 600 kg/m³. The mercury in the left-hand limb of the manometer is at a height 1.9 m below the static pressure tapping in the water pipe, and in the right-hand limb at a height 0.5 m below the static pressure tapping in the oil pipe (Figure 2.25). What is the static pressure difference between the water and the oil?

If the static pressure difference between the water and the oil increases by 6 kPa, what will be the percentage change in the height difference of the mercury either side of the U-tube?

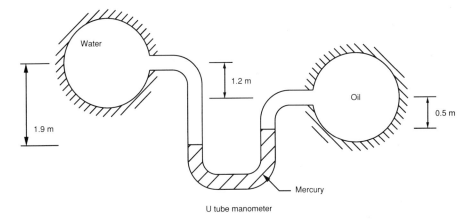

Fig. 2.25 *Example 2.6 — a U-tube manometer is used to measure the static pressure difference between water in one pipe and oil in another pipe.*

Solution
The difference in height of the mercury Z either side of the U-tube is

$$Z = 1.9 - 1.2 - 0.5 = 0.2 \text{ m}$$

Utilising Equation (2.8), if the absolute static pressure of the water is $\{p\}_w$ and the absolute static pressure of the oil is $\{p\}_{oil}$, the pressure difference for equal pressures at level XX is given by

$$\{p\}_w + \{\rho\}_w \times 9.81 \times 1.9 = \{p\}_{oil} + \{\rho\}_{oil} \times 9.81 \times 0.5 + \{\rho\}_m \times 9.81 Z$$

$$\therefore \{p\}_w + 1000 \times 9.81 \times 1.9 = \{p\}_{oil} + 800 \times 9.81 \times 0.5 + 13\,600 \times 9.81 \times 0.2$$

$$\therefore \{p\}_w - \{p\}_{oil} = 12 \text{ kPa}$$

If this pressure difference increases by 6 kPa, the new pressure difference, identified by superscript ′, is

$$\{p'\}_w - \{p'\}_{oil} = 18 \text{ kPa}$$

If the pressure difference increases, the mercury in the left-hand limb will move downwards through a vertical height Z', and the mercury in the right-hand limb must move vertically upwards through the same height because the mercury is incompressible and the U-tube has a constant diameter. Equalising the pressures either side of the U-tube in the same manner as before but at the new level of the mercury in the left-hand limb of the manometer, gives

$$\{p'\}_w + \{\rho\}_w g(1.9 + Z') = \{p'\}_{oil} + \{\rho\}_{oil} g(0.5 - Z') + \{\rho\}_m g(0.2 + 2Z')$$

$$\therefore 18 \times 10^3 + 1000 g(1.9 + Z') = 800 g(0.5 - Z') + 13\,600 g(0.2 + 2Z')$$

$$\therefore Z' = 24.2 \text{ mm}$$

The new difference in height of the mercury either side of the manometer Z'' is given by

$$Z'' = Z + 2Z' = 0.2 + 2 \times 0.0242 = 248 \text{ mm}$$

The method is useful in situations where differential pressures are important.

EXAMPLE 2.7

Water of density 1000 kg/m³ flows through a horizontal pipe. Oil of density 800 kg/m³ flows through another horizontal pipe located at the same level. The left-hand limb of an inverted U-tube manometer is connected to a static pressure tapping in the side of the water pipe, and the right-hand limb is connected to a static pressure tapping in the side of the oil pipe. The fluid in the inverted U-tube is air, assumed to have a constant density of 1.2 kg/m³. The water in the left-hand limb of the manometer rises to a height of 500 mm, and the oil in the right-hand limb rises to a height of 300 mm (Figure 2.26). Determine the static pressure difference between the water and the oil.

Solution

The difference in height of the air Z either side of the U-tube is

$$Z = 0.5 - 0.3 = 0.2 \text{ m}$$

Utilising Equation (2.8), if the absolute static pressure of the water in its pipe is $\{p\}_w$ and the absolute static pressure of the oil in its pipe is $\{p\}_{oil}$, the pressure difference for equal pressures at level XX is given by

$$\{p\}_w - \{\rho\}_w \times 9.81 \times 0.5 = \{p\}_{oil} - \{\rho\}_{oil} \times 9.81 \times 0.3 - 1.2 \times 9.81 Z$$

$$\therefore \quad \{p\}_w - 1000 \times 9.81 \times 0.5 = \{p\}_{oil} - 800 \times 9.81 \times 0.3 - 1.2 \times 9.81 \times 0.2$$

$$\therefore \quad \{p\}_w - 4905 = \{p\}_{oil} - 2354.4 - 2.35$$

$$\therefore \quad \{p\}_w - \{p\}_{oil} = 2.55 \text{ kPa}$$

Note that the pressures decrease going upwards, hence the negative signs, in accordance with Equation (2.8).

Also note that the pressure term associated with the height of air in the manometer is fairly negligible compared to the other pressures. Therefore, in cases where high accuracy is not required, it does not matter that the air is compressible because any changes in its density will have an insignificant effect upon the pressure difference.

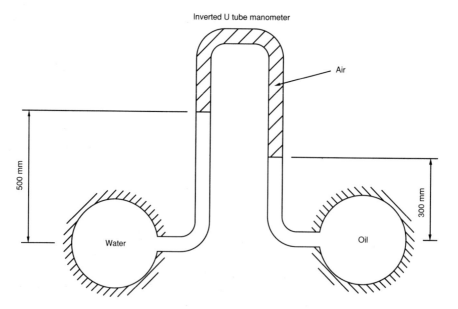

Fig. 2.26 *Example 2.7—an inverted U-tube manometer is used to measure the static pressure difference between water in one pipe and oil in another pipe.*

2.3.7 Pressure gauges and pressure transducers

Almost anything which changes its condition somehow under the effects of fluid pressure can be used for pressure measurement, provided it is calibrated with a manometer. For a practical device, the change must be consistent with time so, in reality, there are only a few instruments which are used commercially. The most common gauge is the Bourdon gauge which relies upon a circular arc of elliptical tubing changing shape to a circular cross-section under the effects of fluid pressure.

All manometers and gauges, such as the Bourdon, require the fluid pressure to be steady for a reasonable length of time. Pressure transducers are devices which can register very rapid changes in pressure. They come in a number of different forms and mostly provide an electrical output. There are transducers based upon the effects of capacitance, rates of change of strain, voltage effects in a piezoelectric crystal and magnetic properties. All have to be calibrated and the only calibration possible is against a manometer under steady conditions, even though they are most likely to be used under dynamic conditions.

2.4 THE PROPERTY TEMPERATURE

The accepted definition of temperature is embodied in the 'zeroth law of thermodynamics', so called because it was expounded after the first and second laws of thermodynamics were in being, but should have come before them. It is that: 'two systems are said to have equal temperatures when they suffer no physical change upon being brought into contact'. This indicates that temperature is an intensive fluid property, like pressure. It is easier, though, to consider the temperature of a fluid as being the 'degree of hotness'. In other words, if a fluid is hot to the touch, it can be said to have a higher temperature than that of a fluid which is cold to the touch. Physically, the value of temperature is related to the vibration energy of the molecules in that they vibrate more as the temperature is raised. The scale of temperature employed can cause some confusion. Temperature is given the symbol T and is measured in degrees Kelvin.

$$\text{Temperature: } T \text{ (units K)}$$

Another common scale is the degree Celsius which is in regular usage because it gives the value of $0\,°C$ to the temperature of a block of ice melting at 101.325 kPa, and a value of $100\,°C$ to the temperature of water boiling to become steam at the same pressure.

Temperature, though, can be measured in a number of ways by devices called thermometers. For example, it is known that the volume of a fluid will expand when heated, thereby raising its temperature. This is the principle of the mercury in glass thermometer shown in Figure 2.27.

Mercury is contained in a bulb at the base of the thermometer. When the bulb is heated, the mercury expands and rises up the stem, which can be calibrated to give a reading of temperature. There is a further bulb at the top of the thermometer in case too much heat is applied and the mercury goes off the top of the scale. Also stamped on the stem is the immersion distance for which the thermometer has been calibrated, in other words the distance which the stem and bulb must be immersed in the fluid whose temperature is required in order to obtain a correct reading. This allows for the expansion of the glass which contains the mercury.

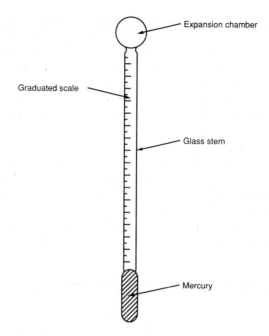

Fig. 2.27 *In a mercury in glass thermometer, the expansion of the mercury when heated by a fluid is used as a measure of the fluid temperature.*

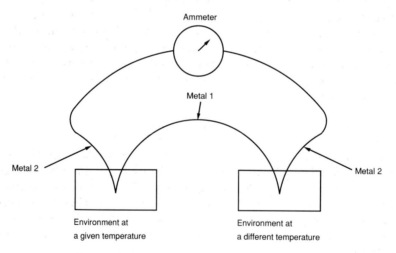

Fig. 2.28 *In a thermocouple, the current produced in a circuit of two metals whose junctions are at different temperatures is used as a measure of the temperature difference of the fluids surrounding the junctions.*

Another way of measuring temperature is through a knowledge of the Seebeck effect. In 1821, Seebeck discovered that if two different metals in a circuit were kept at different temperatures, a small current would flow. This is the basis of the thermocouple thermometer, as shown in Figure 2.28.

Table 2.1 *Common thermocouple pairs.*

Thermocouple pair	Approximate sensitivity (V/K)	Maximum temperature (K)
Chromel/alumel	0.04×10^{-3}	1 500
Iron/constantan	0.06×10^{-3}	1 300
Copper/constantan	0.05×10^{-3}	900
Platinum/platinum-rhodium	0.01×10^{-3}	2 000
Tungsten/molybdenum	0.03×10^{-3}	3 000
Iridium/tungsten	0.02×10^{-3}	2 300

If one of the metal junctions is maintained at a constant given temperature, and the other at a variable higher or lower temperature, a current, which is dependent upon the temperature difference of the two junctions, flows around the circuit. The current can be measured and calibrated to record the temperature difference. The current is actually very small and must be amplified in a practical thermometer. The constant temperature junction is usually built into the back of the measuring instrument, and the variable temperature junction is set into a probe which can be inserted easily into the fluid whose temperature is required to be known. There are many metal pairs that can be used to make a thermocouple depending upon the temperature range, the cost and the accuracy of measurement. Table 2.1 indicates a few of the common pairs, as indicated in manufacturer's catalogues.

Both the mercury in glass thermometer and the thermocouple can be calibrated against recognised fixed points of temperature, for example the freezing- and boiling-points of water under particular atmospheric conditions. Using the mercury in glass thermometer as an illustration, it can be placed in a block of melting ice in an atmosphere where the pressure is 101.325 kPa, and the point to which the mercury rises in the stem be marked off and named 0 °C. Then it can be placed in water boiling to become steam at the same pressure and the point to which the mercury rises in the stem this time named 100 °C. The distance between the two marks can be evenly divided into increments of temperature. The same procedure can be applied to the thermocouple, only now it is a reading of current that is indicative of the temperature. Unfortunately, it is not necessarily so that, at any intermediate value of temperature between the two fixed points, the freezing- and boiling-points of water, the two thermometers will give the same reading of temperature. This is because one is dependent upon the way that mercury expands with temperature, and the other is dependent upon the way that a current is generated by a temperature difference in two metals joined together. In other words, any thermometer that is calibrated will depend upon the medium which varies somehow with a heat input. What is needed is a thermometer which is independent of any physical medium but, unfortunately, no such instrument exists as yet.

Also, what is the significance of the scale? The degree Celsius scale has been referred to, but it is the degree Kelvin scale that is in the SI system of units. This specifies the freezing-point of water at 101.325 kPa as being 273.15 degrees Kelvin, and the boiling-point of water at 101.325 kPa as 373.15 degrees Kelvin. These values have been arrived at by using the most 'accurate' thermometer possible, a constant volume gas thermometer. It is possible with gases to determine temperature levels through a knowledge of the pressure and volume of the gas, two properties that can be found without compromise, and this is the basis of the constant volume gas thermometer. Unfortunately the equation which links the three properties of the gas, the equation of state, itself has limitations imposed upon it, as discussed in Chapter 6. However, an absolute temperature scale is proposed, independent of any medium, as a consequence of the second law of thermodynamics, in Section 12.1.4. and there is sufficient experimental evidence with gases to support it, which enables temperatures to be predicted with confidence.

In practice, because it is difficult to measure temperature exactly, the relationship between the Kelvin and Celsius scale of temperature, for the purposes of this book, is taken as

$$\text{degrees Kelvin} = \text{degrees Celsius} + 273$$

This provides sufficient accuracy for most purposes. Degrees Kelvin should be used in all calculations except for those involving temperature differences where it does not matter. Zero temperature in degrees Kelvin at the bottom of the scale, where the molecules are stationary, is envisaged and this situation has been approached by research workers.

The temperature measured by a thermometer placed into a fluid is the total or stagnation temperature, and it is this temperature that is used in all the calculations. In fact, in a moving fluid, there is also a static temperature. The relationship between the two is expressed by a particular form of the steady flow energy equation, which is introduced in Chapter 8, and it includes the kinetic energy of the fluid in appropriate units, in much the same way that Equation (2.3) defines the total pressure in terms of the static and dynamic pressures. To measure the static temperature of a fluid, the thermometer employed would have to travel at the same velocity as the fluid, which is difficult to organise. While it is the static temperature that should really be used in most of the formulae derived, it is fortunate that there is little difference between the static and total temperatures of a fluid, except at very high velocities. Such situations are not considered in this book and the total temperature is utilised at all times.

Other types of thermometer that are often used are the platinum resistance thermometer and the thermistor. In these, the changes in the electrical resistance with temperature are recorded. In the first case it is the resistance of the metal platinum, and in the second case the resistance of a semiconductor material.

2.5 THE PROPERTY VISCOSITY

The viscosity of a fluid is a property which indicates how 'sticky' it is, in other words, how resistant it is to being moved from place to place, or how resistant it is to deformation. As such, it is of most significance when the fluid is flowing. A useful concept which is used to describe a fluid in motion is the Streamline.

2.5.1 Streamline flow

The molecules which make up a fluid vibrate about a mean free point which moves when the fluid is in motion. When the fluid flows through a pipe or duct for example, it is possible both for the molecules and 'lumps' or particles of fluid, not just to adopt the direction of motion, but also to change their relative position in the fluid stream. In order to describe the flow of the fluid in some easily identifiable way, it is necessary to introduce the concept of a streamline, which may be defined as 'an imaginary line in the fluid across which no fluid is flowing at a given instant in time'. Such a concept implies that the point velocity

Fig. 2.29 *A streamline is an imaginary line in the fluid across which no fluid is flowing at a given instant in time.*

vector of every particle of fluid on the line is at a tangent to the line. It becomes possible to imagine a fluid stream to be composed of a number of streamlines and for the pattern that they produce to give a good indication of the direction of flow of the fluid. The physical boundaries through which the fluid is passing may also be considered as streamlines and they often, but not always, determine the path the fluid will take (Figure 2.29).

2.5.2 The dynamic and kinematic viscosity

As the viscosity of a fluid is a measure of how resistant it is to deformation, consider the two-dimensional case of a fluid contained between two parallel plates, as shown in Figure 2.30. The lower plate is stationary while the upper plate is moving with a velocity C_s.

The fluid is affected by the presence of the two plates. It can be shown by experiments with dyes that a layer of fluid next to any solid surface adopts the same velocity as the surface. Therefore, in Figure 2.30, the layer of fluid next to the stationary plate adopts a zero velocity and the layer of fluid next to the moving plate adopts a velocity of C_s. A layer of fluid may be thought of as a streamline, but it is very much a conceptual idea because it is not possible to say how thick the layer is or, indeed, how many layers or streamlines make up the total flow. Physically it is conceivable that various molecules of the fluid next to the plate will get trapped in the undulations of its surface. In Figure 2.30, because the

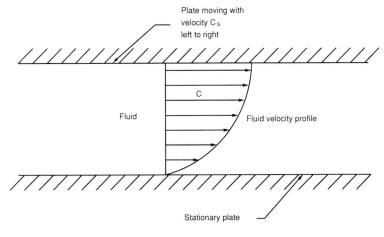

Fig. 2.30 *A velocity gradient is produced in a fluid contained between two parallel plates one of which is stationary, the other moving with velocity C_s.*

layer or streamline at the top is moving with velocity C_s, it exerts a force on the next layer down and so on through a cross section of fluid. At the same time, the layer of fluid at the bottom which is stationary, exerts a retarding force upon the layer of fluid just above it. Therefore, the velocity of the fluid will vary from C_s at the top to zero at the bottom. Intermediate layers of fluid will slip past each other with velocities related to their distance from the plate at rest and a velocity profile will result in the fluid. As each layer has a different velocity, shear stresses τ will develop between the layers. The shear stresses give rise to a force, called a Viscous Force, which is transmitted by the fluid. The existence of the force can be witnessed in the resistance to the motion of the upper plate. The fluid viscous force exerts a Friction Force on the plate which depends upon the fluid shear stress exerted at the surface.

Newton showed that the shear stress in the fluid is proportional to the fluid velocity gradient. In Figure 2.30, if the velocity of the fluid at any height z is C, the shear stress τ is given by

$$\tau \propto \frac{dC}{dz} \tag{2.12}$$

For many fluids, the equation can be written

$$\tau = \mu \frac{dC}{dz} \tag{2.13}$$

where μ is the Dynamic Viscosity. Some fluids, such as blood, do not obey this relationship, but the principal ones used in thermal systems, namely air, water, steam and refrigerant, certainly do and are called Newtonian fluids. The dynamic viscosity is a temperature-dependent property of the fluid. In circumstances where the temperature does not vary greatly, it is a reasonable assumption that the dynamic viscosity is constant. In practice, there are restrictions upon the use of Equation (2.13), as discussed in Section 2.5.6.

The Kinematic Viscosity ν is the dynamic viscosity divided by the density of the fluid. Thus:

Dynamic viscosity: μ (units kg/m s)

Kinematic viscosity: ν (units m^2/s)

where

$$\nu = \frac{\mu}{\rho} \tag{2.14}$$

Real fluids have viscosity and suffer from a retarding friction force whenever they flow from one place to another against a solid surface. The friction force is a similar idea to that which exists when two solid surfaces are rubbing together but, in the case of a fluid which can change its shape, this determines the magnitude of the force. Sometimes it is necessary to imagine an ideal fluid without viscosity and, therefore, no velocity gradient in the flow, no shear stresses, and no retarding friction force.

2.5.3 Molecular explanation of viscosity

The viscous force that arises in a flowing fluid must be due to the rate of change of momentum of the molecules across a streamline and to the force of attraction between the molecules.

In liquids, the molecules are closely packed together, the intermolecular forces are dominant and try to resist the shear stresses established when the liquid is made to move. As the temperature is raised, the intermolecular forces are weakened because the mean free path of the molecules is increased. It becomes easier for the liquid to flow and its viscosity is decreased. For example, the dynamic viscosity of saturated water at 100 °C is 27.8×10^{-5} kg/m s, and 13.3×10^{-5} kg/m s at 200 °C.

In gases, the intermolecular forces are not so dominant as in liquids because the distance between the molecules is so much greater. Instead, it is the momentum exchange that determines how easily the gas will flow. As the temperature of the gas is increased, the molecules become more excited and collide more frequently, increasing the momentum transfer. There is a greater resistance to flow generated in the gas and the viscosity increases. For example, the dynamic viscosity of air at 100 °C is 2.2×10^{-5} kg/m s, whereas at 200 °C it is 2.6×10^{-5} kg/m s.

2.5.4 Laminar and turbulent flow

Reynolds carried out a series of experiments which established the existence of two distinct types of flow. The experiments involved injecting dye into the flow of various fluids under different conditions of pressure and temperature and observing what happened to it. He discovered that the dye either travelled in the fluid in orderly straight lines or it became totally mixed. He called the first type of flow, shown in Figure 2.31, Laminar and the second type of flow, shown in Figure 2.32, Turbulent.

In laminar flow, the particles of fluid move in an orderly manner in straight lines, or layers, even though the velocity with which the particles move along one layer is not necessarily the same as that along another. Only molecules pass from one layer of fluid to another, not particles of fluid. The layers may be considered as streamlines and laminar flow is often called Streamlined Flow.

In turbulent flow, the direction taken by particles of fluid is disorderly. They are constantly crossing each other in following an erratic three-dimensional path, resulting in the thorough

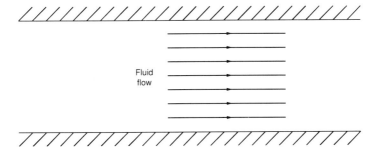

Fig. 2.31 *In the laminar flow of a fluid in a duct, the particles of fluid move in an orderly manner.*

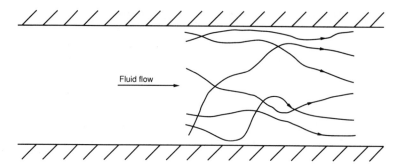

Fig. 2.32 *In the turbulent flow of a fluid in a duct, the particles of fluid move in a random manner.*

mixing of the fluid. Only the average motion of the fluid is in the direction dictated by the system boundary. Sometimes the particles establish themselves in circular motions superimposed upon the general direction of flow and these are called Vortices or Eddies. They usually dissipate themselves at some stage during which their kinetic energy of rotation is converted into another form of energy in the fluid, called internal energy, which is considered in more detail in Chapter 5. This introduces an inefficiency into the fluid flow because it is much more difficult to transfer the energy in the opposite direction.

In his experiments, Reynolds was able to determine whether a fluid flow would be laminar or turbulent by a consideration of the forces that are acting. He derived a dimensionless number, called the Reynolds Number, which could be used to determine the type of flow.

2.5.5 The Reynolds number

Reynolds reasoned that thorough mixing of a fluid and, hence, turbulent flow, results when the fluid inertia forces are dominant. The inertia forces would tend to maintain the velocity and momentum of the fluid. On the other hand, laminar flow is likely when the viscous forces are dominant and can impose a sense of order upon the particles of fluid. This is because the effect of the viscous forces is to try to slow down the motion of the fluid. The Reynolds number, therefore, is proportional to the ratio of the inertia to viscous forces.

The magnitude of the inertia force is expressed in Newton's second law of motion, namely that the force is equal to a mass of fluid multiplied by its acceleration and, hence, a volume of fluid multiplied by its density and its acceleration. However, in flowing fluids, it is more realistic to talk of the volume flow rate of fluid and the rate of change of momentum of the fluid. The inertia force $F_{inertia}$ becomes equal to the volume flow rate multiplied by its density and a typical velocity, as follows:

$$F_{inertia} = \text{(volume flow rate)} \times \text{(density)} \times \text{(velocity)}$$

But the volume flow rate can be further broken down into the cross sectional area of fluid flow multiplied by its velocity:

$$F_{inertia} = \text{(density)} \times \text{(cross sectional area)} \times \text{(velocity}^2\text{)}$$

THE PROPERTY VISCOSITY

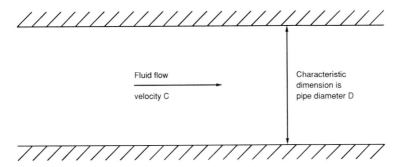

Fig. 2.33 *The characteristic dimension in the Reynolds number for the flow of a fluid in a pipe is the pipe diameter.*

The viscous force $F_{viscous}$ is given by Equation (2.14) as

$$F_{viscous} = (\text{dynamic viscosity}) \times (\text{surface area}) \times (\text{velocity gradient})$$

The ratio of the inertia force to the viscous force is the Reynolds number Re:

$$Re = \frac{\text{Inertia force}}{\text{Viscous force}} = \frac{F_{inertia}}{F_{viscous}}$$

Because it is a ratio of two forces, Re is dimensionless. Further, the ratio of the velocity to the velocity gradient has the dimensions of length and may be thought of as representing a characteristic dimension of the system. While the surface area and cross-sectional area are not the same, their ratio is dimensionless and not a function of any particular system. The Reynolds number, therefore, becomes

$$Re = \frac{\text{Fluid density} \times \text{fluid velocity} \times \text{characteristic dimension}}{\text{Fluid dynamic viscosity}}$$

When a fluid is flowing through a pipe or duct, the characteristic dimension is the diameter of the pipe or the equivalent diameter of the duct D (Figure 2.33). The Reynolds number for pipe or duct flow is

$$Re = \frac{\rho C D}{\mu} \tag{2.15}$$

When a fluid is flowing externally over a body, the characteristic dimension depends upon the application. It might be the length of surface L that the fluid maintains contact with (Figure 2.34), in which case the Reynolds number would be

$$Re = \frac{\rho C L}{\mu} \tag{2.16}$$

2.5.5.1 The critical Reynolds number

Given the mathematical expressions for the Reynolds number in Equations (2.15) and (2.16), clearly in flows where the friction forces dominate (low density, velocity or characteristic

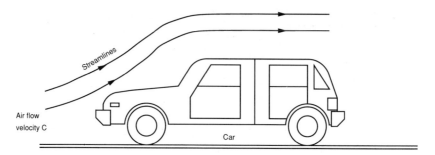

Fig. 2.34 *The characteristic dimension in the Reynolds number for the flow of a fluid over a surface is the length of the surface that the fluid is in contact with.*

length, and high viscosity), the Reynolds number will be low and the flow will be laminar in nature. In flows where the inertia forces dominate (high density, velocity or characteristic length, and low viscosity), the Reynolds number will be high and the flow will be turbulent.

The Critical Reynolds Number is the value when the flow changes from laminar to turbulent in a given physical situation. It is usually expressed as a range because there is no fixed point for the change, rather a transition from one type of flow to the other.

For a pipe or duct, the critical Reynolds number is between $Re > 2000$ and $Re < 3500$. For the external flow over a body, the critical Reynolds number, as defined in Equation (2.16), is between $Re > 10^5$ and $Re < 2 \times 10^6$. When the Reynolds number is within these ranges, the fluid is said to be in the Transition Region. Things like surface roughness and pipe bends accelerate the transition to turbulent flow.

EXAMPLE 2.8

A fluid is flowing through a horizontal pipe of diameter 0.1 m with a velocity of 1 m/s. Determine the Reynolds number of the flow if the fluid is oil of density 800 kg/m³ and dynamic viscosity 0.1 kg/m s.

If the fluid is air of density 1.2 kg/m³ and dynamic viscosity 1.846×10^{-5} kg/m s, what is the Reynolds number of the flow?

If the fluid is water of density 1000 kg/m³ and dynamic viscosity 54.4×10^{-5} kg/m s, what is the Reynolds number of the flow?

What would be the diameter of pipe required to produce a Reynolds number of 1500 and so induce laminar flow when the fluid is water?

Solution

For oil in the pipe, the Reynolds number Re using Equation (2.15) is

$$Re = \frac{\{\rho\}_{oil}\{C\}_{oil}D}{\{\mu\}_{oil}} = \frac{800 \times 1 \times 0.1}{0.1} = 800$$

As it is less than 2000, the flow is laminar. In fact, oil flows in pipes are often laminar.

For air in the pipe, the Reynolds number Re using Equation (2.15) is

$$Re = \frac{\{\rho\}_{air}\{C\}_{air}D}{\{\mu\}_{air}} = \frac{1.2 \times 1 \times 0.1}{1.846 \times 10^{-5}} = 6\,500.5$$

As it is greater than 3500, the flow is turbulent. In fact, most airflows in pipes are turbulent.

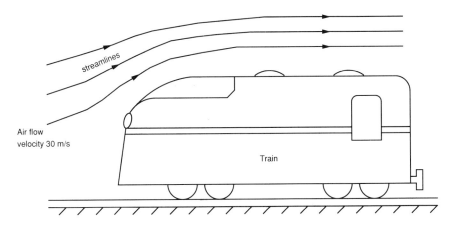

Fig. 2.35 *Example 2.9 — the value of the Reynolds number of the airflow over a train determines whether the flow is laminar or turbulent.*

For water in the pipe, the Reynolds number Re using Equation (2.15) is

$$Re = \frac{\{\rho\}_w \{C\}_w D}{\{\mu\}_w} = \frac{1000 \times 1 \times 0.1}{54.4 \times 10^{-5}} = 183\,823.5$$

As it is greater than 3500, the flow is turbulent. Again, most water flows in pipes are turbulent. To give a Reynolds number of 1500 in the water flow, the diameter of the pipe D would have to be, using Equation (2.15):

$$D = \frac{Re\{\mu\}_w}{\{\rho C\}_w} = \frac{1500 \times 54.4 \times 10^{-5}}{1000 \times 0.1} = 8.16 \text{ mm}$$

This is a fairly small diameter for a typical water pipe, and the water velocity is quite low.

EXAMPLE 2.9

A train is moving through still air with a velocity of 30 m/s (Figure 2.35). The air has a kinematic viscosity of 1.5×10^{-5} m²/s. At what distance from the front of the train will the relative Reynolds number of the air have reached a value of 2×10^6, and so make the airflow turbulent?

Solution

If the distance from the front of the train at which the relative Reynolds number of the air Re reaches 2×10^6 is L, this is given by Equation (2.16) as

$$L = \frac{Re\{\nu\}_{air}}{\{C\}_{air}} = \frac{2 \times 10^6 \times 1.5 \times 10^{-5}}{30} = 1 \text{ m}$$

The air soon becomes turbulent and, as the train is so long, may as well be considered turbulent from the front. In fact, the flow of a fluid over a surface requires special treatment and is dealt with in some detail in Chapter 9.

2.5.6 Shear stresses in laminar and turbulent flow

In laminar Newtonian flow, the simple relationship between the shear stresses in the fluid and the dynamic viscosity and velocity gradient, embodied in Equation (2.13), holds true.

SOME THERMOFLUID PROPERTIES

In other words, the laminar shear stresses τ can be written:

$$\tau = \mu \frac{dC}{dz} \qquad (2.17)$$

However, in turbulent Newtonian flow, with particles of fluid moving erratically from layer to layer, the development of the resultant shear stresses τ is considerably more complex and Equation (2.13) is no longer valid. Instead, it becomes necessary to write

$$\tau = \epsilon \frac{dC}{dz} \qquad (2.18)$$

Here ϵ is called the eddy viscosity. It is a function of many factors and cannot be regarded as a simple temperature-dependent constant. It also has a value considerably greater than the dynamic viscosity and this is very important because it means that the resultant friction forces will be greater in turbulent flow than in laminar flow. This can have an effect upon the design of some thermal systems in which fluid is being transported from one place to another.

Because ϵ in Equation (2.18) is so complex, the analysis of turbulent flows is difficult and recourse is made to a number of other relationships, some of them empirical. One such defines the Friction Factor as 'the ratio of the shear stress to the dynamic pressure of the fluid'. It is given the symbol f and is dimensionless as both the shear stress and the dynamic pressure have the units of Pa. Hence:

Friction factor: f (dimensionless)

where

$$f = \frac{\tau}{0.5\rho C^2} \qquad (2.19)$$

The friction factor is very useful parameter for understanding turbulent flows in particular. For example, the higher the value of f for a pipe, the rougher the surface of the pipe and the greater the friction force opposing the motion of the fluid. As it happens, f is also dependent upon the Reynolds number of the flow and the empirical relationship between the two for the flow under consideration needs to be known for an analysis to proceed. The friction factor is developed further in Chapters 9 and 11.

2.5.7 The measurement of viscosity

There are many ways to measure the dynamic viscosity of a fluid. A device called a Viscometer, described below, is just one of them, but it is reasonably easy to understand. It consists of a drum sited in an outer cylinder and separated from the cylinder by a thin film of the fluid whose viscosity is required to be known, as shown in Figure 2.36.

Either the inner drum can be rotated and a torque measured on the outer cylinder, or the outer cylinder can rotate and the torque be measured on the inner drum. The torque is produced because the fluid adheres to each surface, one is stationary and the other rotating. A velocity gradient is established in the fluid, which generates a friction force on the stationary surface which is measured as a torque.

THE PROPERTY VISCOSITY

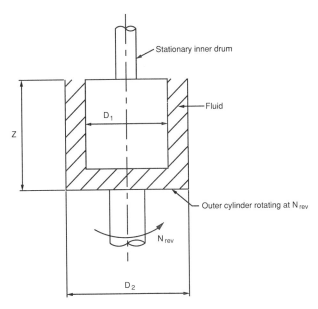

Fig. 2.36 *A viscometer measures the viscosity of a fluid from a knowledge of the torque generated by the viscous forces in a fluid separating a stationary drum from a rotating cylinder.*

Suppose that the outer cylinder of diameter D_2 is rotated at N_{rev}, as in Figure 2.36, and that the torque TQ is measured on the stationary inner drum of diameter D_1. The torque is the product of the force on the inner drum and the radius of the inner drum, where the force is given by the shear stress on the inner drum multiplied by the surface area of the inner drum. If the flow is assumed to be laminar, Equation (2.13) can be used to substitute for the shear stress on the inner drum (taking the velocity gradient in the radial r direction). The torque TQ for a drum height Z is given as

$$TQ = 0.5 \mu \pi D_1 \, ZD_1 \frac{dC}{dr}$$

If the velocity profile may be assumed to be linear in the first instance, then:

$$TQ = \frac{\mu \pi D_1 ZD_1 \, \pi D_2 N_{\text{rev}}}{D_2 - D_1} \qquad (2.20)$$

If the torque on the inner cylinder is measured and its value substituted into Equation (2.20), all the other values in the equation are known except for the dynamic viscosity which may be determined.

The above analysis does not consider the effects of the friction force on the bottom surface of the drum, often called the end effects. To account for these, consider an elemental ring at radius r and of width dr in the fluid separating the bottom surface of the drum from the cylinder. Let the vertical separation distance be Z', as shown in Figure 2.37.

If the additional torque exerted due to the end effects is TQ', this is given by the integral of the frictional force acting on the elemental ring multiplied by the radius of the ring. The force is acting on the area of the ring and the speed of rotation is still N_{rev} acting at radius r.

SOME THERMOFLUID PROPERTIES

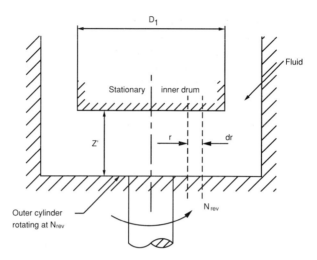

Fig. 2.37 Additional torque results from the viscous force in the fluid separating the bottom of the stationary drum from the rotating outer cylinder in a viscometer.

Thus, utilising Equation (2.14):

$$TQ' = \int \frac{2\pi r \, dr \mu \, 2\pi N_{rev} r r}{Z'}$$

Integrating over diameter D_1 gives

$$TQ' = \frac{N_{rev} \mu \pi^2 D_1^4}{16 Z'} \qquad (2.21)$$

The above analysis can be made for a velocity profile which is not linear but which varies with the radius, as shown in Example 2.11.

In Figure 2.36, the outer cylinder is rotating. In doing so, it will suffer a retarding friction force due to the atmosphere in which the measurement is being made. If the dynamic viscosity of the fluid in the atmosphere, presumably air, is considerably less than the dynamic viscosity of the fluid whose viscosity is being measured, the extra friction force is irrelevant. Alternatively, the problem could be overcome by making the inner drum rotate inside a stationary cylinder, and this is usually done. Equation (2.20) must be modified to allow for the fact that it is now the inner drum rotating at N_{rev} and that the torque is applied on the outer cylinder. Equation (2.20) becomes

$$TQ = \frac{\mu \pi D_2 Z D_2 \, \pi D_1 N_{rev}}{D_2 - D_1} \qquad (2.22)$$

Equation (2.21) for the end effect torque remains the same.

EXAMPLE 2.10

In a simple viscometer, an outer cylinder of diameter 50 mm is made to rotate at 100 rev/s. An inner drum of diameter 40 mm and height 80 mm is held stationary and the torque required to do this is measured as being 0.24 N m when the annular space between the cylinder and drum is filled with a liquid whose viscosity is required to be known. If the friction forces acting on the bottom surface of the drum can be ignored and the velocity profile generated in the liquid is linear, what is the dynamic viscosity of the liquid assuming laminar conditions?

If the friction forces on the lower surface of the drum were to be accounted for, what would be the additional torque required to hold the inner drum stationary, given that the drum is maintained at a vertical height of 10 mm above the cylinder and that the dynamic viscosity of the liquid remains the same?

Solution

Ignoring end effects and assuming laminar conditions and a linear velocity profile, the torque on the inner cylinder TQ is given by Equation (2.20) as

$$TQ = \frac{\mu \pi D_1 Z D_1 \, \pi D_2 N_{\text{rev}}}{D_2 - D_1}$$

The following data are known: $N_{\text{rev}} = 100$ rev/s, $D_2 = 0.05$ m, $D_1 = 0.04$ m, $(D_2 - D_1) = 0.01$ m, $Z = 0.08$ m, $TQ = 0.24$ N m.

Substituting the known values into Equation (2.20) allows the dynamic viscosity of the liquid to be determined, as follows:

$$0.24 = \frac{\mu \pi \times 0.04 \times 0.08 \times 0.04\pi \times 0.05 \times 100}{0.01}$$

$$\therefore \quad \mu = 0.38 \text{ kg/m s}$$

Accounting for the friction force on the bottom of the drum, given that Z' is 0.01 m, Equation (2.21) for the extra torque TQ' is

$$TQ' = \frac{N_{\text{rev}} \mu \pi^2 D_1^4}{16 Z'} = \frac{100 \times 0.38 \pi^2 \times 0.04^4}{16 \times 0.01}$$

$$\therefore \quad TQ' = 0.006 \text{ N m}$$

The total torque on the inner drum is the sum of TQ and TQ', which is

$$TQ + TQ' = 0.24 + 0.006 = 0.246 \text{ N m}$$

In practice, a viscometer will record the torque accounting for the friction forces on the bottom surface. The example is used here to evaluate their contribution of the end effects which, in this case, amounts to approximately 2.5%.

EXAMPLE 2.11

A rotary viscometer consists of a fixed cylinder within which a cylindrical drum rotates. The inside diameter of the outer cylinder is 75 mm and the annular space between the cylinder and drum, which is filled with a particular liquid, is 2.5 mm. When the inner drum is rotated at 24 rev/min, the torque required to hold the outer cylinder stationary is measured as 0.3 N m per metre length of cylinder. Determine the dynamic viscosity of the liquid assuming laminar conditions and that the velocity profile in the liquid is linear. Ignore end effects.

What is the percentage change in the dynamic viscosity of the liquid if the velocity profile is integrated over the distance separating the inner cylinder from the drum?

Solution

In this example, the outer cylinder is stationary and the inner drum rotates. Given that the velocity profile may be assumed linear, that laminar conditions prevail and that end effects may be ignored, Equation (2.22) is valid. If the

SOME THERMOFLUID PROPERTIES

liquid has a dynamic viscosity μ, the torque on the outer cylinder TQ is given by

$$TQ = \frac{\mu \pi D_2 Z D_2 \, \pi D_1 N_{\text{rev}}}{D_2 - D_1}$$

The following data are known: $D_2 = 0.075$ m, $D_1 = 0.07$ m, $(D_2 - D_1) = 0.005$ m, $N_{\text{rev}} = 24/60 = 0.4$ rev/s, $TQ/Z = 0.3$ N m/m.

Substituting in the above equation for TQ allows the dynamic viscosity of the liquid μ to be determined as follows:

$$0.3 = \frac{\mu \pi \times 0.075^2 \pi \times 0.07 \times 0.4}{0.005}$$

$$\therefore \quad \mu = 0.965 \text{ kg/m s}$$

When the velocity profile may not be assumed linear, it must be integrated over the distance separating the inner drum from the outer cylinder. Consider an elemental ring of radius r and height Z in the liquid, as in Figure 2.38.

The torque on the element TQ is the product of the friction force and the radius of the inner drum, where the force is given by the shear stress multiplied by the surface area of the element. If the flow is assumed to be laminar, Equation (2.13) can be used to substitute for the shear stress (taking the velocity gradient in the radial r direction). If the liquid now has a dynamic viscosity μ', the torque TQ for an element height Z is given as

$$TQ = -\mu' 2\pi r Z r \frac{dC}{dr}$$

The negative sign is present because the velocity increases with decreasing radius. Integrating between the inner and outer cylinder diameters, D_1 and D_2 respectively, gives

$$\frac{D_2 - D_1}{D_2 D_1} = \frac{\mu' 2\pi Z (C_2 - C_1)}{2TQ}$$

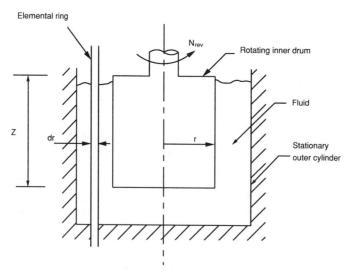

Fig. 2.38 An elemental ring of radius r and height Z in the liquid in a viscometer.

The following data are known: $TQ/Z = 0.3$ N m/m, $D_2 = 0.075$ m, $D_1 = 0.07$ m, $N_{rev} = 0.4$ rev/s, $C_1 = \pi D_1 N_{rev} = \pi \times 0.07 \times 0.4 = 0.088$ m/s, $C_2 = 0$. Hence, the dynamic viscosity of the liquid μ' in this case is

$$\frac{0.07 - 0.075}{0.075 \times 0.07} = \frac{\mu' \pi (0 - 0.088)}{0.3}$$

$$\therefore \mu' = 1.03 \text{ kg/m s}$$

The percentage difference in the value of the dynamic viscosity of the liquid obtained by the two methods $(\%\mu)$ is

$$(\%\mu) = \frac{\mu' - \mu}{\mu} = \frac{1.03 - 0.965}{0.965} = 6.7\%$$

In certain applications this may not be a significant difference and the assumption of a linear profile may be appropriate.

EXAMPLE 2.12

An apparatus used to determine the viscosity of liquids consists of an inner drum of diameter 100 mm held stationary inside an outer cylinder of diameter 101 mm, which rotates at 10 rev/s. The height of liquid in the device can be varied. This allows end effects to be accounted for because they may be considered as an additional height of liquid. For a particular oil, the measured torque on the drum is 0.35 N m when the height of the oil is 60 mm, and the torque is 0.65 N m when the height is 120 mm. If laminar conditions may be assumed to exist and the velocity profile in the oil is linear, determine the dynamic viscosity of the oil.

Solution

Considering Equation (2.20), which is applicable when laminar conditions prevail and the velocity profile is laminar, the different torque measurements are achieved by varying the height of the oil whose viscosity is being measured. All the terms in the equation except TQ and Z are constant. It can be concluded that the torque is proportional to the height of oil. Hence: $TQ \propto Z$. But end effects can be accounted for by assuming an additional height of oil. In this example, the end effects remain the same whatever the height of the oil when the torque is measured. If the additional height of oil to account for end effects is Z'', then $TQ \propto Z + Z''$. Substituting in the values gives

$$\frac{0.35}{0.65} = \frac{0.06 + Z''}{0.12 + Z''}$$

$$\therefore Z'' = 0.01 \text{ m}$$

Equation (2.20) is

$$TQ = \frac{\mu \pi D_1 Z D_1 \pi D_2 N_{rev}}{D_2 - D_1}$$

For the first height of oil, the following data are known: $TQ = 0.35$ N m, $D_2 = 0.101$ m, $D_1 = 0.1$ m, $(D_2 - D_1) = 0.001$ m, $Z = 0.06 + Z'' = 0.06 + 0.01 = 0.07$ m, $N_{rev} = 10$ rev/s.
Substituting into Equation (2.20) allows the viscosity of the oil $\{\mu\}_{oil}$ to be determined as follows:

$$0.35 = \frac{\{\mu\}_{oil} \pi \times 0.1^2 \times 0.07\pi \times 0.101 \times 10}{0.001}$$

$$\therefore \{\mu\}_{oil} = 0.05 \text{ kg/m s}$$

This is a convenient way of accounting for the end effects and easy to carry out in practice on a typical viscometer.

3
Work and Heat Transfer in a Thermal System

With sufficient properties now defined to be able to determine the condition of a fluid at any state, it is instructive at this stage to introduce the idea of a simple generalised thermal system, which could represent a car engine, a thermal power plant or any of the systems considered in Chapter 1. Within a given boundary, it must contain a fluid which undergoes a process. With reference to Figure 3.1, the fluid at state 1 can be described by reference to the value of three of its main properties there, namely the pressure p_1, the temperature T_1 and the volume V_1. Similarly, the fluid at state 2 by p_2, T_2 and V_2.

The analysis of any work and heat transfer as a consequence of something happening to the fluid in the system is what is of interest. In Figure 3.1, the work and heat transfer are shown to occur only when the fluid is changing its state, undergoing a process, and they may be either into or out of the system. This reflects what happens in practice. For example, in a piston cylinder reciprocating engine, when the fuel is burned and there is a heat transfer into the system, the temperature of the fluid increases. When the piston is forced downwards producing a work output at the crankshaft, the fluid is expanding and increasing its volume.

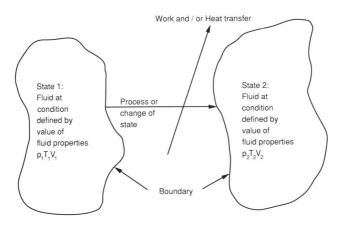

Fig. 3.1 *A generalised thermal system in which a fluid undergoes a process and changes its properties from state 1 to state 2, giving rise to work and heat transfer.*

It could be concluded that it is because there is a work or heat transfer into or out of the system that the fluid changes some of its property values, those relevant to the particular energy transfer taking place. Alternatively, it could also be concluded that it is because some of the fluid property values at state 1 are different to those at state 2 that energy in the form of work or heat is realised. In other words, the amount of work or heat transfer that becomes apparent during a change of state is dependent upon the way that the properties change their value and vice versa. The question becomes, which changes in property value affect the work transfer and which affect the heat transfer?

It now becomes important to consider the type of process followed by the fluid, because two are possible and it is necessary to distinguish between them. The fluid undergoes a Non-flow Process when it remains entirely within the boundary during a change of state. Such is the case in a reciprocating piston cylinder engine or compressor when the valves are closed and the piston moves up and down, compressing or expanding the fluid. The fluid undergoes a Flow Process when it flows into and out of the boundary during a change of state, as in the combustion chamber of a gas turbine plant. The type of process refers to any change of state that the fluid is subjected to, whereas the terms open and closed system, introduced in Chapter 1, are applicable to the system as a whole. It is possible for a fluid to undergo flow processes and yet be in a closed system, depending upon where the boundary is drawn. For example, in a thermal power station, the fluid flows into and out of each component, undergoing flow processes as a result. But for the power station as a whole, the fluid remains entirely within the boundary of the system, namely the walls of each component and the interconnecting pipework. It is continually recycled within a closed system.

In order to examine the work and heat transfer more closely, consider in the first instance the fluid being in a non-flow process. The most common example is when it is trapped inside a reciprocating piston cylinder mechanism and the valves are closed, as in Figure 3.2(a), (b) and (c). Whenever the piston is moving or heat is being added due to the combustion of a fuel, the fluid is experiencing a non-flow process. Is it possible to derive some appropriate formulae in terms of the fluid properties which will allow any associated work and heat transfer to be determined in a theoretical manner?

There are some other examples of non-flow processes, such as the heating and cooling of fluids in vessels of fixed capacity, but it is the intention here to concentrate upon the reciprocating piston cylinder mechanism alone as the most relevant application. In Figure 3.2(a), (b) and (c), some of the terminology associated with this technology is identified. The fluid is contained in a boundary which is represented by the space in the cylinder above the piston. The piston is free to move up and down inside the cylinder, but a piston ring prevents the fluid from escaping between the piston and the walls of the cylinder to the surroundings. The piston is attached via a connecting-rod to a crankshaft which transfers the reciprocating motion of the piston into a rotating motion of the crankshaft, or vice versa. The fluid can enter and leave the cylinder via the two valves. The boundary is not fixed with time. When the piston is at the top of its travel, called top dead centre (TDC), the fluid occupies what is called the Clearance Volume. When the piston has moved to the bottom of its travel, called bottom dead centre (BDC), the fluid occupies the whole of the Cylinder Volume. The boundary has changed from the clearance volume to the cylinder volume during the change of state. The volume through which the piston moves is called the Swept Volume, the diameter of the cylinder the Bore, and the distance between TDC and BDC the Stroke.

WORK AND HEAT TRANSFER IN A THERMAL SYSTEM

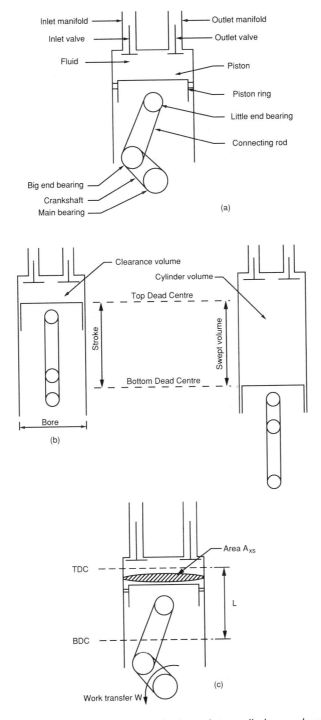

Fig. 3.2 (a), (b), (c) *Work transfer in a piston cylinder mechanism.*

Energy in the fluid, which may arise from a heat input, is converted into a physical movement of the crankshaft, thereby producing a work output. Alternatively, a work input to the crankshaft can be converted into energy in the fluid. When a fluid causes a net amount of work to be done at the crankshaft, the piston cylinder mechanism is behaving as an engine. But when the net work is done by the crankshaft on the fluid, the piston cylinder mechanism is behaving as a compressor or pump. If the fluid, after a number of movements of the piston, is returned to its original condition, it has completed a cycle. Piston cylinder thermal systems such as the Stirling engine have been devised in which the fluid remains entirely within the cylinder and completes a cycle. The fluid can be said to undergo a number of non-flow processes and to be in a closed system. In most practical reciprocating piston cylinder mechanisms, the valves are opened at certain stages and the fluid is induced into the cylinder or exhausted from it.

3.1 WORK TRANSFER IN A NON-FLOW PROCESS

Work transfer was defined in Chapter 1 as an energy that arises as a force is applied over a distance. It is given the symbol W. Thus:

Work transfer: W (units J)

In Figure 3.2(b), let the fluid in the piston cylinder mechanism undergo a change of state from state 1 at TDC, to state 2 at BDC, and assume that it is a gas or vapour. This could be achieved if the pressure of the fluid at state 1 is somewhat higher than the atmospheric pressure. Work will be realised at the crankshaft because the piston will be forced downwards, which turns the shaft. The work transfer at the crankshaft is equal to the force times the distance moved, as in Chapter 1. But that work is produced by changes in the property values of the fluid and, in thermofluids, it is necessary to define the work transfer by what happens to the fluid itself. In the case of the fluid in the piston cylinder mechanism, the force is provided by the pressure difference Δp between the fluid pressure at state 1 and the fluid pressure after expansion at state 2, multiplied by the perpendicular area upon which the pressure acts, namely the surface area of the piston A_s. If the piston moves through a distance L between TDC and BDC, then

$$W = \Delta p A_s L$$

But $A_s L$ is the swept volume of the cylinder V. Therefore, mathematically, the work transfer between states 1 and 2 is written

$$W_{12} = -\int_1^2 p \, dV$$

The negative sign is required for the sign convention adopted for work and heat transfer, because some work transfer may be into a system and some may be out of a system, and it is necessary to distinguish between them. The sign convention is explained in detail in Section 3.3.4. Because volume is a mass dependent or extensive property, there is also a specific work transfer w. Thus:

Specific work transfer: w (units J/kg)

where $W = mw$.

WORK TRANSFER IN A NON-FLOW PROCESS

The equation for the work transfer in a non-flow process becomes

$$W_{12} = mw_{12} = -\int_1^2 p\, dV = -\int_1^2 mp\, dv \qquad (3.1)$$

This is the conventional way of writing the formula for the work transfer in a non-flow process. Physically, it means that both the pressure and volume of the fluid can change between state 1 and state 2, and the work transfer is the sum of the infinite number of changes in pV during the process.

If a graph is drawn of the pressure of the fluid against its volume as it goes from state 1 to state 2, the area under the graph is the work transfer during the change of state. When the fluid undergoes a cycle, the net work transfer is the area of the graph (Figure 3.3). This is the same as the sum of the work transfer in each non-flow process that makes up the cycle.

Alternatively, the equation may be integrated in three ways. Firstly, if the non-flow process is a constant volume one and $dV = 0$, the work transfer is also zero as follows:

$$W_{12} = -\int_1^2 p\, dV = -\int_1^2 p \times 0 = 0$$

Secondly, if the non-flow process is a constant pressure one, in which case:

$$W_{12} = -\int_1^2 p\, dV = -p(V_2 - V_1)$$

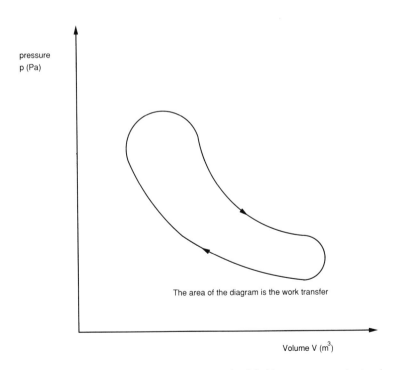

The area of the diagram is the work transfer

Fig. 3.3 *Work transfer is the area of a graph of fluid pressure against volume.*

And thirdly, if a relationship between the pressure and volume is known, for example by experiment. In fact, just such a relationship has been found for many gases, as mentioned in Chapter 2 and dealt with in Chapter 6, which is

$$pV^n = \text{const.}$$

where n usually has a fixed value. In this case the integration is

$$W_{12} = -\int_1^2 p\, dV = -\text{const.} \int_1^2 \frac{dV}{V^n} = -\frac{\text{const.}\left(V_2^{1-n} - V_1^{1-n}\right)}{1-n}$$

EXAMPLE 3.1

When the piston in a reciprocating piston cylinder mechanism is at TDC, the fluid contained in the clearance volume has a pressure of 1 MPa and a volume of 0.05 m³. As the piston moves to BDC, the pressure and volume changes are continually monitored and it is found that the fluid expands according to a law $pV^{1.4} = \text{const.}$ When the piston is at BDC, the fluid volume is 0.1 m³ (Figure 3.4). What is the work transfer to the crankshaft during this non-flow process?

Solution

Let the fluid go from state 1 to state 2. From Equation (3.1), the work transfer W_{12} in a non-flow process is given by

$$W_{12} = -\int_1^2 p\, dV$$

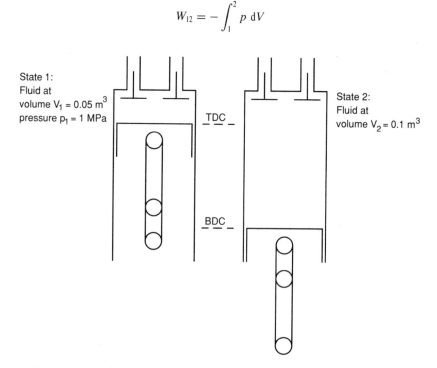

Fig. 3.4 *Example 3.1—determination of the work transfer in a piston cylinder mechanism as the fluid undergoes a non-flow process.*

WORK TRANSFER IN A NON-FLOW PROCESS

As both the pressure and volume of the fluid are varying, it is not possible to integrate the equation. But, in this example, the relationship between the pressure and volume has been determined. Substituting in Equation (3.1) for p gives

$$W_{12} = -\text{const.} \int_1^2 \frac{dV}{V^{1.4}}$$

$$\therefore \quad W_{12} = -\frac{\text{const.} \cdot \left(V_2^{1-1.4} - V_1^{1-1.4}\right)}{1 - 1.4}$$

But the constant can be determined from

$$\text{const.} = p_1 V_1^{1.4} = p_2 V_2^{1.4}$$

At state 1 the constant has a value of

$$\text{const.} = 10 \times 10^5 \times 0.05^{1.4} = 150\,85.4$$

Therefore, the work transfer W_{12} is

$$W_{12} = -\frac{150\,85.4 \times \left(0.1^{-0.4} - 0.05^{-0.4}\right)}{1 - 1.4} = -30.3 \text{ kJ}$$

The answer being negative does mean something, but an explanation of its significance must await the introduction of the sign convention adopted in Section 3.3.4. Application of Equation (3.1) has allowed an estimate of the work transfer to be made during one stroke of the piston in a piston cylinder mechanism.

EXAMPLE 3.2
It is proposed to build an engine, based upon a reciprocating piston cylinder mechanism, in which the fluid will undergo the following four non-flow processes and complete a cycle:

1. State 1 to state 2—constant pressure expansion at a pressure of 1 MPa from a volume of 0.02 m³ to a volume of 0.09 m³.
2. State 2 to state 3—constant volume drop in pressure to a pressure of 0.1 MPa.
3. State 3 to state 4—constant pressure compression at a pressure of 0.1 MPa from a volume of 0.09 m³ to a volume of 0.02 m³.
4. State 4 to state 1—constant volume rise in pressure to the initial state where the pressure is 1 MPa.

Sketch the cycle on a graph of fluid pressure against volume, determine the work transfer in each non-flow process, and the net work transfer when the fluid completes one cycle.

Solution
The fluid pressure against volume graph for the cycle is shown in Figure 3.5. The work transfer as the fluid goes from state 1 to state 2 W_{12} is given by Equation (3.1) as

$$W_{12} = -\int_1^2 p \, dV$$

As the pressure remains constant during the non-flow process, the equation can be integrated to give

$$W_{12} = -p_1(V_2 - V_1) = -10 \times 10^5 \times (0.09 - 0.02) = -70 \text{ kJ}$$

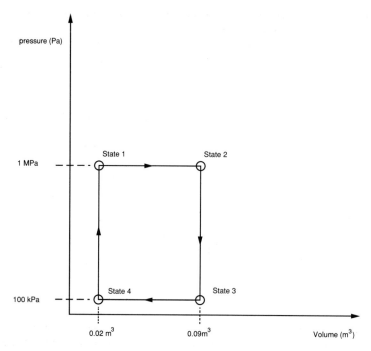

Fig. 3.5 *Example 3.2—a graph of pressure against volume for the fluid undergoing a cycle made up of four non-flow processes.*

The work transfer as the fluid goes from state 2 to state 3, W_{23}, is also given by Equation (3.1). But, in this non-flow process, the volume remains constant, so $dV = 0$. Therefore, the work transfer is zero. The same applies to the work transfer as the fluid goes from state 4 to state 1, W_{41}. Thus:

$$W_{23} = W_{41} = 0$$

The work transfer as the fluid goes from state 3 to state 4, W_{34}, is again given by Equation (3.1). The non-flow process is a constant pressure one, so the equation can be integrated to give:

$$W_{34} = -p_3(V_4 - V_3) = -1 \times 10^5 \times (0.02 - 0.09) = +7 \text{ kJ}$$

The fact that this work transfer is positive shows the sign convention in action. The net work transfer in the cycle $\sum(W)_{\text{cyc}}$ is given by the sum of the work transfers in each non-flow process, as follows:

$$\sum(W)_{\text{cyc}} = W_{12} + W_{23} + W_{34} + W_{41} = -70 + 0 + 7 + 0 = -63 \text{ kJ}$$

The net work transfer in the cycle $\sum(W)_{\text{cyc}}$ is also given numerically by the area of the graph:

$$\sum(W)_{\text{cyc}} = (10 - 1) \times 10^5 \times (0.09 - 0.02) = 63 \text{ kJ}$$

where the difference in sign arises as a consequence of the sign convention. To measure the area and obtain the net work transfer is easy enough in this example in which the fluid follows a rectangular cycle on a p against V diagram but, for more complex shapes of graph, it is necessary to adopt the analytical approach. By applying

Equation (3.1) to the fluid as it undergoes each identifiable non-flow process that makes up the cycle, an estimate has been made of the net work transfer of the proposed engine.

By convention, cycles like this one which go round in a clockwise direction are associated with engines in which there is a net work transfer out of the fluid, and cycles which go in an anti-clockwise direction are associated with compressors which require a net work input.

3.2 HEAT TRANSFER IN A NON-FLOW PROCESS

Work transfer is a reasonably easy concept to understand in the sense that it is an energy that is realised when a force is applied through some distance. Heat transfer, on the other hand, is less easy to describe. In Chapter 1, it was defined as an energy that arises due to a temperature difference. It is given the symbol Q and, like work transfer, it is mass dependent giving rise to a specific heat transfer q. Thus:

Heat transfer: Q (units J)

Specific heat transfer: q (units J/kg)

It seems reasonable that, if work transfer in terms of the fluid can be defined mathematically from a consideration of mechanics as above, so heat transfer should be able to be similarly defined. Both are energies that arise in a thermal system. In fact, work and heat transfer are inextricably linked in the first and second laws of thermodynamics, one of which will be considered in Chapter 4 and the other in Chapter 12. However, at this stage, it has been shown that work transfer is a product of two fluid properties, namely pressure and volume. Obviously heat transfer is an energy associated with temperature. Also, it can be demonstrated physically that heat flows from a body at a given temperature to a body at a lower temperature (Figure 3.6). For example, if a Bunsen burner is applied to one end of a metal bar, it is not long before the other end starts to rise in temperature. In other words, heat as energy, has been transferred by intermolecular collisions along the bar.

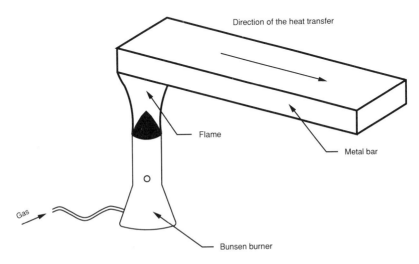

Fig. 3.6 *Heat flows from a body at a higher-temperature to a body at a lower temperature.*

It seems clear, therefore, that temperature should be one of the fluid properties that will make up the mathematical formula for heat transfer in a non-flow process. Temperature is a property like pressure that was defined in Chapter 2 as being intensive. The mathematical formula for heat Q_{12} realised when a fluid undergoes a non-flow process from state 1 to state 2, will be

$$Q_{12} = \int_1^2 T \, d?$$

What is the other property which will complete the formula? A check of its units reveals them to be joules/degree Kelvin in the SI system. At this stage, no fluid property has been identified with these units, so the approach taken is to imagine that such a property exists. It is given the name Entropy and the symbol S. It must be an extensive property like volume and be mass dependent, giving rise to a specific entropy s. Thus:

Entropy: S (units J/K)

Specific entropy: s (units J/kg K)

where $S = ms$.

Inventing a fluid property in this way might seem a dubious approach. However, the existence of entropy can be justified when the second law of thermodynamics is discussed in Chapter 12. For now, the fact that it is needed to write down a mathematical formula for heat transfer, in the same format as work transfer, is sufficient. It is a property like volume, and can be treated in the same way, even if it is not possible to sense it, or indeed measure it. Unlike volume, though, true values of the entropy of a fluid cannot be found. All that can be determined is the entropy change in a fluid as it undergoes a process and this can be achieved from a knowledge of the other fluid properties that can be measured. Fortunately, this is sufficient to analyse a thermal system. In fact, determination of the entropy change of a fluid is often not even necessary for the calculation of work and heat transfer. This does not diminish its significance. Rather, the concept of fluid entropy is vitally important to the understanding of the efficiency of a thermal system.

The formula, therefore, for heat transfer when a fluid changes from state 1 to state 2 in a non-flow process is

$$Q_{12} = mq_{12} = \int_1^2 T \, dS = \int_1^2 mT \, ds \qquad (3.2)$$

The heat transfer formula of Equation (3.2) is given a positive sign by the sign convention adopted. As with work transfer, the heat transfer during a change of state can be determined from the area under a graph of temperature against entropy (Figure 3.7), or from the area of the graph if the fluid completes a cycle (also given by the sum of the heat transfer in each non-flow process that makes up the cycle).

Again, as with work transfer, the equation may be integrated in three ways. Firstly, if the non-flow process is a constant entropy one and $dS = 0$, the heat transfer is also zero as follows:

$$Q_{12} = \int_1^2 T \, dS = \int_1^2 T \times 0 = 0$$

HEAT TRANSFER IN A NON-FLOW PROCESS

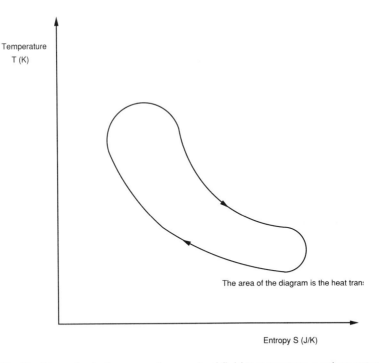

Fig. 3.7 *Heat transfer is the area of a graph of fluid temperature against entropy.*

Secondly, if the non-flow process is a constant temperature one, in which case:

$$Q_{12} = \int_1^2 T \, dS = T(S_2 - S_1)$$

And thirdly, if a relationship between the temperature and entropy is known, for example from experiment. But, as it is not possible to measure entropy directly, no such relationship exists and this integration cannot proceed.

It might seem strange that there can be a heat transfer into or out of a fluid while its temperature remains constant. In fact, this is what happens during the phase change of a fluid. For a fluid to change from the liquid to vapour phase, known as Boiling or Evaporating, there must be a heat supply to the fluid, and for the fluid to change from the vapour to the liquid phase, known as Condensing, there must be a heat rejection from the fluid. The heat transfer at constant temperature in either direction during a phase change is called the Latent Heat. Heat transfer associated with a temperature difference is called Sensible Heat.

EXAMPLE 3.3

A kettle containing 2 kg of water is placed upon an electric stove in a room where the atmospheric pressure is 100 kPa. As the heat from the stove is transferred to the water, the temperature of the water rises. When the temperature of the water reaches 99.6 °C, as further heat is supplied, it is noticed that the temperature remains constant until all the water is converted to steam, in other words, while the water is changing phase (Figure 3.8). If the further heat supply from the stove is measured as being 4516 kJ (this can be done through measuring the

WORK AND HEAT TRANSFER IN A THERMAL SYSTEM

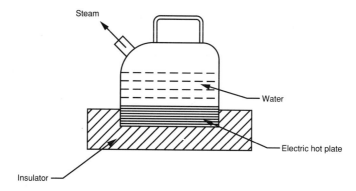

Fig. 3.8 *Example 3.3—determining the entropy change of water in a kettle as it changes phase to steam at constant temperature.*

volts and the amps of the electrical supply), what is the change of total and specific entropy of the water during the change of phase?

Solution

Let the water go from state 1 to state 2. The heat transfer to the water Q_{12} is given by Equation (3.2) as

$$Q_{12} = \int_1^2 \{mT\ ds\}_w$$

But during the change of phase the temperature of the water remains constant. Therefore, the equation can be integrated to give

$$Q_{12} = \{mT(s_2 - s_1)\}_w$$

The change of specific entropy of the water during the phase change is (noting that 99.6 °C must be converted to 372.6 K)

$$\{s_2 - s_1\}_w = \frac{4516 \times 10^3}{2 \times 372.6} = 6.06\ \text{kJ/kg K}$$

The change in entropy is

$$\{S_2 - S_1\}_w = 6.06 \times 2 = 12.12\ \text{kJ/K}$$

It can be seen from this example that values of entropy change can be obtained, albeit for the limited number of processes where the temperature remains constant. However, it is an indication that, although the property entropy has been invented to suit the proposed theory, it does not have to be a restriction on the development of the theoretical analysis. If entropy values are known, they can be used to estimate the heat transfer in a non-flow process and the net heat transfer in a cycle.

3.3 SPECIAL CHARACTERISTICS OF WORK AND HEAT TRANSFER IN A NON-FLOW PROCESS

3.3.1 Work and heat transfer must cross a boundary

The formulae for work and heat transfer in a non-flow process contain the properties pressure, volume, temperature and entropy. It might be thought that any changes in these

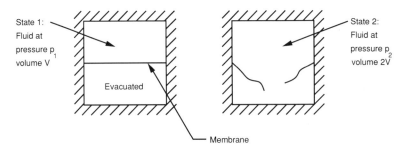

Fig. 3.9 *A container of volume $2V$ divided in half by a membrane, the upper half containing a fluid and the lower half evacuated.*

properties during a process will give rise to work and heat transfer. However, consider the following example. A container of volume $2V$ is divided in half by a membrane, as shown in Figure 3.9.

In state 1, fluid at a high pressure p_1 sits in the top half and occupies a volume V. The state 1 boundary is the inside volume of the top half of the container, the bottom half being evacuated. If the membrane is now broken, the high-pressure fluid in the top half will expand into the lower half. In state 2, the fluid has a volume $2V$ and pressure p_2. The state 2 boundary is now the inside volume of the whole container. There has been a change of pressure as a result of the process, from p_1 to p_2, and a change of volume, from V to $2V$. The formula for W_{12} implies that some work transfer has taken place. But this work is only realisable if it can be transferred out of the fluid, across the boundary and into some physical mechanism. This can be achieved by having a paddle-wheel, situated in front of the hole in the membrane, which turns as the fluid expands through it, provided the shaft of the paddle-wheel crosses the boundary and takes the work out of the system where use can be made of it.

The same principle applies to heat transfer, but is more difficult to describe because the fluid property entropy cannot be measured in the same way as pressure, volume and temperature at this stage. In the particular case of the combustion of a fuel in a fluid containing oxygen, as in an engine, the heat is generated internally. But to be useful, that heat must be transferred across the boundary, either in the form of work at a crankshaft or as heat in another fluid.

3.3.2 The formulae for work and heat transfer only apply to reversible non-flow processes

Concentrating upon the example of work transfer again, consider the piston cylinder mechanism in Figure 3.2. As the piston moves from TDC to BDC, work is realised at the crankshaft. The fluid in the cylinder at state 1, when the piston is at TDC, has a pressure p_1 and volume V_1. After expansion of the fluid to state 2, when the piston is at BDC, the fluid has a pressure p_2 and volume V_2. The amount of work transfer at the crankshaft due to the changes in the fluid pressure and volume during the process can be measured. Let there be 10 J of work.

It is now possible for the piston to go back from BDC to TDC and for the fluid to be returned to its original state of pressure p_1 and volume V_1. In other words, for the process to be reversed. To achieve this, work must be put into the crankshaft when the fluid is being compressed. But if only 10 J of work is put into the crankshaft, it would not be possible, in practice, to return the fluid to state 1 and the piston to TDC because, in real fluids undergoing a process, some of the energy is converted into a form that is not useful and this must be accounted for. The energy that is no longer available has been used to overcome the effects of friction which manifests itself in a number of ways. Whenever one solid body slides over another, a friction force exists which retards the motion. This is what happens when the piston slides against the cylinder walls. Unfortunately, this retarding force acts against the movement of the piston when it is going from TDC to BDC and also in the reverse direction when it is going from BDC to TDC. In addition, further friction forces arise when the fluid expands and is compressed in the cylinder. Fluids suffer friction against solid surfaces in the same sort of way that other solids do, the main difference being that they do not maintain their shape, as discussed in Section 2.5. The effect, though, is similar in that fluid friction forces oppose the motion of the piston both when it is going from TDC to BDC and vice versa.

Therefore, in the piston cylinder arrangement of Figure 3.2, if there is no friction, the work output as the piston moves down from TDC to BDC will be greater than when friction is present, and might have a numerical value of say 12 J. Hence, the losses due to friction are of the order of 2 J. Similarly, with no friction, it will only require 12 J of work to be put into the piston to make it go back to TDC, for the fluid to be returned to its original condition at state 1 and for the process to be reversed. But if friction is present, it will now require 14 J of work, the 12 J assuming no friction plus the extra 2 J required to overcome the friction forces which always oppose the motion (Figure 3.10).

As far as the formula for work transfer is concerned, if the fluid goes from state 1 with a pressure of p_1 and volume V_1, to state 2 where the pressure is p_2 and volume V_2, and back to state 1 again, the value of W will be the same because the changes in pressure and volume are the same in each process. This is only true for the case of the fluid expanding

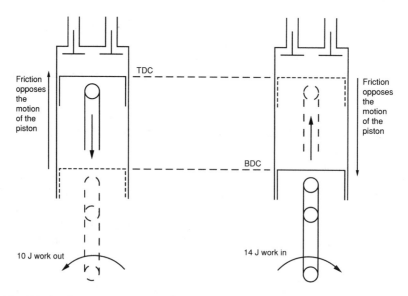

Fig. 3.10 *Friction forces oppose the motion of a piston in a cylinder whenever it is moving.*

and being compressed assuming no friction, an ideal case. In real fluids and real processes, friction forces which oppose the motion exist. The conclusion must be that the formula for work transfer in a non-flow process only applies to the ideal case of no friction.

For the case of heat transfer, the fluid will go from state 1 where the temperature is T_1 and entropy S_1 and the piston is at TDC, to state 2 where the temperature is T_2 and the entropy S_2 and the piston is at BDC, back to state 1 again. Assuming that the piston in the cylinder is made to move from TDC to BDC by virtue of the fluid having previously received energy as heat, which implies that T_1 is at a higher temperature than T_2, the heat transfer will take place from the fluid to the surroundings. But when the piston is returning from state 2 back to state 1 and the fluid temperature is increasing from T_2 to T_1, the required heat transfer from the surroundings into the fluid cannot occur because it is against the temperature gradient of the fluid and heat can only be transferred from a higher to a lower temperature regime. Ideally it is possible to suggest that the heat could be transferred through an infinite number of intermediate stages between state 2 and state 1 in which there is an infinitesimal temperature difference between the surroundings and the fluid.

The ideal case is called Reversible (Figure 3.11) in thermofluids because the fluid is made to return to its original condition, and the amount of work and heat transfer is the same when the piston goes from TDC to BDC in the expansion process as when it returns from BDC to TDC in the compression process, albeit in the opposite (reverse) direction. Both processes are referred to as Reversible Processes and, as the fluid actually undergoes a cycle, it is referred to as a Reversible Cycle.

Reversibility is an ideal state of affairs because all real fluids suffer from the effects of friction and heat transfer gradients. Real processes are condemned to be irreversible. Fortunately, making the assumption of reversible processes and being able to use the formulae for work and heat transfer can be justified because the answers achieved are reasonably well matched to the amounts of work and heat transfer that are realised in practice. In other words, if a piston cylinder reciprocating engine was set up in the laboratory, and the work

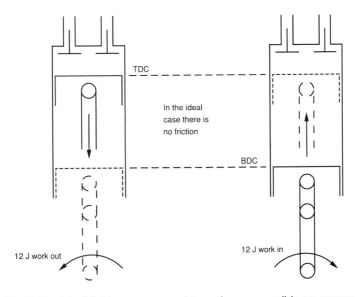

Fig. 3.11 *Ideal fluids are assumed to undergo reversible processes.*

transfer measured from the area of a graph of the fluid pressure versus the fluid volume and compared to the work transfer predicted by the formula for W, the two answers would be similar to each other. In that sense, the formulae for work and heat transfer can be used with some confidence for predicting the performance of reciprocating machinery. In effect, the reversible formulae for work and heat transfer in non-flow processes, Equations (3.1) and (3.2) respectively, describe the infinite number of changes that occur in the fluid property values of pressure and volume, or temperature and entropy, as the fluid goes from state 1 to state 2. The exact shape of the graph is known and the area under the graph is determined accordingly. In real processes, it is only possible to identify by measurement a limited number of intermediate fluid conditions between state 1 and state 2 and the exact shape of the graph cannot be determined. The better the ideal reversible graph follows the actual graph, whatever it is, the more accurately it predicts the work or heat transfer in the process.

3.3.3 Work and heat transfer are not properties of the fluid

The properties of the fluid are used to describe its condition in any given state. For work and heat transfer, the properties pressure, temperature, volume and now entropy have been introduced. It might be thought that, as the reversible work and heat transfer are described in the formulae by the product of two properties, they might also be properties of the fluid, but this is not the case.

Take the example of a skier on a mountain. This is not an exact analogy because a skier is not a thermal system, but it is useful to demonstrate the point. At the bottom of the mountain (state 1), the air pressure is 100 kPa and air temperature 280 K. At the top of the mountain (state 2), the air pressure is 90 kPa and air temperature 270 K. When the skier moves from the bottom of the mountain to the top, he or she will undergo a change of state. To get to the top, the skier could walk, which would require a certain amount of work to be done. Alternatively, the skier could catch the ski lift which would require no work on his or her behalf. The skier has gone through the same change of state as defined by the values of the properties at the bottom and the top of the mountain but, in choosing two different ways to get to the top, he or she has utilised different amounts of work. By the same argument the skier has transpired different amounts of heat. Similarly on the downward route. The skier could come straight down the mountain, which would be exciting and require little work to be done, or the skier could take a much longer route through a gradual decline which would require considerably more work to be expended and more heat to be transpired. Again, in both cases, the skier has moved from state 2 to state 1 as defined by the pressure and temperature, but has done differing amounts of work and transferred different quantities of heat (Figure 3.12).

In addition, it is noticeable from the example of the skier that work is only done and heat transferred during the change of state, not when he or she is either at the bottom of the mountain in state 1, or at the top of the mountain in state 2. Hence, work and heat are energies that only appear at the boundary during a change of state in a thermal system.

Therefore, the conclusion is that work and heat transfer are not properties of the fluid even though, mathematically, they consist of the integral of the product of two properties in a non-flow process. They only exist during a change of state and the amount of work and heat transfer depends upon the way the state is changed, in other words on the non-flow process undergone by the fluid or, in the case of the skier, on the route that is taken. This is important for two reasons. Firstly, further properties of the fluid will be encountered as the theory is developed and it can now be seen that it will be possible to recognise and identify

SPECIAL CHARACTERISTICS OF WORK AND HEAT TRANSFER

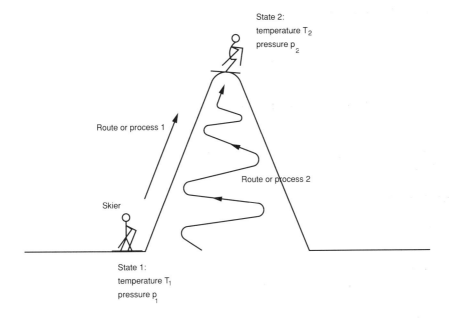

Fig. 3.12 *Fluid property values do not depend upon the route taken or process.*

a property if it can be used to describe the condition of the fluid at any state and for its change in value not to depend upon the process during a change of state, just like pressure, volume, temperature and entropy. Secondly, as the amount of work and heat transfer does depend upon the process, it must be possible to define various processes in which the work and/or heat transfer can be maximised or minimised, depending upon the application. This is the job of an engineer, to design a thermal system such as, for example, an engine that will produce the most work for the least heat input, or a compressor which will require the least work input and the smallest heat losses for the largest increase in fluid pressure.

It should also be noted again that it is not necessary for all the property values to change to realise work and heat transfer during a non-flow process. For example, it is possible to envisage that, in the case of the skier, the temperature at the top of the mountain could be the same as at the bottom because the sun is shining brightly at the top, whereas the bottom is in the shade. Further, in the case of the skier going up the mountain in a chair lift, no work is done whether the skier goes all the way to the top or gets off half-way up. In the latter case, the skier would have changed his or her state from state 1 to state 3, at whatever the conditions of the pressure and temperature were half-way up the mountain.

In a mathematical analysis of a thermal system, it is often necessary to differentiate and integrate the equations derived to describe the system. Note that this is possible for the reversible work and heat transfer formulae of Equations (3.1) and (3.2) respectively. The convention used will be as follows. For a property like pressure p, as the fluid goes from state 1 to state 2, the differential and integral will be written

$$\text{Differential: } dp$$

$$\text{Integral: } p_2 - p_1$$

74 WORK AND HEAT TRANSFER IN A THERMAL SYSTEM

For work and heat transfer which are not properties:

$$\text{Differential: } dW, \ dQ$$

$$\text{Integral: } W_{12}, \ Q_{12}$$

3.3.4 Sign convention for work and heat transfer

A piston cylinder mechanism produces a work transfer when the fluid expands and the piston moves from TDC to BDC, but requires a work input when the fluid is compressed and the piston moves from BDC to TDC. Similarly heat can be transferred into or out of a system. It is necessary to distinguish between them and the sign convention that will be used in this book is as follows:

1. Work transfer into a thermal system — positive.
2. Work transfer out of a thermal system — negative.
3. Heat transfer into a thermal system — positive.
4. Heat transfer out of a thermal system — negative.

The convention is based upon the principle that all energy gains into a system are considered a good idea, and that all energy outputs from a system are a bad idea.

Signs for work and heat transfer can cause confusion when carrying out thermal analyses. But if the following rules are obeyed, there should be no problems:

Sign rule A. If numerical values for W and Q are determined from formulae or equations, there is no need to worry about the signs, they will take care of themselves, provided the usual convention is applied whereby any difference is written as final value minus initial value. In the case of a piston moving from TDC where the volume is V_1, to BDC where the volume is V_2, the change in volume is $(V_2 - V_1)$.

EXAMPLE 3.4

A piston in a cylinder is at TDC where the fluid pressure is 5 MPa and volume 0.01 m³. A fuel is burned in the fluid which raises its temperature and pressure but, at the same time, the piston moves downwards expanding the fluid and so reducing its pressure. The fuel is burned at such a rate that the pressure rise due to combustion is balanced by the pressure fall due to expansion, creating a constant pressure process. The combustion is terminated when the fluid volume has reached 0.02 m³ (Figure 3.13). Determine the work transfer during the non-flow process assuming it is reversible.

Solution

Let the fluid go from state 1 to state 2. Equation (3.1) for the work transfer in a non-flow process gives

$$W_{12} = -\int_1^2 p \, dV$$

This equation can be integrated because the pressure of the fluid is constant at 5 MPa. Therefore:

$$W_{12} = -p(V_2 - V_1) = -50 \times 10^5 \times (0.02 - 0.01) = -50 \text{ kJ}$$

SPECIAL CHARACTERISTICS OF WORK AND HEAT TRANSFER

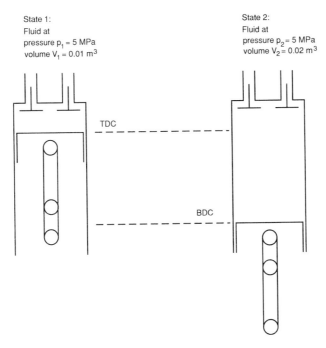

Fig. 3.13 *Example 3.4—work transfer due to the expansion of a fluid in a piston cylinder mechanism.*

The fact that the answer is −50 kJ shows that the work transfer is a work output from the system.

Sign rule B. If actual numerical values for W and Q are known, the sign convention must be applied before they are substituted into formulae or equations.

EXAMPLE 3.5

In a reciprocating piston cylinder mechanism, the piston is half-way between BDC and TDC where the fluid pressure is 1 MPa and volume 0.05 m³. The piston is to be made to move upwards because 10 kJ of work is put into the crankshaft. But the exhaust valve is left open, so the fluid is delivered at a constant pressure of 1 MPa to a tank connected to the exhaust manifold (Figure 3.14). What is the volume of the fluid in the cylinder after the work has been done on the piston assuming a reversible non-flow process?

Solution

Let the fluid go from state 1 to state 2. Because the work transfer is a work input and it is a numerical value, it must be given an appropriate sign which is positive in this case. Equation (3.1) for the work transfer in a non-flow process gives

$$W_{12} = -\int_1^2 p \, dV$$

This can be integrated because the pressure of the fluid is constant. Therefore:

$$W_{12} = -p(V_2 - V_1)$$

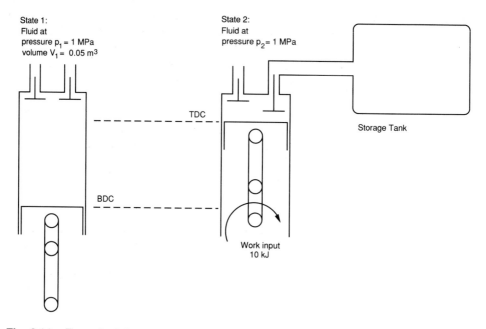

Fig. 3.14 *Example 3.5—a work input is required to deliver compressed fluid in a piston cylinder mechanism to a store.*

Substituting in the numerical values gives

$$+10 \times 10^3 = -10 \times 10^5 (V_2 - 0.05)$$
$$\therefore V_2 = 0.04 \text{ m}^3$$

Here V_2 is smaller than V_1 which it should be because the piston is moving towards TDC. If the negative sign in Equation (3.1) had been forgotten, V_2 would have turned out to be

$$10 \times 10^3 = 10 \times 10^5 (V_2 - 0.05)$$
$$\therefore V_2 = 0.06 \text{ m}^3$$

Here V_2 is larger than V_1 which is obviously impossible when the piston is on its upward stroke. The practical implications of any numerical solution must always be considered, especially when the sign convention is being applied.

Sign rule C. If work and/or heat is transferred from one fluid to another, as in a heat exchanger, or from one component to another, as in a motor driving a compressor, use only the numerical value for the energy transferred, do not give it a sign. If the energy transferred is later to be used in a formula or equation, the appropriate sign must be applied to it then, in line with sign rule B.

In the case of a heat exchanger, for example, there is a transfer of energy from a hot fluid to a cold fluid. The heat is transferred out of the hot fluid and into the cold fluid. In any transfer of work and/or heat between components or fluids, the energy will need to change signs depending upon whether it is being transferred into or out of a system. Sign rule C is designed to get round this problem by not allocating any sign during the energy transfer. Instead the energy transfer is considered as a numerical exchange of energy alone and the appropriate sign is only allocated when that energy is later used in a formula or equation and sign rule B applies.

In most cases, the transfer of energy is associated with some of the energy not ending up in the intended destination. In the case of the heat exchanger, for example, it is most likely that not all the heat in the hot fluid will find its way into the cold fluid. Instead some will be transferred into the atmosphere. In this case, it is necessary to introduce an energy transfer efficiency and this is dealt with in more detail in Section 4.2.2.

EXAMPLE 3.6

Consider a reciprocating compressor being driven by an electric motor (Figure 3.15). Ignoring losses in the drive system, if the electric motor produces 10 kJ of work, what is the work transferred to the reciprocating compressor?

If the work transferred to the compressor is used to move the piston from BDC position under constant pressure conditions of 200 kPa, what will be the volume change of the fluid in the cylinder if the non-flow process is reversible?

Solution

Neglecting any losses between the two components, the energy transfer will be such that the work output from the electric motor W_{EM} is equal to the work input to the compressor W_C.

Using sign rule C which stipulates that no sign is attached to the energy transfer, the work input to the compressor W_C is

$$W_C = W_{EM} = 10 \text{ kJ}$$

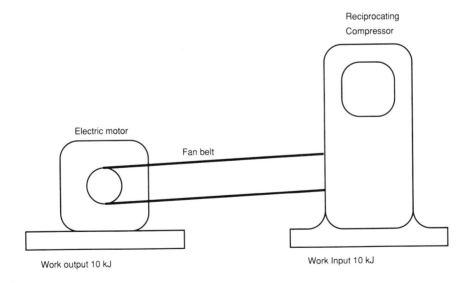

Fig. 3.15 *Example 3.6—an electric motor providing a work input to a reciprocating piston cylinder compressor.*

If the value of W_C is to be used in a formula, the appropriate sign must be given to it. As it is a work input, it must be positive. For the fluid in the reciprocating compressor going from state 1 to state 2, the work transfer for the non-flow process is, from Equation (3.1):

$$W_C = W_{12} = -\int_1^2 p \, dV$$

As the fluid in the cylinder is maintained at constant pressure, this equation can be integrated:

$$\therefore \quad W_C = W_{12} = -p(V_2 - V_1)$$

Substituting in the numerical values and following sign rule B, gives:

$$+10 \times 10^3 = -2 \times 10^5 (V_2 - V_1)$$
$$\therefore \quad V_2 - V_1 = -0.05 \text{ m}^3$$

Hence V_2 is smaller than V_1 which it should be because the piston is moving from BDC position.

3.4 WORK AND HEAT TRANSFER IN A FLOW PROCESS

Clearly the formulae derived above for W and Q in Equations (3.1) and (3.2) are based upon an analysis of a piston cylinder mechanism and only apply to non-flow processes. It is not possible, at this stage, to identify a formula for work and heat transfer in the same way for flow processes. This might seem as if flow processes are more difficult to analyse but, as will be seen in Chapter 8 where flow processes are considered in detail, it is not necessary to do so.

The existence of the property entropy, though, is equally relevant to flow processes as non-flow processes. Its existence may have been assumed from the necessity to write down an equation for the heat transferred in a non-flow process, but it is a property just like pressure and must be treated in the same way. In the case of the skier, in going up the mountain, he or she goes from a pressure p_1 to a pressure p_2. Equally, he or she must go from an entropy S_1 to an entropy S_2.

Also relevant to flow processes as much as to non-flow processes is the fact that work and heat transfer are not properties of the system, that work and heat transfer must cross the boundary to become effective, and their sign convention. It will be seen in Chapter 8 that it is also relevant to speak of reversible flow processes in open systems.

3.4.1 Time-dependent values

In most thermal systems, a fluid is made to undergo a process more than once and time becomes an important consideration. Clearly, in flow processes, the fluid is entering and leaving a component at a certain rate but, even in non-flow processes, the process may be repeated, or a number of non-flow processes may make up a cycle which the fluid goes through so many times per second.

The convention adopted when the rate is relevant is to place a dot upon the mass of fluid m (as in Section 2.1), the volume of fluid V (as in Section 2.2), the work transfer W and the heat transfer Q. Only these four need be considered as they are sufficient to identify

WORK AND HEAT TRANSFER IN A FLOW PROCESS   79

that something is happening per second to the fluid. When it is a rate of work transfer, it is usually referred to as a power, as stated in Section 1.3. Thus:

Fluid mass:	m (units kg)
Fluid mass flow rate:	$\dot{m}$ (units kg/s)
Fluid volume:	V (units m^3)
Fluid volume flow rate:	$\dot{V}$ (units m^3/s)
Work transfer:	W (units J)
Work transfer/second (power):	$\dot{W}$ (units J/s = W)
Heat transfer:	Q (units J)
Heat transfer/second (heat transfer rate):	$\dot{Q}$ (units J/s = W)

EXAMPLE 3.7

It is proposed to build an engine in which a fluid undergoes the following four non-flow reversible processes and completes a cycle:

1. State 1 to state 2—heat addition of 36.6 kJ at a constant temperature of 500 K during which the entropy rises.
2. State 2 to state 3—constant entropy drop in temperature until the temperature is 300 K.
3. State 3 to state 4—heat rejection at a constant temperature of 300 K as the entropy decreases through the same amount as it increased when going from state 1 to state 2.
4. State 4 to state 1—constant entropy rise in temperature back to 500 K.

Sketch the cycle on a graph of fluid temperature against entropy.
What is the change of entropy as the fluid goes from state 1 to state 2?
How much heat is transferred as the fluid goes from state 2 to state 3, and from state 4 to state 1?
What is the heat rejected as the fluid goes from state 3 to state 4?
If the fluid completes the cycle 10 times per second, what is the net heat transfer in the appropriate units?

Solution

The cycle drawn on a graph of fluid temperature against entropy is as shown in Figure 3.16. From Equation (3.2), the change in entropy as the fluid goes from state 1 to state 2 is given by

$$Q_{12} = \int_1^2 T \, dS$$

As the temperature remains constant during the non-flow process, the equation can be integrated to give

$$Q_{12} = T_1(S_2 - S_1)$$

The heat transfer Q_{12} is a heat addition which is positive according to sign rule B. The change of fluid entropy is

$$S_2 - S_1 = \frac{36.6 \times 10^3}{500} = 73.2 \text{ J/K}$$

As the fluid goes from state 2 to state 3, the heat transfer Q_{23} is again given by Equation (3.2):

$$Q_{23} = \int_2^3 T \, dS$$

WORK AND HEAT TRANSFER IN A THERMAL SYSTEM

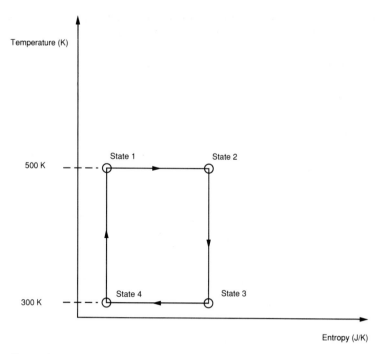

Fig. 3.16 *Example 3.7—a graph of temperature versus entropy for the fluid as it undergoes a cycle.*

But as the non-flow process is one of constant entropy:

$$dS = 0 \quad \therefore \quad Q_{23} = 0$$

Hence, there is no heat transfer between state 2 and state 3. There is also no heat transfer between state 4 and state 1 which is another constant entropy non-flow process. Therefore $Q_{41} = 0$.

When the fluid goes from state 3 to state 4, the heat transfer is again given by Equation (3.2) as

$$Q_{34} = \int_3^4 T \, dS$$

As before, this is a constant temperature non-flow process so the equation can be integrated to give

$$Q_{34} = T_3(S_4 - S_3)$$

But the decrease in fluid entropy in going from state 3 to state 4 is numerically the same as the increase when going from state 1 to state 2. Thus:

$$S_4 - S_3 = -(S_2 - S_1) = -73.2 \text{ J/K}$$

Substituting gives the heat transfer Q_{34} as

$$Q_{34} = 300(-73.2) = -22 \text{ kJ}$$

The fact that the answer is negative shows that it is a heat rejection according to the sign convention. The net heat transfer in the cycle $\sum(Q)_{cyc}$ is given by the sum of all the heat transfers in each non-flow process:

$$\sum(Q)_{cyc} = Q_{12} + Q_{23} + Q_{34} + Q_{41} = 36.6 + 0 - 22 + 0 = 14.6 \text{ kJ}$$

The same answer could have been obtained by determining the area of the diagram. In other words

$$\sum(Q)_{cyc} = (500 - 300) \times 73.2 = 14.6 \text{ kJ}$$

If the cycle is repeated 10 times per second, the net heat transfer per second in the cycle $\sum(\dot{Q})_{cyc}$ is

$$\sum(\dot{Q})_{cyc} = \sum(Q)_{cyc} \times 10 = 14.6 \times 10 = 146 \text{ kW}$$

The determination of the change in entropy as the fluid goes from state 1 to state 2 has been used to find the heat transfer that is realised as the fluid goes from state 3 to state 4. Although the significance of the amount of entropy change in each non-flow process is perhaps not yet apparent, it does not prevent the property entropy being used to evaluate the net heat transfer in the cycle, which is the objective of the analysis.

As in Example 3.2, because the cycle goes in a clockwise direction, the system is an engine producing a net work output from a net heat transfer input.

4
The First Law of Thermodynamics

Chapter 3 developed two theoretical equations applicable when a fluid is undergoing a reversible non-flow process, one for predicting the work transfer and one for predicting the heat transfer. The formulae are of the same format because work and heat transfers are both energies that arise only when the fluid is changing its state, and they also share a number of other characteristics which distinguish them from the properties of the fluid. However, it is difficult to proceed further with the equations as they stand and it is necessary to approach the analysis of a thermal system from a different line of enquiry.

As a typical purpose of a thermal system is to allow the conversion of work transfer to heat transfer or vice versa, it might be fair to assume that there must exist a relationship between the work and heat transfers, and this is just what is expressed in the first and second laws of thermodynamics.

A zeroth law of thermodynamics has already been expounded in Section 2.4. This was the definition of the temperature of the fluid. The first and second laws are the foundations upon which the study of thermodynamics is based. But they are not laws in the strict meaning of the word. Instead, they are statements based upon observation which are accepted because they apply to all practical thermal systems. They can be shown to be invalid in situations where there are only a few molecules under consideration, but this is not a general application. The first law is dealt with in this chapter and the second law in Chapter 12.

The first law of thermodynamics expresses a relationship between the work and heat transfer realised during the changes of state of a fluid in a thermal system when the fluid undergoes a cycle, and may be quoted as follows: 'when any closed system goes through a cycle, the net work delivered to the surroundings is proportional to the net heat taken from the surroundings'.

Originally, heat and work transfer were given different units and the constant of proportionality was called the Joule Equivalent of Heat. Now that heat and work transfer are both recognised as energy forms there is no need for the constant, and mathematically, the first law of thermodynamics is written:

$$\sum(Q)_{\text{cyc}} + \sum(W)_{\text{cyc}} = 0 \qquad (4.1)$$

There are three words in the law which need highlighting. Firstly, the system must complete a cycle which means that the fluid properties must return to their original condition, however loosely this is interpreted. Secondly, it is the net work and net heat transfer that is of interest, in other words some of the work and heat transfer may be positive and some negative, in keeping with the sign convention proposed in Section 3.3.4. And thirdly, it applies to a

THE FIRST LAW OF THERMODYNAMICS

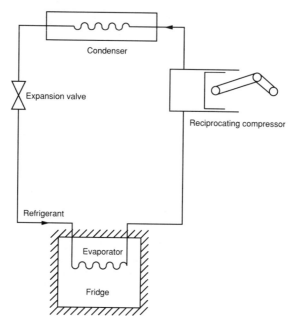

Fig. 4.1 *A refrigerator is a closed system made up of components in which the fluid follows flow processes.*

closed system. This might seem rather restrictive, but in fact is not the case. The system must be closed in that the fluid must go round a circuit and complete a cycle. But in doing so, it can pass through a number of components which are open in that the fluid enters in one condition but leaves in another. If the system boundary is the walls of each component and the interconnecting pipework, the fluid never crosses the boundary and the system must be considered closed. Therefore, the first law can be applied to systems such as refrigerators which consist of four open components, but in which the refrigerant fluid remains entirely within the system and completes a cycle (Figure 4.1). Hence, Equation (4.1) can also be written in terms of the net work transfer per second and the net heat transfer per second as follows:

$$\sum (\dot{Q})_{cyc} + \sum (\dot{W})_{cyc} = 0$$

Note that in Figure 4.1, following the convention for a cycle, the refrigerant is shown to follow an anticlockwise direction around the system, indicating a net work input.

The law of the conservation of energy implies that energy can neither be created or destroyed, but merely changed from one form to another. The first law of thermodynamics is, in effect, a particular statement of that law in that it is applied to a thermal system under specified conditions such that only the work and heat transfer need be accounted for.

4.1 THE FIRST LAW OF THERMODYNAMICS IN ACTION

The first law can be observed, and so proved as such, by applying it to a variety of closed cycles, adding together the positive and negative heat and work transfers, and confirming

4.1.1 Methods for determining work transfer

Work transfer can be readily measured on a device called a dynamometer, as shown in Figure 4.2. A dynamometer consists of a wheel which is made to rotate, for example by the crankshaft of a piston cylinder mechanism or the output shaft of a rotary turbine. A load can be applied to the lever which will produce a friction force on the wheel, with the force F being measured on the spring balance. If the analogy is drawn with a motor car, the load effectively simulates whether the car is going uphill or downhill. Knowledge of the force F, the distance at which it is applied L and the rotational speed of the wheel N_{rev} allows the work transfer, known as the Brake Power $\dot{W}_{bp}$, to be calculated as follows:

Work transfer/second = brake power = torque × speed = force × distance × speed

$$\therefore \dot{W}_{bp} = -2\pi F L N_{rev} \qquad (4.2)$$

where the negative sign is introduced again for the sign convention. Work input, for example, to a compressor, is usually provided by an engine, a reciprocating device such as a piston cylinder mechanism or a rotary device such as a turbine but, in either case, the dynamometer can be used on the drive to discover its output. If the drive is an electric motor, its power can be found from a knowledge of the amps and the volts.

EXAMPLE 4.1

A reciprocating petrol engine is running in the laboratory and the power output is being measured on a dynamometer. Unfortunately, it is an old dynamometer and the measurements of load are in pounds! The following readings of engine speed and dynamometer load are taken, as shown in Table 4.1. Determine the brake power outputs of the engine if the load is measured at a distance of 0.5 m.

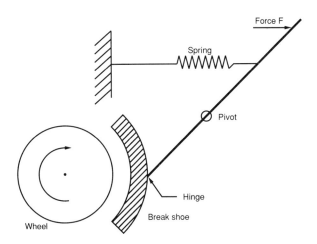

Fig. 4.2 *A dynamometer for determining the brake power of an engine.*

86   THE FIRST LAW OF THERMODYNAMICS

Table 4.1 *Example 4.1—readings of engine speed and dynamometer load.*

Engine speed (rev/min)	1000	2000	3000	4000	5000	6000
Dynamometer load (lb)	64	58	57	53	47	37

Table 4.2 *Example 4.1—readings of engine speed and brake power.*

Engine speed (rev/min)	1000	2000	3000	4000	5000	6000
Brake power $\dot{W}_{bp}$ (kW)	−14.9	−27.0	−39.8	−49.4	−54.8	−51.7

Solution

The dynamometer load in pounds can be converted to newtons by multiplying by a factor of 4.45. The rev/min can be converted to rev/s by dividing by 60. Utilising Equation (4.2), the brake power output of the engine $\dot{W}_0$ at the respective engine speeds is as shown in Table 4.2. The negative sign merely indicates that it is a power output according to the sign convention. The power generated by a reciprocating engine often declines at higher speeds due to the mechanical limitations of the valve operation which serves to restrict the mass flow of air entering the cylinder in particular, and friction effects in the inlet manifold which heat up the air, reducing its density and, hence, mass flow rate.

However, the work output from a reciprocating engine determined from a dynamometer is not the same as the work transfer predicted by the non-flow formula for W, as in Equation (3.1). This is because there are friction forces in the system between the fluid and the final output, friction both in the fluid and in the mechanical drive system to the dynamometer.

For the fluid in a piston cylinder mechanism undergoing non-flow processes, the mathematical formula for W means that it is the area under the graph of fluid pressure p against volume V during a change of state and, in a cycle, is the area of the diagram. A typical p against V plot for the fluid in a reciprocating engine is shown in Figure 4.3.

Values for the work transfer realised in the fluid in a reciprocating engine or compressor can be obtained if a plot of the fluid pressure and volume changes is made. This can be achieved by placing a pressure transducer in the clearance volume at the top of the cylinder to measure the pressure, and recording the volume changes in the cylinder by means of a magnetic pick-up attached to the crankshaft. If both signals are fed into an oscilloscope, the graph of p against V is obtained, and the determination of its area A_{pV}, after suitable calibration, is the work transfer per cycle. It is called an area but its units are those of pressure × volume which is of energy. If the piston cylinder mechanism is doing N_{cyc} cycles per second, the work transfer is expressed as a power, called the Indicated Power $\dot{W}_{ip}$, and given by

$$\dot{W}_{ip} = \pm A_{pV} N_{cyc} \qquad (4.3)$$

where the negative sign is for an engine and the positive sign for a compressor. It is not necessarily so that the number of cycles per second performed by an engine N_{cyc} is the same as the number of revolutions per second of the crankshaft N_{rev}. In fact, in a two-stroke engine, $N_{cyc} = N_{rev}$, but in a four-stroke engine, $N_{cyc} = 0.5 N_{rev}$.

THE FIRST LAW OF THERMODYNAMICS IN ACTION 87

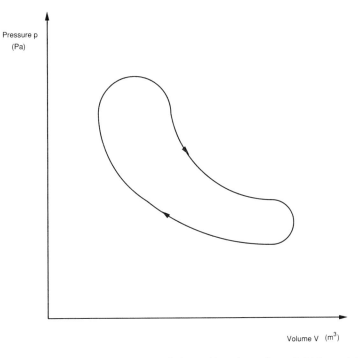

Fig. 4.3 *A typical graph of the pressure variation with volume for a fluid in a piston cylinder reciprocating engine.*

In other words, the crankshaft must complete two revolutions before one engine cycle is completed and it takes four strokes of the piston to complete the cycle.

The difference between the indicated power $\dot{W}_{ip}$ and the brake power $\dot{W}_{bp}$ is the Friction Power $\dot{W}_F$:

$$\dot{W}_F = \dot{W}_{ip} - \dot{W}_{bp}$$

The friction power is, in effect, the power required to overcome friction between the fluid in the piston cylinder and the final drive (Figure 4.4). This manifests itself as a heat transfer $\dot{Q}_F$ out of the system. In other words there is an energy transfer from the friction to the heat. According to sign rule C, the numerical value for the friction power is equated with the heat produced $\dot{Q}_F$ and, if the heat is used in a further equation or formula, it must be given an appropriate sign. Thus:

$$\dot{W}_F = \dot{W}_{ip} - \dot{W}_{bp} = \dot{Q}_F \tag{4.4}$$

The heat transfer arises because the effects of the friction forces opposing the motion are to raise the temperature, which causes heat to be transferred to the surroundings. In an engine, the friction forces are not just those of the fluid in the cylinder and the piston moving against the cylinder walls, as described in Chapter 3, but also those arising in the drive mechanism to the dynamometer. Equation (4.4) is really the first law of thermodynamics applied to the limited case of the conversion of friction power to heat.

This technique can be applied to a piston cylinder mechanism whether it is an engine or a compressor but, in the case of a compressor, $\dot{W}_{ip}$ will be less than $\dot{W}_{bp}$ because the work

88   THE FIRST LAW OF THERMODYNAMICS

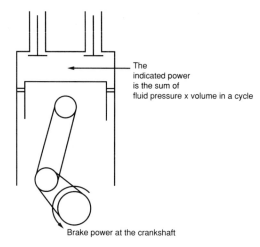

Fig. 4.4 *The friction power is the difference between the indicated power and the brake power of an engine.*

has to be put into the system. The friction power is as described above except that $\dot{W}_{ip}$ and $\dot{W}_{bp}$ are the other way round.

For an open system such as a rotary turbine or compressor, the indicated power is not given by Equation (4.3). The energy in the fluid can still be determined from a knowledge of the change in the fluid properties, but the appropriate formula depends upon the application.

4.1.2 Methods for determining heat transfer

It is useful to introduce another fluid property when considering heat transfer, namely the Specific Heat c. This is because the heat transfer Q is defined mathematically in terms of the entropy S in a non-flow reversible process. But this is not a measurable quantity and so a different approach needs to be taken to that adopted for W above. The specific heat is defined as 'the heat required to raise the temperature of unit mass of a substance by one degree'.

$$\text{Specific Heat: } c \text{ (units J/kg K)}$$

For a mass of fluid m, temperature rise dT, the specific heat c is given in terms of the heat requirement dQ by

$$c = \frac{dQ}{m\, dT} \tag{4.5}$$

If c can be assumed constant, the equation can be integrated to give

$$Q_{12} = mc(T_2 - T_1) \tag{4.6}$$

which is a very useful equation in thermofluids. In a flow process, the equation becomes

$$\dot{Q}_{12} = \dot{m}c(T_2 - T_1)$$

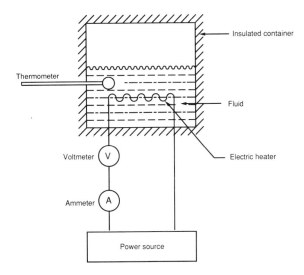

Fig. 4.5 *Determining the specific heat of a fluid.*

The heat transfer associated with the temperature difference of Equation (4.6) is called the sensible heat transfer as stated in Section 3.2. To obtain values of c for different substances, solids and fluids, the heat may be applied electrically and evaluated from a knowledge of the volts and amps. Measuring the temperature rise $(T_2 - T_1)$ will enable the specific heat to be calculated (Figure 4.5). It will be seen later that there are two main types of specific heat, but at this stage all that is necessary is to know that it can be measured and used as a means of determining heat transfer in a thermal system.

Heat transfer often occurs in thermal systems due to the combustion of a fuel and this gives rise to the Calorific Value CV of the fuel which is defined as 'the heat released when unit mass of fuel is burned under normal temperature and pressure conditions'.

$$\text{Calorific value: } CV \text{ (units J/kg)}$$

The calorific value of fuels can be measured in devices such as the bomb calorimeter and gas fuel calorimeter. Essentially, they make use of Equation (4.6) in that the fuel is burned in air or oxygen in a container surrounded by a jacket of water. The heat released by the fuel is transferred to the water and, knowing the mass of water $\{m\}_w$, its specific heat $\{c\}_w$ and by measuring its temperature rise $\{T_2 - T_1\}_w$, the heat transfer is calculated. Knowing the mass of fuel $\{m\}_{fuel}$ gives the calorific value, as follows. The heat available in the fuel Q_{comb} is

$$Q_{comb} = \{m\}_{fuel} CV \tag{4.7}$$

From Equation (4.6), the heat gained by the water is

$$Q_{12} = \{mc(T_2 - T_1)\}_w = Q_{comb} \tag{4.8}$$

Equation (4.8) allows the calorific value of the fuel to be calculated (Figure 4.6). When there is a continuous combustion of a fuel, Equation (4.7) becomes

$$\dot{Q}_{comb} = \{\dot{m}\}_{fuel} CV$$

90 THE FIRST LAW OF THERMODYNAMICS

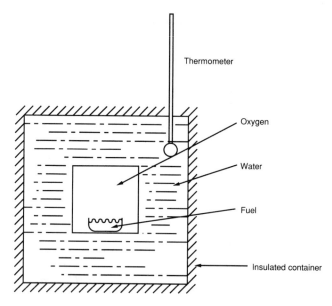

Fig. 4.6 *Determining the calorific value of a fuel.*

Some approximate values of CV for different fuels are:

$$\begin{aligned}
\text{Natural gas:} \quad & CV = 47\,000 \text{ kJ/kg} \\
\text{Petrol:} \quad & CV = 43\,000 \text{ kJ/kg} \\
\text{Black coal:} \quad & CV = 32\,000 \text{ kJ/kg} \\
\text{Brown coal:} \quad & CV = 20\,000 \text{ kJ/kg}
\end{aligned}$$

4.2 EFFICIENCY OF A THERMAL SYSTEM

4.2.1 Overall thermal efficiency

Interest lies in producing thermal systems which have a required energy output for a minimum energy input. In Chapter 1, the efficiency term was introduced as a method of assessing the performance of a thermal system. It is now possible to be more specific and to define an Overall Thermal Efficiency η as follows: 'the overall thermal efficiency η is the ratio of the useful net energy output to the energy input into a thermal system which must be paid for'.

For example, in the case of an engine, the useful net energy out is the work output at the crankshaft. The energy input that must be paid for is the fuel which undergoes combustion producing heat (Figure 4.7). For an electric-driven compressor, the useful net energy out is in the form of high-pressure fluid, expressed in the units of energy, and the energy input that has to be paid for is the electricity required to power the electric motor to drive the compressor (Figure 4.8). The overall thermal efficiency, therefore, is a very useful measure of the performance of a thermal system and relating it to an energy input that has to be costed gives it a significance which reflects real life.

EFFICIENCY OF A THERMAL SYSTEM

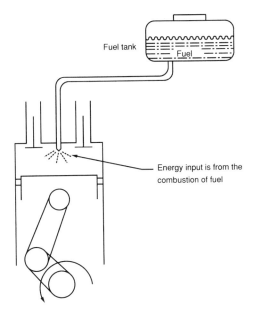

Fig. 4.7 *The overall thermal efficiency of an engine.*

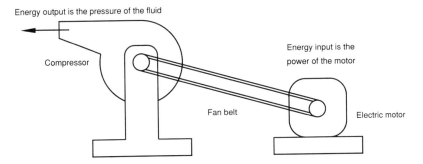

Fig. 4.8 *The overall thermal efficiency of an electric-driven compressor.*

There are systems in which the energy input does not have to be paid for, for example those that benefit from solar energy or, it might be argued, a hydroelectric scheme. In such cases, it is appropriate to use the energy transfer efficiency below.

4.2.2 Energy transfer efficiency

None of the components that make up a thermal system can transfer energy perfectly. There are always 'losses' in the system, in other words, energy that does not end up where it is supposed to. In that sense, because energy is never 'lost', it is energy that is transferred into a form that is not required or cannot be made use of. For example, in the case of a compressor

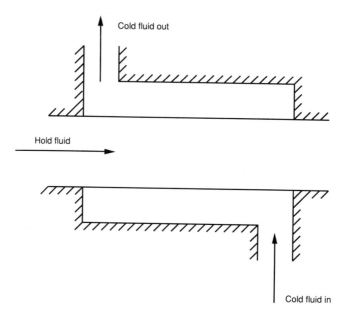

Fig. 4.9 *The energy transfer efficiency of a heat exchanger.*

driven by an electric motor, the work output from the motor is greater than the work input to the compressor due to friction 'losses' in the drive shaft connecting the two components. In fact, the energy is not 'lost', merely that the work done in overcoming friction appears as heat which is not what is required at the compressor. Therefore, it is more pertinent to talk about energy transfer than energy loss. In general, the Energy Transfer Efficiency η_e is defined as 'the ratio of the energy received to the energy that has been delivered'.

This is a fairly broad definition and requires some common sense to be applied when utilising it. It is not only relevant to energy transfers between components but also to fluids. In a heat exchanger which is transferring heat from a hot fluid to a cold fluid, the energy received is that by the cold fluid, and the energy delivered is that by the hot fluid. Because some of the heat is 'lost' to the atmosphere, the energy conversion efficiency is not 100% (Figure 4.9).

Use must be made of sign rule C in the application of the energy transfer efficiency whereby no sign is allocated to the numerical values for the energy transfer.

EXAMPLE 4.2

An electric motor has a power output of 100 kW. It is to be used to drive a compressor. The energy input to the compressor is measured as being 80 kW (Figure 4.10). What is the energy transfer efficiency of the drive system?

The drive is replaced by a new system which has an energy transfer efficiency of 90%. What is now the work input into the compressor?

Solution

The work output from the electric motor $\dot{W}_{EM}$ must have a negative sign according to sign rule B. Thus:

$$\dot{W}_{EM} = -100 \text{ kW}$$

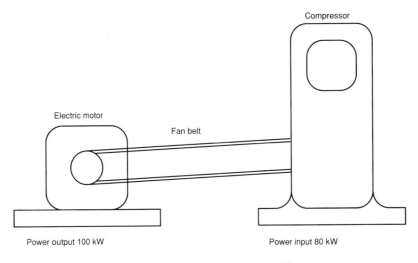

Fig. 4.10 *Example 4.2—an electric motor driving a compressor.*

The work input to the compressor $\dot{W}_C$ must have a positive sign according to sign rule B. Thus:

$$\dot{W}_C = +80 \text{ kW}$$

But when the energy transfer from the electric motor to the compressor is being considered, no signs are allocated according to sign rule C. As the energy received is by the compressor and the energy delivered is by the electric motor, the energy transfer efficiency η_e is given by

$$\eta_e = \frac{\dot{W}_C}{\dot{W}_{EM}} = \frac{80}{100} = 80\%$$

If the energy transfer efficiency is improved to 90%, the work input to the compressor, using sign rule C, is given by

$$\eta_e = 0.9 = \frac{\dot{W}_C}{100} \quad \therefore \quad \dot{W}_C = 90 \text{ kW}$$

If this value is to be used in a further equation or formula, it must be given a positive sign according to sign rule B. The level of the drive efficiency can often have an important influence upon the overall efficiency of a system.

EXAMPLE 4.3

In an oil to water heat exchanger, the hot oil from an engine enters the heat exchanger with a flow rate of 0.5 kg/s and a temperature of 80 °C, and leaves at a temperature of 60 °C. The cold water enters with a flow rate of 0.15 kg/s and at a temperature of 20 °C, and leaves at a temperature of 55 °C (Figure 4.11). If the oil and water specific heats are 2.5 and 4.2 kJ/kg K respectively, determine the energy transfer efficiency of the heat exchanger.

Due to changing circumstances the oil flow rate is to be increased to 0.6 kg/s, but the temperature drop of the oil has to be maintained. A modification is made which allows more heat transfer within the fixed size of the heat exchanger. The mass flow rate of water and inlet water temperature remain at their initial values. If the energy transfer efficiency of the heat exchanger is also unchanged, determine the new outlet temperature of the water.

94   THE FIRST LAW OF THERMODYNAMICS

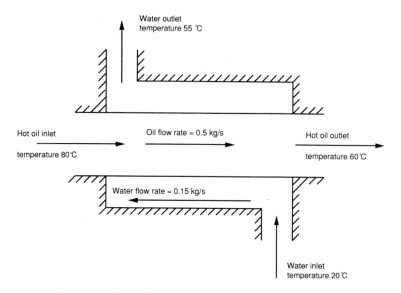

Fig. 4.11 *Example 4.3—an oil to water heat exchanger.*

Solution

Assuming that the oil goes from state 1 to state 2, using Equation (4.6), the heat transferred from the oil $\dot{Q}_{12}$ is

$$\dot{Q}_{12} = \{\dot{m}c(T_2 - T_1)\}_{\text{oil}} = 0.5 \times 2.5 \times (60 - 80) = -25 \text{ kW}$$

Assuming that the water goes from state 3 to state 4, using Equation (4.6), the heat gained by the water $\dot{Q}_{34}$ is

$$\dot{Q}_{34} = \{\dot{m}c(T_4 - T_3)\}_{\text{w}} = 0.15 \times 4.2 \times (55 - 20) = 22.1 \text{ kW}$$

Not all the heat from the oil is transferred to the water, some is transferred to the atmosphere. Using sign rule C, the energy transfer efficiency η_e is

$$\eta_e = \frac{22.1}{25} = 88.4\%$$

If the oil flow rate is increased to 0.6 kg/s, the heat transferred from the oil $\dot{Q}'_{12}$ is now:

$$\dot{Q}'_{12} = 0.6 \times 2.5 \times (80 - 60) = -30 \text{ kW}$$

The energy transfer efficiency remains the same, so the heat gained by the cold water $\dot{Q}'_{34}$ is now

$$\eta_e = 0.884 = \frac{\dot{Q}'_{34}}{\dot{Q}'_{12}} = \frac{\dot{Q}'_{34}}{30}$$

$$\therefore \quad \dot{Q}'_{34} = 26.5 \text{ kW}$$

Before determining the outlet water temperature from Equation (4.6), the appropriate sign must be applied to $\dot{Q}'_{34}$. As it is a heat gained by the cold water it must be positive. The new outlet water temperature $\{T'_4\}_{\text{w}}$ is

$$\dot{Q}'_{34} = +26.5 = 0.15 \times 4.2(\{T'_4\}_{\text{w}} - 20)$$

$$\therefore \quad \{T'_4\}_{\text{w}} = 62.1\,°\text{C}$$

In practice, it might be possible, and perhaps easier, to increase the water flow rate to cope with the additional heat being dissipated by the oil.

4.2.3 Efficiencies and costs

Efficiencies, as defined above, are a valuable way of assessing the performance of a thermal system but, unfortunately, they do not tell the full story. Costs are of vital importance too, but only the running costs and not the capital costs of any thermal system can be dealt with here. Costs do vary from year to year, in most cases they increase, and from country to country, which can result in cost calculations being inappropriate, but the importance of relating thermal performance to some monetary value outweighs any disadvantages.

To carry out an analysis of the running cost of a thermal system requires a knowledge of the price of any energy that is being consumed. Energy is such big business throughout the world that most of the energy sectors cost their particular fuel in a unit that is seen as being appropriate to them, rather than conforming to some agreed standard. Oil is usually costed in US$ per barrel and natural gas in a local currency per m^3. It is not realistic to cover all the possible variations so, in order to evaluate how much money has to be spent operating a thermal system, all energies will be priced in $/kWh. To obtain an Annual Running Cost $\$_{ARC}$, the energy transfer that has to be paid for E in kW, the number of hours of operation per year N_{hrs} of the plant, and the cost of the energy $\$_{kWh}$ in $/kWh must be known. Then

$$\$_{ARC} = E N_{hrs} \$_{kWh} \tag{4.9}$$

where E is always given a positive value. The Simple Payback Period (SPP) is defined as the ratio of 'the extra capital cost that is incurred in making a change to a thermal system to the saving in the running cost due to that change'. While it is not the intention to deal with capital costs, if an SPP for an investment is stated, the amount of capital that can be spent in making a change to a thermal system can be calculated.

The price of oil, which is traded and transported throughout the world, can vary quite wildly depending upon trouble erupting in various areas, particularly in oil-producing countries. A cartel consisting of a number of nations tries to set the price, but with only limited success. The price of natural gas does not vary so much because it is tied to a region due to the difficulties and costs encountered in transporting it. Coal, being a less user-friendly fuel, has to compete with the other two and its price is subject to their variations, unless long-term contracts have been signed between coal producers and consumers. Oil, natural gas and coal are primary fuels. Electricity is a secondary energy or fuel because it is produced following the combustion of one of the primary fuels. Therefore it is subject to the overall thermal efficiency of the power station. Electricity at night-time is generally half the price of electricity during the daytime in order to encourage greater consumption during the night, because the power station cannot be simply turned off when the demand drops. However, although the price of fuels may vary somewhat from country to country, it is still common to find that the price of natural gas is typically one-third of the price of daytime electricity.

EXAMPLE 4.4

A 70% energy transfer efficiency boiler uses coal at a cost of 0.04 $/kWh, runs for 6000 hours per year and produces 500 kW of steam (Figure 4.12). What is the annual running cost of the boiler?

THE FIRST LAW OF THERMODYNAMICS

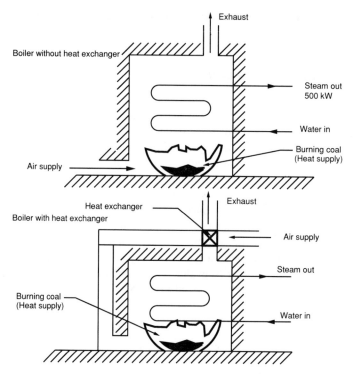

Fig. 4.12 *Example 4.4— reducing the annual running cost of a boiler by the inclusion of a heat exchanger in the exhaust to preheat the air for combustion.*

If a heat exchanger could be fitted to the exhaust to preheat the air before combustion, how much money could be spent on it if an SPP of three years is specified and it is estimated that the boiler energy transfer efficiency would increase to 80%?

Solution

At an energy transfer efficiency of 70%, the energy in the fuel $\dot{Q}_{comb}$, using sign rule C, is given by

$$\eta_e = 0.7 = \frac{500}{\dot{Q}_{comb}}$$

$$\therefore \dot{Q}_{comb} = \frac{500}{0.7} = 714.3 \text{ kW}$$

From Equation (4.9), the annual running cost $\$_{ARC}$ of the boiler is

$$\$_{ARC} = E N_{hrs} \$_{kWh} = 714.3 \times 6000 \times 0.04 = \$171\,432 \text{ per year}$$

If the boiler energy transfer efficiency is improved to 80%, the improved annual running cost $\$'_{ARC}$ is

$$\$'_{ARC} = \frac{500 \times 6000 \times 0.04}{0.8} = \$150\,000 \text{ per year}$$

The saving in annual running cost is

$$\$_{ARC} - \$'_{ARC} = \$21\,432 \text{ per year}$$

Therefore, the amount of capital that could be spent for a 3-year SPP is:

$$\text{Amount of capital} = 3 \times 21\,432 = \$64\,296$$

Heat exchangers which are used to preheat the air before combustion in a boiler are usually good investments with a reasonably short pay-back period.

4.3 EXAMPLES OF THE FIRST LAW OF THERMODYNAMICS

With the methods described above, it is now possible to obtain measurements of work and heat transfer in a thermal system and to justify the first law of thermodynamics. All the thermal systems explored would appear in the energy triangle of Figure 1.3. The reciprocating piston cylinder engine, the steam plant and the gas turbine all convert heat input into work output. The heat pump is a little different in that it converts a work input into a heat output. Note that the first law must be applied to the fluid undergoing a cycle.

4.3.1 Energy balance on a reciprocating engine

It is possible to apply the first law of thermodynamics to a reciprocating piston cylinder engine although the terms closed system and cycle may require a fairly liberal interpretation. A cycle is effectively completed because at any time fresh air at a given state is always being induced into the cylinder and products of combustion exhausted to the atmosphere, so the processes are continually being repeated. It could be a closed system if the boundary is drawn wide enough to allow the products of combustion at exhaust to be recycled as air at inlet! It is easier to consider it from the point of view of the conservation of energy, the energy input into the engine must equal the energy output, and all the energies take the form of work or heat transfer. Therefore, an energy balance on a reciprocating engine should reveal that the first law of thermodynamics is valid. A similar analysis could also be carried out on a reciprocating compressor although, in this case, the energy output would be in the form of high-pressure fluid.

With reference to Figure 4.13, the energy input to the engine is in the form of heat transfer due to the combustion of a fuel. If the calorific value of the fuel and the mass flow rate of fuel into the engine are known, the heat input $\dot{Q}_{comb}$ can be determined. Assuming that the combustion of the fuel is complete, the heat input into the engine $\dot{Q}_i$ will be equal to the heat produced by the combustion of the fuel $\dot{Q}_{comb}$.

The energy output takes a number of forms. Firstly, there is the brake power produced by the engine $\dot{W}_{bp}$ which can be measured on a dynamometer. Then there are a number of heat outputs. Some of the heat produced by the fuel is transferred to the cooling water which circulates through the cylinder walls. This is necessary to prevent the engine becoming too hot, thereby distorting the engine block. Assuming that it goes from state 1 to state 2, the heat transfer in the cooling water $\dot{Q}_{12}$ can be determined by measuring its mass flow rate, specific heat and temperature difference. Some heat is transferred into the exhaust. Assuming

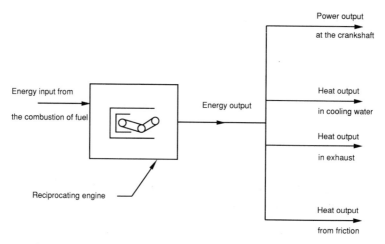

Fig. 4.13 The energy transfer in a reciprocating engine.

that it goes from state 3 to state 4, this quantity of heat transfer $\dot{Q}_{34}$ can also be determined by measuring the mass flow rate of exhaust, its specific heat and temperature rise, which is the difference in temperature between the exhaust and the air at inlet to the engine. The remainder of the energy is used in overcoming the effects of friction in the engine, in other words the friction power $\dot{W}_F$, expressed as a heat transfer out of the system, $\dot{Q}_F$.

If the first law of thermodynamics is valid, then

$$\sum(\dot{Q})_{cyc} + \sum(\dot{W})_{cyc} = 0$$

and hence:

$$\dot{Q}_i + \dot{Q}_{12} + \dot{Q}_{34} + \dot{Q}_F + \dot{W}_{bp} = 0$$

According to the sign convention adopted, $\dot{Q}_{12}$, $\dot{Q}_{34}$, $\dot{Q}_F$ and $\dot{W}_{bp}$ will all be negative.

EXAMPLE 4.5

A performance test on a single-cylinder, four-stroke, reciprocating internal combustion petrol engine produced the following data:

Petrol flow rate into engine	$= 2 \times 10^{-6}$ m³/s
Density of petrol	$= 740$ kg/m³
Calorific value of petrol	$= 43\,000$ kJ/kg
Airflow rate into engine	$= 0.025$ m³/s
Air density	$= 1.2$ kg/m³
Air temperature at engine inlet	$= 285$ K
Exhaust temperature at engine outlet	$= 775$ K
Specific heat of exhaust	$= 1.01$ kJ/kg K
Cooling water flow rate	$= 0.12$ kg/s
Cooling water inlet temperature	$= 285$ K
Cooling water outlet temperature	$= 325$ K
Specific heat of cooling water	$= 4.2$ kJ/kg K

Load on dynamometer = 320 N
Dynamometer load applied at distance = 0.3 m
Engine speed = 2000 rev/min
Area of $p-V$ diagram = 1.668 kJ

Carry out an energy balance on the engine to show that the first law of thermodynamics is valid and determine the overall thermal efficiency of the engine.

Solution

From Equation (4.7), the heat input to the engine $\dot{Q}_{comb}$ is

$$\dot{Q}_{comb} = \{\dot{m}\}_{fuel} CV = 2 \times 10^{-6} \times 740 \times 43\,000 = 63.6 \text{ kW}$$

Assuming that the combustion of the fuel is complete, the heat input into the engine $\dot{Q}_i$ will be equal to the heat produced by the combustion of the fuel $\dot{Q}_{comb}$.

$$\therefore \quad \dot{Q}_i = \dot{Q}_{comb} = 63.6 \text{ kW}$$

From Equation (4.6), the heat gained by the cooling water $\dot{Q}_{12}$ in going from state 1 to state 2 is

$$\dot{Q}_{12} = \{\dot{m}c(T_2 - T_1)\}_w = 0.12 \times 4.2 \times (325 - 285) = 20.2 \text{ kW}$$

Noting that the mass flow rate of the exhaust is the sum of the mass flow rate of the air and the mass flow rate of the fuel, the heat gained by the exhaust $\dot{Q}_{34}$ in going from state 3 to state 4, from Equation (4.6), is

$$\dot{Q}_{34} = \{\dot{m}c(T_4 - T_3)\}_{ex} = (0.025 \times 1.2 + 2 \times 10^{-6} \times 740) \times 1.01 \times (775 - 285)$$
$$\therefore \quad \dot{Q}_{34} = 15.6 \text{ kW}$$

From Equation (4.2), the brake power output $\dot{W}_{bp}$ at the dynamometer is

$$\dot{W}_{bp} = -FL2\pi N_{rev} = -\frac{320 \times 0.3 \times 2\pi \times 2000}{60} = -20.1 \text{ kW}$$

Noting that it is a four-stroke engine and that the number of cycles per second is half the number of revolutions per second, the indicated power $\dot{W}_{ip}$ from Equation (4.3) is

$$\dot{W}_{ip} = -A_{pV} N_{cyc} = -\frac{1.668 \times 2000}{60 \times 2} = -27.8 \text{ kW}$$

From Equation (4.4), the friction heat loss $\dot{Q}_F$ is

$$\dot{Q}_F = \dot{W}_F = \dot{W}_{ip} - \dot{W}_{bp} = -27.8 - (-20.1) = -7.7 \text{ kW}$$

It is negative from the calculation because it is a heat output from the system. Using sign rule B, the heat transfers out of the system in the cooling water and the exhaust are also negative. Therefore, the sum of the heat transfers is

$$\sum(\dot{Q})_{cyc} = \dot{Q}_i + \dot{Q}_{12} + \dot{Q}_{34} + \dot{Q}_F = 63.6 - 20.2 - 15.6 - 7.7 = 20.1 \text{ kW}$$

and the sum of the work transfers is

$$\sum(\dot{W})_{cyc} = \dot{W}_{bp} = -20.1 \text{ kW}$$

$$\therefore \quad \sum(\dot{Q})_{cyc} + \sum(\dot{W})_{cyc} = 20.1 - 20.1 = 0$$

100 THE FIRST LAW OF THERMODYNAMICS

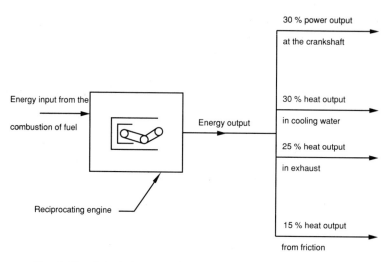

Fig. 4.14 *A typical energy balance on a reciprocating engine.*

The first law of thermodynamics is obeyed (Figure 4.14).

To carry out an energy balance in percentage terms, and making use of sign rule C:

$$\text{Energy input in power output} = \frac{\dot{W}_{bp}}{\dot{Q}_i} = \frac{20.1}{63.6} = 31.6\%$$

$$\text{Energy input in cooling water} = \frac{\dot{Q}_{12}}{\dot{Q}_i} = \frac{20.2}{63.6} = 31.8\%$$

$$\text{Energy input in exhaust} = \frac{\dot{Q}_{34}}{\dot{Q}_i} = \frac{15.6}{63.6} = 24.5\%$$

$$\text{Energy input in friction} = \frac{\dot{Q}_F}{\dot{Q}_i} = \frac{7.7}{63.6} = 12.1\%$$

The overall thermal efficiency η is

$$\eta = \frac{\dot{W}_{bp}}{\dot{Q}_{comb}} = \frac{20.1}{63.6} = 31.6\%$$

This is quite a high value for the overall efficiency of a petrol engine, but is possible for a stationary unit operating at a fixed speed. Unfortunately, it shows that nearly 70% of the energy input is not converted into useful energy output. Some of the energy in the cooling water could be classified as useful if it is circulated through a heat exchanger inside the passenger compartment in a car in order to keep the occupants warm. But this is only applicable in the winter. As stated in Chapter 1, the overall thermal efficiency of the engine is low, but this is not a fault in the design of the engine. Over 25% of the energy input is dissipated in the exhaust. In Example 6.7, actual fluid temperatures and pressures are determined for a typical engine cycle, and it will be seen that it is not possible to reduce the exhaust temperature to atmospheric conditions because of limitations on the exhaust pressure, which cannot go below atmospheric or the exhaust gas will not be forced out of the exhaust valve.

4.3.2 Energy balance on a steam plant

A simplified steam plant, or thermal power station, producing electricity consists of four main components, as shown in Figure 4.15, where the fluid is shown to follow a clockwise path around the system in keeping with the convention for a net work output.

In the boiler, liquid water at high pressure is converted to steam vapour at approximately the same pressure. A heat input $\dot{Q}_i$ is required to boil a fluid and this is provided by the combustion of a fuel in the boiler. As there is an energy transfer efficiency in the boiler, only some of the heat of combustion $\dot{Q}_{comb}$ actually ends up in the working fluid. The steam is expanded in a rotary turbine, producing a work output $\dot{W}_T$ which drives the alternator for electricity. In the expansion, the steam pressure is converted to kinetic energy which drives the rotor, as described in Section 1.2. Therefore, the steam leaves the turbine as a low-pressure vapour. In the condenser, the steam is cooled and changes phase, becoming liquid water at approximately the same low-pressure with which it enters the condenser. Heat $\dot{Q}_c$ is given out by the steam condensing, which must be dissipated, and this can be achieved by passing cold water from a nearby river through the condenser. Finally, in the feed pump, the low-pressure water is pumped up to become the high-pressure water required in the boiler, the feed pump behaving like a rotary compressor. The feed pump, therefore, requires a work input $\dot{W}_{FP}$, which is usually supplied by bleeding some steam from the high-pressure part of the circuit and passing it through a small separate turbine dedicated to running the feed pump.

Although each of the four main components of the system, the boiler, turbine, condenser and feed pump, are very definitely open-flow devices in which the fluid enters and leaves

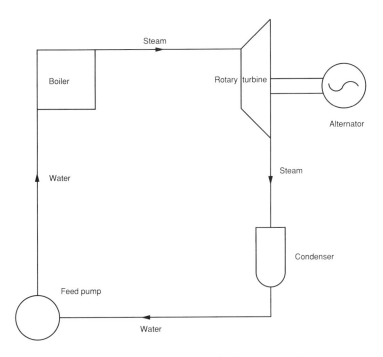

Fig. 4.15 *The energy transfer in a steam plant.*

THE FIRST LAW OF THERMODYNAMICS

each component in turn, the water/steam completes a closed cycle by continually passing round the system. Thus, it should demonstrate that the first law of thermodynamics is valid and that

$$\sum(\dot{Q})_{cyc} + \sum(\dot{W})_{cyc} = 0$$
$$\therefore \quad \dot{Q}_i + \dot{Q}_c + \dot{W}_T + \dot{W}_{FP} = 0$$

According to the sign convention, $\dot{Q}_c$ is a heat output from the system and should be negative, and $\dot{W}_T$ is a work output from the system which should also be negative.

EXAMPLE 4.6

In a coal-fired steam power station, the following performance data was recorded:

Mass flow of coal consumed in boiler	= 2.46 kg/s
Calorific value of coal	= 35 000 kJ/kg
Energy transfer efficiency of the boiler	= 70%
Power output from the turbine	= 21.2 MW
Mass flow rate of cooling water into condenser	= 248.8 kg/s
Temperature of cooling water into condenser	= 280 K
Temperature of cooling water out of condenser	= 310K
Specific heat of cooling water	= 4.2 kJ/kg K
Energy transfer efficiency of condenser	= 80%
Power input to feed pump	= 105 kW
Energy transfer efficiency of feed pump drive	= 95%

Show that the first law of thermodynamics is valid and determine the overall thermal efficiency of the system.

Solution

In the boiler, fuel is burned producing heat. From Equation (4.7), the heat transfer available from the fuel $\dot{Q}_{comb}$ is

$$\dot{Q}_{comb} = \{\dot{m}\}_{fuel} CV = 2.46 \times 35\,000 = 86.1 \text{ MW}$$

At an energy transfer efficiency η_e of 70%, the energy input to the working fluid $\dot{Q}_i$ is

$$\dot{Q}_i = \dot{Q}_{comb}\eta_e = 86.1 \times 0.7 = 60.3 \text{ MW}$$

The work output from the turbine $\dot{W}_T$ is 21.2 MW. Thus:

$$\dot{W}_T = -21.1 \text{ MW}$$

In the condenser, the heat gained by the cooling water $\dot{Q}_{12}$ in going from state 1 to state 2 is given by Equation (4.6) as

$$\dot{Q}_{12} = \{\dot{m}c(T_2 - T_1)\}_w = 248.8 \times 4.2 \times (310 - 280) = 31\,348.8 \text{ kW}$$

But there is an energy transfer efficiency of 80% in the condenser. Therefore, the actual heat given out by the steam condensing $\dot{Q}_c$ is

$$\dot{Q}_c = \frac{31\,348.8}{\eta_e} = \frac{31\,348.8}{0.8} = 39\,186 \text{ kW} = 39.2 \text{ MW}$$

In the feed pump there is a power output at the drive system of 105 kW, so with an energy transfer efficiency of 95%, the power input to the feed pump $\dot{W}_{FP}$ is

$$\dot{W}_{FP} = \eta_e \times 105 = 0.95 \times 105 = 99.75 \text{ kW} = 0.1 \text{ MW}$$

According to sign rule B, the heat transferred out of the working fluid in the condenser is negative. The sum of the heat transfers is

$$\sum (\dot{Q})_{cyc} = \dot{Q}_i + \dot{Q}_c = 60.3 - 39.2 = 21.1 \text{ MW}$$

The sum of the work transfers is

$$\sum (\dot{W})_{cyc} = \dot{W}_T + \dot{W}_{FP} = -21.2 + 0.1 = -21.1 \text{ MW}$$

$$\therefore \quad \sum (\dot{Q})_{cyc} + \sum (\dot{W})_{cyc} = 21.1 - 21.1 = 0$$

The first law of thermodynamics is obeyed (Figure 4.16). The overall thermal efficiency of the system η, making use of sign rule C and noting that both the feed pump work input and the fuel for the boiler must be paid for, is given by

$$\eta = \frac{\sum (\dot{W})_{cyc}}{\dot{Q}_{comb} + \dot{W}_{FP}} = \frac{21.2}{86.1 + 0.105} = 24.6\%$$

The overall thermal efficiency is quite low because the steam plant is rather small. Modern large coal- and natural-gas-fired steam plants of capacity around 600 MW are capable of reaching efficiencies approaching 40%. However, even at such a high level, 60% of the input energy does not appear as useful energy output. With reference to Chapter 1, this is not a design fault in the power station. In order to recycle the fluid, it must be condensed in the condenser from the steam vapour state to the liquid water state before returning via the feed pump to the boiler as high-pressure water. Much energy is dissipated in cooling towers or rivers in order to achieve the phase change.

It might be thought that the steam at exit from the turbine could be directly fed into the feed pump before the boiler. However, compressing a vapour like this requires considerably more work input than that needed to compress a liquid through the same pressure range. Therefore, the net work output from the plant would decrease. In addition, the net work transfer achieved in the cycle is given by the area of the steam pressure against volume diagram. As will be seen in Chapter 7, unless the steam is condensed to become liquid water in the condenser, the area of the diagram for practical values of steam pressure and temperature is relatively tiny, again indicating a small net work output from the system.

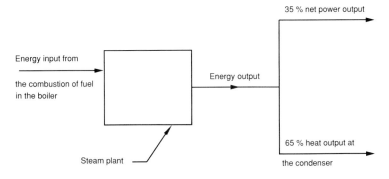

Fig. 4.16 *A typical energy balance on a steam plant.*

4.3.3 Energy balance on a refrigerator/heat pump

The four main components that make up an air-to-air vapour compression refrigerator or heat pump are shown in Figure 4.17. The fridge and the heat pump operate on the same cycle but perform different functions. In both, as in Figure 4.1, the refrigerant follows an anticlockwise path around the system, in keeping with the convention for a net work input.

In the compressor, low-temperature and low-pressure refrigerant vapour is compressed to a higher pressure and temperature. The compressor must be driven by an external power source, for example an electric motor, and so requires a power input $\dot{W}_C$. In the condenser, the refrigerant is made to condense at approximately constant pressure to become a liquid, thereby giving out heat $\dot{Q}_c$. To make it condense, the refrigerant must be cooled by the air at a lower temperature. In the expansion valve, the refrigerant liquid is expanded to a low-pressure and low-temperature. There is no work input or output and any heat transfer effects are small and can be neglected. In the evaporator, the refrigerant liquid is made to become a vapour at the conditions required for entry to the compressor. In order for the refrigerant to boil, it requires a heat input $\dot{Q}_{ev}$ which it receives from the air passing over the evaporator coils. For this to be achieved, the air must be at a higher temperature than the refrigerant.

The system is referred to as being air to air because the heat transferred into the refrigerant at the evaporator is from an air source, and that delivered by the refrigerant in the condenser is to an air sink. Vapour compression refers to the compression of the refrigerant vapour in the compressor. The duty of a refrigerator is to cool down the air which surrounds the evaporator in the fridge. The duty of a heat pump is to heat up the air which surrounds the condenser, but otherwise the two plants behave similarly.

As with the steam plant, although the four main components are open, the system as a whole is closed, the refrigerant completes a cycle and the system should obey the first law

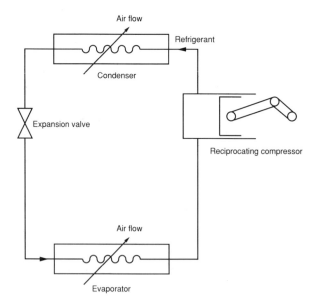

Fig. 4.17 *The energy transfer in a refrigerator/heat pump.*

of thermodynamics. Hence:

$$\sum (\dot{Q})_{\text{cyc}} + \sum (\dot{W})_{\text{cyc}} = 0$$

$$\therefore \quad \dot{Q}_{\text{ev}} + \dot{Q}_{\text{c}} + \dot{W}_{\text{C}} = 0$$

According to the sign convention, $\dot{Q}_c$ is a heat output from the condenser and should be negative. The heat pump is an example of a system which converts a work input into a heat output.

EXAMPLE 4.7

A domestic air-to-air vapour compression refrigerator has a work input to the refrigerant of 0.8 kW at the compressor. If the heat transferred from the refrigerant to the surroundings at the condenser is 1.7 kW, determine the heat transferred into the refrigerant at the evaporator assuming that the first law of thermodynamics is valid. If the electric motor driving the compressor has an energy transfer efficiency of 75%, determine the annual running cost of the refrigerator if it is in operation 5000 hours per year and the electricity for the motor costs 0.12 $/kWh.

If the drive energy transfer efficiency is improved to 90% by a simple modification, what are the savings in the annual running cost?

Solution

If the first law of thermodynamics is valid, Equation (4.1) is

$$\sum (\dot{Q})_{\text{cyc}} + \sum (\dot{W})_{\text{cyc}} = 0$$

If the heat transferred into the refrigerant at the evaporator is $\dot{Q}_{\text{ev}}$, the heat transferred out of the refrigerant at the condenser $\dot{Q}_c$, and the work input at the compressor $\dot{W}_C$, Equation (4.1) can be written

$$\dot{Q}_{\text{ev}} + \dot{Q}_c + \dot{W}_C = 0$$

According to sign rule B, $\dot{Q}_c$ is negative. Thus:

$$\dot{Q}_{\text{ev}} - 1.7 + 0.8 = 0$$

$$\therefore \quad \dot{Q}_{\text{ev}} = +0.9 \text{ kW}$$

As it is positive, it is a heat input to the refrigerant. At an energy transfer efficiency of 75%, the power required at the electric motor $\dot{W}_{\text{EM}}$, using sign rule C, will be

$$\dot{W}_{\text{EM}} = \frac{\dot{W}_c}{\eta_e} = \frac{0.8}{0.75} = 1.1 \text{ kW}$$

From Equation (4.9), the annual running cost $\$_{\text{ARC}}$ is

$$\$_{\text{ARC}} = E N_{\text{hrs}} \$_{\text{kWh}} = 1.1 \times 5000 \times 0.12 = \$660 \text{ per year}$$

With the modification to the electric drive, the new energy transfer efficiency η'_e is 90%. Therefore, the new electric motor power input $\dot{W}'_{\text{EM}}$ is

$$\dot{W}'_{\text{EM}} = \frac{\dot{W}_C}{\eta'_e} = \frac{0.8}{0.9} = 0.9 \text{ kW}$$

Using Equation (4.9), the new annual running cost $\$'_{ARC}$ is

$$\$'_{ARC} = 0.9 \times 5000 \times 0.12 = \$540 \text{ per year}$$

The saving in running cost is

$$\$_{ARC} - \$'_{ARC} = 660 - 540 = \$120 \text{ per year}$$

There are few alternatives to a refrigerator for achieving low temperatures, but it is important to reduce the running costs to as low a level as possible.

EXAMPLE 4.8

A domestic air-to-air vapour compression heat pump uses a 3 kW electric motor to drive the compressor. The drive has an energy transfer efficiency of 75%. Air of density 1.2 kg/m^3 and specific heat 1.005 kJ/kg K passes at the rate of 0.96 m^3/s over the evaporator coils, suffering a 4°C difference in temperature as it does so. The evaporator has an energy transfer efficiency of 80%. How much heat can be delivered by the refrigerant to the air passing over the condenser? Assume that the first law of thermodynamics is valid.

Given that the condenser has an energy transfer efficiency of 80%, determine the overall thermal efficiency of the heat pump.

Solution

For the air passing over the evaporator from state 1 to state 2, the heat transferred from the air $\dot{Q}_{12}$, from Equation (4.6), is

$$\dot{Q}_{12} = \{\dot{m}c(T_2 - T_1)\}_{air} = 0.96 \times 1.2 \times 1.005 \times (-4) = -4.6 \text{ kW}$$

Note that $\{T_2 - T_1\}_{air}$ is (-4) because the air temperature decreases. Applying sign rule C, and with an energy transfer efficiency of 80%, the energy in the refrigerant at the evaporator $\dot{Q}_{ev}$ is

$$\dot{Q}_{ev} = \frac{\dot{Q}_{12}}{\eta_e} = \frac{4.6}{0.8} = 5.8 \text{ kW}$$

Note that the energy transfer efficiency is defined as above because the refrigerant has to have more cooling energy than the air it is trying to lower the temperature of in order to counteract any heat gains to the air as its temperature falls.

At an energy transfer efficiency of 75%, the work input in the refrigerant $\dot{W}_C$ from the electric motor drive $\dot{W}_{EM}$ is

$$\dot{W}_C = \dot{W}_{EM}\eta_e = 3 \times 0.75 = 2.25 \text{ kW}$$

Assuming that the first law of thermodynamics is valid, Equation (4.1) is

$$\sum(\dot{Q})_{cyc} + \sum(\dot{W})_{cyc} = 0$$

If the heat transfer into the refrigerant at the evaporator is $\dot{Q}_{ev}$, the heat transfer from the refrigerant at the condenser is $\dot{Q}_c$, and the work input to the refrigerant from the compressor is $\dot{W}_C$, Equation (4.1) can be written:

$$\dot{Q}_{ev} + \dot{Q}_c + \dot{W}_C = 0$$
$$\therefore \quad +5.8 + \dot{Q}_c + 2.25 = 0$$
$$\therefore \quad \dot{Q}_c = -8.05 \text{ kW}$$

This is as it should be in that the heat transferred out of the refrigerant at the condenser to the surroundings is a negative according to the sign convention, and it has come out to be negative according to sign rule A. At an energy transfer efficiency of 80% at the condenser, and using sign rule C, the heat delivered to the air passing over it as it goes from state 3 to state 4, $\dot{Q}_{34}$, is

$$\dot{Q}_{34} = 8.05 \times 0.8 = 6.4 \text{ kW}$$

In a heat pump, the useful energy out is the heat delivered to the air at the condenser, and the energy input that must be paid for is the electricity for the electric motor. Thus, the overall thermal efficiency η, using sign rule C, is

$$\eta = \frac{\dot{Q}_{34}}{\dot{W}_{EM}} = \frac{6.4}{3} = 213\%$$

Clearly, it is not right to have an efficiency greater than 100%. Therefore, for the case of a heat pump, the overall thermal efficiency is renamed the Coefficient of Performance of the heat pump COP_{hp} where: $COP_{hp} = 2.13$

There is nothing invalid about the result. The first law of thermodynamics has been followed and the analysis is correct. The answer arises from the definition of the efficiency. In effect, because the heat input at the evaporator is free, it need not be included in the term for the efficiency. Physically, the result means that every 1 kW of electricity paid for will produce 2.13 kW of heat. To have a heating device with such a high 'efficiency' is a big advantage. Unfortunately, heat pumps cannot always compete economically on capital and running cost grounds with natural gas heating in particular. Also, depending upon the refrigerant utilised, the heat output occurs at temperatures typically in the range of 50–60 °C, which can restrict the number of applications. High-temperature refrigerants, which allow higher temperature heat outputs, do exist, but they are expensive and render the capital cost of the heat pump unattractive.

Note that the coefficient of performance of a refrigerator COP_{fridge} is defined in terms of the heat transferred out of the air at the evaporator, because that is the purpose of a refrigerator, divided by the work input at the compressor. The COP of a heat pump can be shown, from the first law of thermodynamics, to always have a value greater than 1. This is not so for a refrigerator, although in practice, most refrigerators do have a COP above 1.

4.3.4 Energy balance on a closed cycle gas turbine

A simplified closed cycle gas turbine plant producing electricity consists of four main components, as shown in Figure 4.18, where the gas follows a clockwise path around the system, in keeping with the convention for a net work output.

In the heat exchanger, the working fluid enters at high pressure after being compressed in the compressor. There is a heat input $\dot{Q}_i$ to the working fluid which can be provided by the external combustion of a fuel. The heat exchanger is approximately a constant pressure process so the working fluid enters the turbine at high pressure and high temperature. The energy content is converted to work in the turbine as the fluid expands, the work output $\dot{W}_T$ being used to drive an alternator to produce electricity. The fluid, therefore, leaves the turbine at a lower pressure and temperature. In the cooler, it gives up further heat $\dot{Q}_{cool}$ to a heat sink, perhaps the water from a nearby river. The cooler is approximately a constant pressure process. The low-pressure and low-temperature fluid is compressed in the compressor to a higher pressure and temperature. The compressor requires a work input $\dot{W}_C$ which is usually provided by the turbine through a shaft which connects the two components. This completes the cycle for the fluid as it is now back to its former condition and ready to enter the heat exchanger again. The fluid used is a gas — it could be air but more likely it is helium as that has a better specific heat.

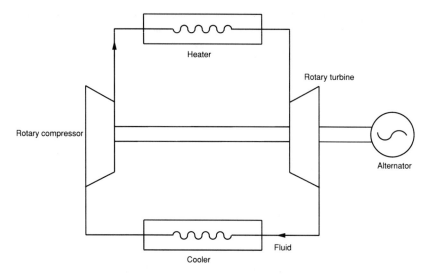

Fig. 4.18 *The energy transfer in a closed cycle gas turbine.*

The cycle should demonstrate that the first law of thermodynamics is valid and, therefore, that

$$\sum (\dot{Q})_{\text{cyc}} + \sum (\dot{W})_{\text{cyc}} = 0$$

$$\therefore \quad \dot{Q}_i + \dot{Q}_{\text{cool}} + \dot{W}_T + \dot{W}_C = 0$$

According to the sign convention, $\dot{Q}_{\text{cool}}$ is a heat output from the system and should be negative, and $\dot{W}_T$ is a work output from the system which should also be negative.

EXAMPLE 4.9

In a closed cycle gas turbine plant, the gas absorbs a heat input of 1000 kW in the heat exchanger. In the cooler, the water from a nearby river rises in temperature by 10 K, has a mass flow rate of 12 kg/s, and a specific heat of 4.2 kJ/kg K. The cooler has an energy transfer efficiency of 80%. The compressor requires a power input of 100 kW. If the compressor power input is provided by the turbine through a shaft connecting the two components, what is the net power output of the turbine?

If the heat input to the heat exchanger is provided by coal burning externally at an energy transfer efficiency of 65%, what is the mass flow rate of coal required if its calorific value is 32 000 kJ/kg?

What is the overall thermal efficiency of the plant?

If the coal costs $0.02/kWh and the plant runs for 5000 hours per year, what is the annual running cost of the plant?

If the alternator has an energy transfer efficiency of 95%, how much electrical power does it produce?

If the electricity produced by the alternator can be sold to consumers at the rate of $0.12/kWh, how much money will this generate, again assuming the plant runs for 5000 hours per year?

Solution

From Equation (4.6), the heat gained by the river water in the cooler as it goes from state 1 to state 2 $\dot{Q}_{12}$ is

$$\dot{Q}_{12} = \{\dot{m}c(T_2 - T_1)\}_w = 12 \times 4.2 \times 10 = 504 \text{ kW}$$

If the cooler has an energy transfer efficiency of 80%, the heat rejected by the fluid $\dot{Q}_{cool}$ in the cooler is

$$\dot{Q}_{cool} = \frac{\dot{Q}_{12}}{\eta_e} = \frac{504}{0.8} = 630 \text{ kW}$$

If the first law of thermodynamics is valid, Equation (4.1) is

$$\sum(\dot{Q})_{cyc} + \sum(\dot{W})_{cyc} = 0$$

If the heat input to the working fluid is $\dot{Q}_i$, the work output from the turbine $\dot{W}_T$, and the work input to the compressor $\dot{W}_C$, Equation (4.1) becomes

$$\dot{Q}_i + \dot{Q}_{cool} + \dot{W}_T + \dot{W}_C = 0$$

But $\dot{Q}_{cool}$ is negative according to sign rule B. Thus:

$$+1000 - 630 + \dot{W}_T + 100 = 0$$
$$\therefore \dot{W}_T = -470 \text{ kW}$$

The work output from the turbine should be negative according to the sign convention. As the turbine drives the compressor, the net power output of the system $\sum(\dot{W})$ is

$$\sum(\dot{W}) = \dot{W}_T + \dot{W}_C = -470 + 100 = -370 \text{ kW}$$

If the coal is burned with an energy transfer efficiency of 65%, the heat released by the coal $\dot{Q}_{comb}$ is:

$$\dot{Q}_{comb} = \frac{1000}{0.65} = 1538.5 \text{ kW}$$

From Equation (4.7), the mass flow rate of coal $\{\dot{m}\}_{fuel}$ required to produce this heat is

$$\{\dot{m}\}_{fuel} = \frac{\dot{Q}_{comb}}{CV} = \frac{1538.5}{32\,000} = 0.048 \text{ kg/s}$$

The overall thermal efficiency of the plant η, using sign rule C, is

$$\eta = \frac{\sum(\dot{W})}{\dot{Q}_{comb}} = \frac{370}{1538.5} = 24.1\%$$

From Equation (4.9), the annual running cost $\$_{ARC}$ of the coal boiler is

$$\$_{ARC} = E N_{hrs} \$_{kWh} = 1538.5 \times 5000 \times 0.02 = \$153\,850 \text{ per year}$$

If the alternator has an energy transfer efficiency of 95%, the power output at the alternator $\dot{W}_{alt}$, using sign rule C, is:

$$\dot{W}_{alt} = \sum(\dot{W})\eta_e = 370 \times 0.95 = 351.5 \text{ kW}$$

If the resultant electricity is sold to the consumer at \$0.12/kWh, the money generated, making use of Equation (4.9), is

$$\text{Money generated} = \dot{W}_{alt} N_{hrs} \$_{kWh} = 351.5 \times 5000 \times 0.12 = \$210\,900 \text{ per year}$$

The overall thermal efficiency of a closed cycle gas turbine plant for electricity generation is somewhat less than an equivalent steam turbine plant at the larger power outputs, and an equivalent diesel engine at the smaller power outputs, and they do not represent a commercial proposition. A considerable proportion of the energy input in the heater is dissipated in the cooler in order for the fluid to complete a cycle. The cooler could be dispensed with, in which case the system becomes an Open Cycle Gas Turbine. There is a saving in capital cost and system weight, but only air can be used as the fluid, whereas the other gases available in a closed cycle gas turbine, such as helium, have advantages in terms of their heat transfer characteristics.

Eliminating the cooler does not solve the problem of the low overall thermal efficiency and the need to dissipate heat to the atmosphere. The first law of thermodynamics may be applied to an open cycle gas turbine (such as an aircraft engine) in the same way that it is applied to a reciprocating engine, with the proviso of a liberal interpretation of the terms closed system and cycle. With no cooler fitted, the hot exhaust goes directly into the atmosphere and dissipates the heat without the need for a physical means of achieving this. The exhaust is hot, but it is not possible to reduce it to the atmospheric temperature because of the limitation imposed upon it by the pressure (the exhaust gas pressure cannot go below atmospheric otherwise the gas cannot be forced out of the turbine). The overall thermal efficiency is as low as for the closed cycle gas turbine, but it is not a design fault of the system.

4.4 COMBINED HEAT AND POWER PLANTS

All the power-producing plants discussed in this chapter—the reciprocating engine, the steam plant, the open and closed cycle gas turbine—have low overall thermal efficiencies because much of the energy input is exhausted to the atmosphere as heat. If this energy as heat can be utilised, it may be included as a useful energy output and the overall thermal efficiency would increase considerably. Such is the basis of a combined heat and power plant.

In a reciprocating engine, it is possible to fit a heat exchanger to provide hot water, for example. The water can first be used as the cooling water for the engine and secondly be heated by the exhaust gases. It is capable of reaching high and useful temperatures and finds many applications in industry.

In a steam plant, it is too expensive to dump the steam at exit from the turbine and continually input fresh water to the boiler via a feed pump. Instead the steam must be condensed to the liquid phase and recycled. If it is to be recycled, the latent heat produced in the condenser could be dumped at roughly the temperature of the available sink, the atmospheric temperature in the case of cooling towers, or the temperature of a nearby river or the sea if they are being used, about 30–40 °C. From Chapter 7, it will be seen that at this temperature, the pressure of the steam required for condensation to take place is approximately 6 kPa. This is a low pressure, but it means that the pressure drop of the steam in the turbine will be the highest achievable, and the power output of the turbine and the overall thermal efficiency of the plant, defined purely in terms of the power output, will be the maximum possible. The heat dissipated at the temperature of 30–40 °C finds little commercial application. However, if the condensation of the steam is made to take place at approximately atmospheric pressure, the heat transfer will be available at a temperature of about 100 °C. This heat can be picked up by a separate water system and used elsewhere. Such is the basis of a district heating scheme. The power output from the turbine is reduced, but the overall thermal efficiency of the plant actually increases provided the power output and the useful heat output (as received by the separate water system) are included in the equation.

The exhaust gas in an open cycle gas turbine is released at a very high temperature. The heat inherent in the gas may be transferred to useful hot water in a waste heat boiler or heat exchanger. Alternatively it may be utilised to generate high-temperature and high-pressure steam which is then further cycled through a steam system. Such is the basis of a combined cycle plant. The same may be applied to a closed cycle gas turbine, replacing the cooler.

All these approaches will improve the overall thermal efficiency of the system, but they are not worth doing if there is not a suitable demand for the heat. Even in a district heating scheme, nobody requires central heating in the summer-time. The heat is still being produced and must be got rid of somehow. It is often not worth, from the financial point of view, contemplating the installation of a combined heat and power plant unless there is a requirement for the heat on a regular basis for a considerable number of hours per year. In the end, it is a commercial decision that must be taken, the technology is available and the thermofluid principles are well established to determine the optimum quantities of heat and power that can be generated.

A heat pump is also a device which is very susceptible to a cost analysis. Unfortunately, the capital cost of a typical domestic heat pump per kilowatt is normally higher than a natural gas boiler, so any saving in running cost that arises which will reduce the payback period incurred through investing in one, is welcome. The disadvantage of an electric-driven heat pump is the cost of the electricity. The heat pump could be run at night, taking advantage of cheaper off-peak electricity rates, and the heat produced placed in a thermal store for use during the day, but the extra capital cost of the store, the greater occupied space, and the added complexity of the system mitigate against this approach. The heat pump could be driven by a natural gas engine, but for a typical small engine of 20 kW output and with a heat pump COP of 3, this gives a system heat output of 60 kW which takes it out of the domestic market. It is fine for the commercial and industrial market where there is a demand for the heat. However, introducing a natural gas engine of efficiency 30%, reduces the overall thermal efficiency of the complete system down to the level of a natural gas boiler, which is much simpler to operate and cheaper to buy. On the other hand, if the waste heat of the engine is also utilised, the system might be complex but attractive from the energy and financial aspects. Applications in hospitals, swimming-pools and hotels are numerous.

5 The Non-flow Energy Equation

Now that the first law of thermodynamics has been justified, it can be developed a little bit further. It is worth persevering with the study only of non-flow processes (as in a reciprocating piston cylinder mechanism when the valves are closed) in the first instance because this yields some useful information. Although most thermal systems incorporate devices in which the fluid undergoes flow processes, by restricting the analysis at this stage to non-flow processes, two other fluid properties can be identified.

Consider a thermal system which is undergoing a cycle but which consists of two non-flow processes only. Suppose that these processes can be described on a graph of fluid property A versus fluid property B, as shown in Figure 5.1.

For the first cycle, the fluid goes from state 1 to state 2 by non-flow process X and back to state 2 by non-flow process Z. In the second cycle, the fluid goes from state 1 to state 2 by non-flow process Y, but also returns to state 2 by non-flow process Z.

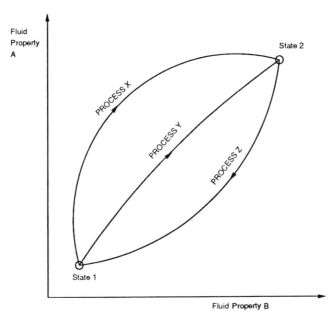

Fig. 5.1 *Graph of fluid property A against fluid property B for a fluid undergoing a cycle made up of two non-flow processes.*

THE NON-FLOW ENERGY EQUATION

The first law of thermodynamics can be written as

$$\sum (Q + W)_{\text{cyc}} = 0$$

For cycle XZ:

$$\sum (Q + W)_{\text{cycXZ}} = \sum (Q + W)_{\text{X}} + \sum (Q + W)_{\text{Z}}$$

And for cycle YZ:

$$\sum (Q + W)_{\text{cycYZ}} = \sum (Q + W)_{\text{Y}} + \sum (Q + W)_{\text{Z}}$$

Obviously, $\sum (Q+W)_{\text{Z}}$ is the same in both equations. Also, the first law of thermodynamics gives

$$\sum (Q + W)_{\text{cycXZ}} = \sum (Q + W)_{\text{cycYZ}} = 0$$

Therefore:

$$\sum (Q + W)_{\text{X}} = \sum (Q + W)_{\text{Y}}$$

In other words, the $\sum (Q + W)$ is independent of whether the fluid goes from state 1 to state 2 by non-flow process X or by non-flow process Y, or by any other non-flow process for that matter. But this fulfils the definition of a property. So the conclusion must be that the term $\sum (Q+W)$ is a property of the fluid, even though it consists of two terms which are not properties, namely Q and W. However, this is not unusual because Q and W themselves consist of the multiplication of two properties.

But what is this new fluid property that has been identified as a consequence of the first law of thermodynamics? It clearly must have the same units as the heat and the work transfer. So it is likely that it is an energy and it is called the Internal Energy U. As for Q and W, it must be mass dependent, giving rise to a specific internal energy u as follows:

Internal energy: U (units J)

Specific internal energy: u (units J/kg)

where $U = mu$.

The first law of thermodynamics may now be applied between state 1 and state 2 to a fluid undergoing a non-flow process in terms of the change in internal energy as follows:

$$Q_{12} + W_{12} = U_2 - U_1 = m(u_2 - u_1) \tag{5.1}$$

This equation is known as the Non-flow Energy Equation (NFEE), because it has been derived from a consideration of non-flow processes and utilises the first law of thermodynamics which is restricted to a closed system. It is a particular form of the principle of the Conservation of Energy in that it is limited to non-flow processes. It does not mean that the property internal energy is only applicable in non-flow processes. Its existence has been established by considering a non-flow process, but properties apply to all processes as they are used to describe the condition of a fluid at any state, independent of the process (Figure 5.2).

To give the internal energy a physical definition, it may be thought of as the inherent energy that a fluid possesses due to the molecules of which it is composed. Each molecule is

THE NON-FLOW ENERGY EQUATION   115

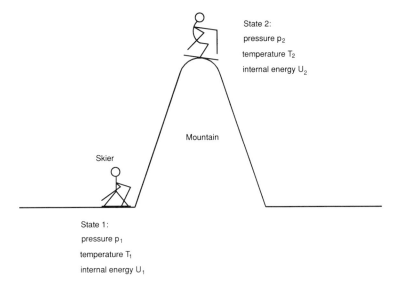

Fig. 5.2 *The internal energy of a fluid is a property.*

considered to be vibrating, rotating and diffusing and the sum of the molecular energies that this represents is the internal energy (Figure 5.3). Rather like entropy, it is another property of the fluid that it is not possible to sense or measure directly, but this need not provide a restriction on the further development of the necessary theory. Provided values of W and Q can be determined for a number of different non-flow processes, it becomes possible to derive a scale of values for the internal energy of a fluid, although it will only be for the internal energy difference as a fluid undergoes a change of state rather than absolute values of the internal energy itself that can be determined, as with the entropy of the fluid.

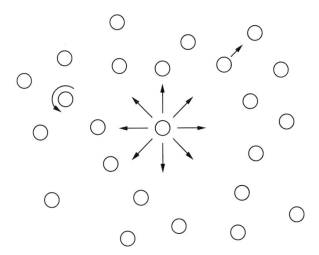

Fig. 5.3 *The internal energy of a fluid is the energy inherent in the molecules of the fluid due to their vibration, rotation and diffusion.*

As the internal energy is a measure of the molecular activity of the fluid, clearly the more active the molecules the higher the internal energy level. By the same token, as the temperature of a fluid is increased, its molecules vibrate more. Hence, raising the temperature of a fluid also increases its internal energy, or vice versa. Sometimes this is done to advantage, for example when heat is supplied to a fluid in an engine cycle, which is later converted into work transfer. At other times, it is difficult to convert the increase in internal energy into a more useful form. For example, when a fluid is made to flow over a solid surface, the friction force acting at the boundary tries to slow down its passage, as described in Section 2.5.2. If the fluid is provided with sufficient energy to maintain a constant flow rate, the energy used to overcome the retarding effect of the friction force at the boundary is converted into internal energy and the temperature of the fluid increases. It is not easy to convert the increase in internal energy into energy of a more useful nature. Indeed it may produce other deleterious effects. In the inlet manifold of a reciprocating piston cylinder engine, the increase in temperature of the air at high revolutions of the engine is enough to decrease the density of the air sufficiently to lower the mass flow rate, leading to a decline in the power output.

5.1 FIVE BASIC NON-FLOW PROCESSES FOR CHANGING THE STATE OF A FLUID

Work and heat transfer have been identified as not being properties of the fluid but rather that the amount of work and heat realised during a change of state is dependent upon the process. It is now necessary to be more specific about the types of non-flow process that may be undergone by the fluid. As the reversible formulae for W and Q are in terms of four fluid properties, it follows that manipulation of these properties will define the non-flow process. Therefore, four non-flow processes can be recognised immediately, namely a Constant Pressure non-flow process, a Constant Volume non-flow process, a Constant Temperature non-flow process, and a Constant Entropy non-flow process. One other non-flow process must be possible, that is when all the four properties vary, and this is called a Polytropic non-flow process. There must be five ways or non-flow processes by which a fluid can change its state and these are now considered.

At this stage, four main formulae have been identified for a fluid going from state 1 to state 2, which may be applicable to each non-flow process. These are

$$W_{12} = mw_{12} = -\int_1^2 p \, dV = -\int_1^2 mp \, dv \qquad (3.1)$$

$$Q_{12} = mq_{12} = \int_1^2 T \, dS = \int_1^2 mT \, ds \qquad (3.2)$$

$$\sum(Q)_{\text{cyc}} + \sum(W)_{\text{cyc}} = 0 \qquad (4.1)$$

$$Q_{12} + W_{12} = U_2 - U_1 = m(u_2 - u_1) \qquad (5.1)$$

The first two formulae apply only to reversible non-flow processes. The intention now is to examine whether some other information can be gleaned following the identification of the five types of process that can be undertaken by a fluid changing from state 1 to state 2 by a non-flow process, but any derivation based upon the definitions of W_{12} and Q_{12} is also restricted by the definition of reversibility.

One or more of the constant pressure, constant volume, constant temperature and constant entropy processes can be found in the main ideal cycles that have been proposed for reciprocating engines, namely the Carnot, Stirling, Otto and Diesel cycles. Although there is quite a bit of difference between the real and theoretical cycles, the latter still provide the basis for practical engines.

5.1.1 Constant pressure non-flow process

This is also known as an Isobaric non-flow process. The formula for W_{12} can be integrated for a reversible process because p is a constant. Therefore, W_{12} becomes

$$W_{12} = -p(V_2 - V_1) = -mp(v_2 - v_1) \qquad (5.2)$$

Also, differentiating Equation (5.1) and substituting for dW assuming a reversible process:

$$\mathrm{d}Q + \mathrm{d}W = \mathrm{d}U \quad \therefore \quad \mathrm{d}Q = \mathrm{d}U + p\,\mathrm{d}V$$

But for a constant pressure process:

$$\mathrm{d}(pV) = p\,\mathrm{d}V + V\,\mathrm{d}p = p\,\mathrm{d}V$$

Substituting gives

$$\mathrm{d}Q = \mathrm{d}U + \mathrm{d}(pV) = \mathrm{d}(U + pV) = \mathrm{d}H$$

Integrating:

$$Q_{12} = H_2 - H_1 \qquad (5.3)$$

Because the fluid properties $(U + pV)$ occur regularly in thermofluids, they are given a separate symbol H which is called the Enthalpy. This is, therefore, another property of the fluid, as it is made up of other properties, and must be mass dependent with a specific enthalpy h. Thus:

Enthalpy: H (units J)

Specific enthalpy: h (units J/kg)

Where

$$H = mh = U + pV = m(u + pv) \qquad (5.4)$$

Although it has been derived assuming a reversible non-flow constant pressure process, enthalpy is a property of the fluid just like volume and must be applicable whatever the process undergone by the fluid. As with the property internal energy though, it will only be possible to determine changes in the enthalpy of a fluid as it undergoes a process, rather than specific values.

Equation (5.3) shows that the enthalpy change of a fluid during a constant pressure non-flow reversible process is equal to the heat transfer. This is important because in a number of non-flow-type thermal systems the heat is transferred under constant pressure conditions.

118 THE NON-FLOW ENERGY EQUATION

For example, when a fluid changes phase and in the non-flow heat supply process of the theoretical Diesel cycle, which is the basis of the practical diesel engine. In the latter, the fuel is injected into the cylinder, which is full of compressed air, and burns which results in the pressure rising. At the same time, the piston is moving downwards from TDC position causing the fluid pressure to fall. If the two effects balance each other, the combustion of the fuel and the heat supply will take place at constant pressure.

As enthalpy is a property of the fluid, it is also applicable to flow processes just like any other property and it is in this regard that enthalpy assumes its major significance. An energy equation for a flow process, similar to the NFEE of Equation (5.1), will be developed in Chapter 8, and it will be seen that the enthalpy of the fluid is the major determinant of the heat transfer in boilers, combustion chambers and heat exchangers, and of the work transfer in turbines and compressors. In other words, it is a very important fluid property although its derivation here might seem somewhat innocuous.

EXAMPLE 5.1

A fluid is contained in the cylinder of a reciprocating piston cylinder mechanism. The piston is at TDC, so the fluid occupies the clearance volume. The fluid pressure is 5 MPa and the fluid volume 0.07 m^3. As the piston moves towards BDC and the fluid expands, a heat supply of 6000 kJ is transferred into it, thereby maintaining the pressure constant at 5 MPa. When the fluid volume has reached 0.1 m^3, the heat addition ceases (Figure 5.4).

Determine the change in enthalpy and the change in internal energy of the fluid, and the work transfer during the process, given that it may be assumed to be reversible.

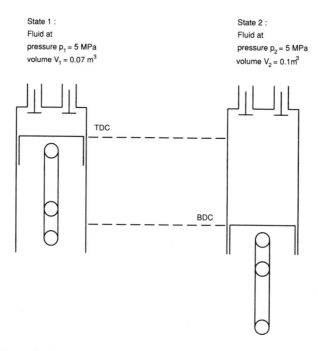

Fig. 5.4 *Example 5.1—a fluid in a piston cylinder mechanism expands in a constant pressure process as the piston moves downwards from TDC due to a heat supply of 6000 kJ.*

Solution

Let the fluid go from state 1 to state 2 in the reversible non-flow constant pressure process. Equation (5.3) allows the change of enthalpy of the fluid to be determined. Thus:

$$Q_{12} = H_2 - H_1$$

The heat transfer is a heat addition which is positive according to sign rule B. Hence:

$$+6000 = H_2 - H_1$$
$$\therefore \quad H_2 - H_1 = +6000 \text{ kJ}$$

The definition of enthalpy enables the change in internal energy of the fluid to be calculated. From Equation (5.4):

$$H_2 - H_1 = (U_2 + p_2 V_2) - (U_1 + p_1 V_1)$$
$$\therefore \quad U_2 - U_1 = 6000 \times 10^3 + 50 \times 10^5 \times 0.07 - 50 \times 10^5 \times 0.1 = 5850 \text{ kJ}$$

The work transfer is given by Equation (5.2) as

$$W_{12} = -p(V_2 - V_1) = -50 \times 10^5 \times (0.1 - 0.07) = -150 \text{ kJ}$$

Being negative, it is a work output. Alternatively, W_{12} could have been calculated from the first law of thermodynamics applied between the states. Using Equation (5.1):

$$Q_{12} + W_{12} = U_2 - U_1$$
$$\therefore \quad W_{12} = -6000 + 5850 = -150 \text{ kJ}$$

Equation (5.1) indicates that more work output could be obtained if the internal energy rise of the fluid could be minimised, which can be achieved by increasing the swept volume of the cylinder (as shown directly in Equation (5.2)).

5.1.2 Constant volume non-flow process

This is also known as an Isochoric non-flow process, in which

$$V_1 = V_2 \quad \therefore \quad dV = 0$$

But the work transfer in a reversible non-flow process is given by Equation (3.1). Hence, under constant volume conditions:

$$W_{12} = -\int_1^2 p \, dV = 0 \tag{5.5}$$

From Equation (5.1):

$$Q_{12} = mq_{12} = U_2 - U_1 = m(u_2 - u_1) \tag{5.6}$$

Therefore, during a constant volume change of state, no work is transferred and the heat transfer is equal to the change in the internal energy of the fluid. If heat is supplied, the internal energy of the fluid increases and its temperature rises, and vice versa. A constant volume non-flow heat supply process is utilised in the theoretical Otto cycle which is the

120 THE NON-FLOW ENERGY EQUATION

basis of the reciprocating petrol engine. The heat transfer into the fluid is then converted into useful work output in another non-flow process. A constant volume non-flow heat rejection process is utilised in both the theoretical Otto cycle and the theoretical Diesel cycle. This occurs when the exhaust valve is opened and the products of combustion rush out through the exhaust manifold.

EXAMPLE 5.2

A vessel of fixed volume 0.5 m³ contains a fluid. Heat is supplied to the fluid through the external combustion of coal. The mass of coal burned is 0.2 kg and its calorific value is 29 000 kJ/kg. There is an energy transfer efficiency of 60% when the heat of combustion is transferred to the fluid (Figure 5.5). What is the rise in internal energy of the fluid during the change of state if the process may be assumed to be reversible?

If the mass of the fluid is 0.9 kg, what is the initial specific volume of the fluid and the change in its value?

Solution

When the coal is burned, the heat released Q_{comb} according to Equation (4.7) is

$$Q_{comb} = \{m\}_{fuel} CV = 0.2 \times 29\,000 = 5800 \text{ kJ}$$

Let the fluid go from state 1 to state 2 in the reversible non-flow constant volume process.
At an energy transfer efficiency of 60%, the heat input to the fluid during the change of state Q_{12} is

$$Q_{12} = \{Q\}_{comb} \eta_e = 5800 \times 0.6 = 3480 \text{ kJ}$$

The change in internal energy of the fluid is given by Equation (5.6) as

$$Q_{12} = U_2 - U_1$$

where Q_{12} is a heat input to the fluid so, according to sign rule B, it must be allocated a positive sign. Therefore:

$$+3480 = U_2 - U_1$$
$$\therefore \quad U_2 - U_1 = +3480 \text{ kJ}$$

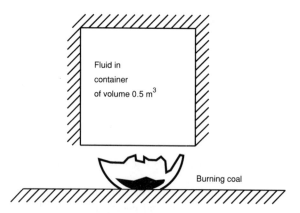

Fig. 5.5 Example 5.2—the internal energy of a fluid increases as heat is supplied in a constant volume process.

FIVE BASIC NON-FLOW PROCESSES FOR CHANGING THE STATE OF A FLUID

The specific volume of the fluid is the ratio of the total volume to the mass. But as the total volume V remains constant at 0.5 m³, and the mass m remains constant at 0.9 kg, the specific volume v also remains constant with a value of

$$v = \frac{V}{m} = \frac{0.5}{0.9} = 0.56 \text{ m}^3/\text{kg}$$

The change in specific volume of the fluid is zero as it is a constant volume process.

5.1.3 Constant temperature non-flow process

This is also known as an Isothermal non-flow process. The formula for Q_{12} can be integrated for a reversible process because T is now a constant. Thus:

$$Q_{12} = mq_{12} = T(S_2 - S_1) = mT(s_2 - s_1) \tag{5.7}$$

Isothermal non-flow reversible heat supply and heat rejection processes are two of the four processes which make up the theoretical Carnot cycle. Although there are not any practical engines working on the Carnot cycle, it is important because it provides a theoretical maximum cycle efficiency against which other cycles can be judged. The Carnot efficiency arises from the second law of thermodynamics and is dealt with in more detail in Chapter 12. Two reversible isothermal non-flow processes are also employed in the theoretical Stirling cycle.

EXAMPLE 5.3

A fluid is contained in the cylinder of a reciprocating piston cylinder mechanism. The piston is at BDC. The piston is made to move to TDC by a work input of 50 kJ. The fluid is compressed but its temperature is maintained constant at a value of 300 K by surrounding the cylinder with cooling water which absorbs the heat of compression. The mass of cooling water is 4.3 kg, its specific heat is 4.185 kJ/kg K, and the temperature rise of the water is measured to be 2.5 K. There is an energy transfer efficiency from the heat of compression to the cooling water of 90%, in other words 10% of the heat of compression is transferred to the atmosphere.

Determine the change in entropy and the change in internal energy of the fluid assuming the process is reversible (Figure 5.6).

Solution

The energy gained by the cooling water in going from state 3 to state 4, Q_{34}, according to Equation (4.6), is

$$Q_{34} = \{mc(T_4 - T_3)\}_w = 4.3 \times 4.185 \times 2.5 = 45 \text{ kJ}$$

At an energy conversion efficiency of 90%, the heat transferred out of the fluid during the non-flow reversible constant temperature compression process Q_{12} between states 1 and 2 is

$$Q_{12} = \frac{Q_{34}}{\eta_e} = \frac{45}{0.9} = 50 \text{ kJ}$$

The change of entropy of the fluid is given by Equation (5.7). Noting that Q_{12} must be negative according to sign rule B:

$$Q_{12} = T(S_2 - S_1)$$

$$\therefore \quad S_2 - S_1 = \frac{-50}{300} = -0.167 \text{ kJ/K}$$

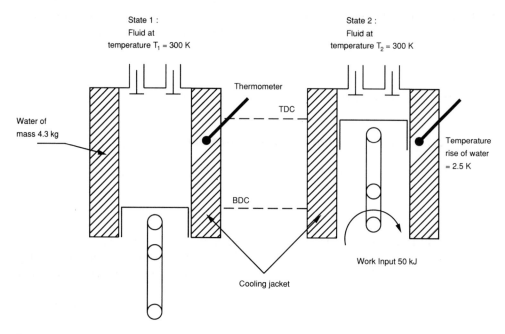

Fig. 5.6 *Example 5.3—a fluid in a piston cylinder mechanism is compressed in a constant temperature process as the piston moves upwards from BDC due to a work input of 50 kJ.*

The change in internal energy of the fluid is given by Equation (5.1). As before, Q_{12} is negative, but the work input term W_{12}, from sign rule B, is positive:

$$Q_{12} + W_{12} = U_2 - U_1$$
$$\therefore \quad U_2 - U_1 = (-50) + 50 = 0$$

Hence, there is no change in the internal energy of the fluid. In fact, this is a characteristic of an isothermal process when the fluid under consideration is a perfect gas and is the basis of what is known as Joule's law, as will be seen in Section 6.1.2.

5.1.4 Constant entropy non-flow process

This is also known as an Isentropic non-flow process in which

$$S_1 = S_2 \quad \therefore \quad dS = 0$$

But the heat transfer in a reversible non-flow process is given by Equation (3.2). Hence, under constant entropy conditions:

$$Q_{12} = \int_1^2 T \, dS = 0 \qquad (5.8)$$

From Equation (5.1):

$$W_{12} = mw_{12} = U_2 - U_1 = m(u_2 - u_1) \qquad (5.9)$$

Therefore, during a constant entropy change of state, no heat is transferred and the work transfer is equal to the change in the internal energy of the fluid.

Another word used to describe what happens to a fluid during a change of state when there is no heat transfer is Adiabatic. Any process with no heat transfer may be adiabatic, but the formula for Q containing entropy is for a reversible non-flow process only. Therefore, an isentropic non-flow process is a Reversible Adiabatic non-flow process, but an adiabatic process need not be reversible. As an isentropic process is the ideal one, this leads to the idea that different adiabatic processes can be categorised or ranked by determining how far from the isentropic case they diverge. An efficiency term can be introduced to quantify the divergence, and this is the approach adopted for a number of components which make up thermal systems. Chapter 8 describes this approach in detail.

The theoretical Carnot, Otto and Diesel cycles all include two reversible isentropic non-flow processes. It is a convenient process for going from one temperature level to another without the need for any heat transfer.

EXAMPLE 5.4

A fluid is contained within the cylinder of a reciprocating piston cylinder mechanism. The piston is at BDC. A work input of 600 kJ is utilised to push the piston to TDC and to compress the fluid. The cylinder is heavily lagged to prevent heat transfer, so the compression is adiabatic (Figure 5.7).

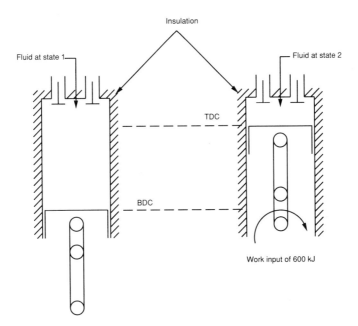

Fig. 5.7 *Example 5.4—a fluid in a piston cylinder mechanism is compressed in a constant entropy non-flow process as the piston moves upwards from BDC due to a work input of 600 kJ.*

What is the change in internal energy of the fluid during the change of state if the compression may be assumed to be reversible?

If the energy in the fluid, when the piston is at TDC, is utilised to push the piston back down to BDC such that the fluid ends up in the same condition as when it started, thereby completing a cycle, what is the work output on the expansion stroke and the change in internal energy of the fluid, again assuming a reversible adiabatic non-flow process?

Solution

Let the fluid go from state 1 to state 2 in the reversible non-flow constant entropy process.

If the cylinder is lagged, the compression is adiabatic. If the non-flow process is also reversible, it is isentropic. Therefore, the change in internal energy of the fluid is given by Equation (5.9) as

$$W_{12} = U_2 - U_1$$

The work transfer of compression W_{12} is a work input which requires a positive sign according to sign rule B. The change in internal energy is

$$+600 = U_2 - U_1 \quad \therefore \quad U_2 - U_1 = +600 \text{ kJ}$$

The internal energy of the fluid increases under compression because the molecules are squeezed into a smaller volume and have a higher activity level, thereby increasing their temperature even though the process is adiabatic. All the work input of compression is converted into internal energy of the fluid. If the fluid completes a cycle, the internal energy must revert to its initial condition. Thus:

$$U_1 - U_2 = -600 \text{ kJ}$$

Equation (5.9) gives the work transfer in the expansion W_{21} (noting that the fluid is now going from state 2 back to state 1) as

$$W_{21} = U_1 - U_2 = -600 \text{ kJ}$$

As W_{21} has a negative value, it is a work output which is what is expected when a piston moves from TDC to BDC. The solution shows that the isentropic non-flow process is indeed reversible but, as discussed in Section 3.3.2, this is the ideal state of affairs because the effects of friction have been ignored.

5.1.5 Polytropic non-flow process

In this non-flow process, all the four properties that go to make up the formulae for W and Q vary. This makes analysis of the effects of the change of state very difficult. Fortunately, it is possible to rely upon some experimental data to ease the mathematics. It can be shown that *some* fluids obey the relationship:

$$pV^n = \text{const.}$$

This formula is one which sometimes describes a polytropic non-flow process. The index of expansion or compression, n, may be assumed to have a constant value in most cases. The formula can be written in a number of ways, such as

$$pV^n = p_1 V_1^n = p_2 V_2^n = \text{const.} \tag{5.10}$$

$$pv^n = p_1 v_1^n = p_2 v_2^n = \frac{p_1}{\rho_1^n} = \frac{p_2}{\rho_2^n} = \text{const.} \tag{5.11}$$

FIVE BASIC NON-FLOW PROCESSES FOR CHANGING THE STATE OF A FLUID

For those fluids that do follow the polytropic relationship, the formula for the work transfer in a reversible non-flow process (Equation (3.1)) can now be integrated by making the substitution for p in terms of V. Thus:

$$W_{12} = -\int_1^2 p \, dV = -\int_1^2 \frac{\text{const.} \, dV}{V^n}$$

$$\therefore \quad W_{12} = -\frac{\text{const.}(V_2^{1-n} - V_1^{1-n})}{1-n}$$

$$\therefore \quad W_{12} = -\frac{p_2 V_2^n V_2^{1-n} - p_1 V_1^n V_1^{1-n}}{1-n}$$

$$\therefore \quad W_{12} = -\frac{p_2 V_2 - p_1 V_1}{1-n} = -\frac{m(p_2 v_2 - p_1 v_1)}{1-n} \qquad (5.12)$$

Many theoretical isothermal and isentropic non-flow processes tend towards being polytropic in practice, and Equation (5.12) is useful for determining the work transfer even though it is derived assuming a reversible change of state. Unfortunately there is no equivalent formula to Equations (5.10) and (5.11) linking the temperature and entropy of the fluid, and it is not possible to obtain an expression for the heat transfer in a polytropic non-flow reversible process in the same way. Instead, the heat transfer can be found from the NFEE applied between the two end states.

For the particular case when the index of expansion or compression $n = 1$, the integration becomes

$$W_{12} = -p_1 V_1 \ln \frac{V_2}{V_1} = -mp_1 v_1 \ln \frac{v_2}{v_1} \qquad (5.13)$$

One type of fluid which obeys the polytropic formula is perfect gases, as defined in Chapter 6. This makes analysis of the work and heat transfer somewhat straightforward. For those fluids which do not follow the polytropic relationship, a different approach needs to be taken. This is the case with water and water changing to steam, as in Chapter 7. When in the vapour phase, some liquids do behave in a polytropic manner in accordance with the formula.

EXAMPLE 5.5

A fluid is contained within a reciprocating piston cylinder mechanism. The piston is at BDC. The pressure of the fluid is 100 kPa and the volume of the fluid is 0.08 m³. The fluid is compressed until the piston reaches TDC where the pressure is 5 MPa. The compression is considered to be a reversible polytropic non-flow process and the fluid follows the polytropic formula with an index of compression of 1.2. The heat transfer out of the fluid during compression is measured as being 10 kJ (Figure 5.8)

Determine the work transfer during the process and the change in the internal energy of the fluid.

Solution

Let the fluid go from state 1 to state 2 in the reversible non-flow polytropic process.

From Equation (5.10), the volume of the fluid V_2 after compression is given by

$$p_1 V_1^{1.2} = p_2 V_2^{1.2}$$

$$\therefore \quad V_2 = \frac{V_1 p_1^{1/1.2}}{p_2^{1/1.2}} = \frac{0.08 \times 1^{1/1.2}}{50^{1/1.2}} = 0.0031 \text{ m}^3$$

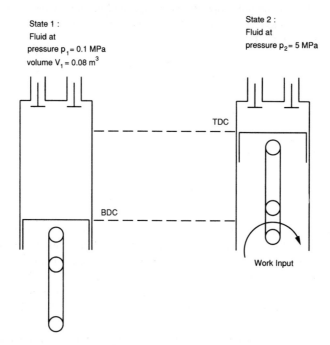

Fig. 5.8 *Example 5.5—a fluid in a piston cylinder mechanism is compressed in a polytropic non-flow process as the piston moves upwards from BDC due to a work input, giving rise to a heat output of 10 kJ.*

From Equation (5.12), the work done in compression of the fluid W_{12} is given by

$$W_{12} = -\frac{m(p_2 v_2 - p_1 v_1)}{1-n}$$

$$\therefore W_{12} = -\frac{50 \times 10^5 \times 0.0031 - 10^5 \times 0.08}{1 - 1.2} = +37.5 \text{ kJ}$$

As W_{12} is positive, it is a work input, as expected for a compression process.

Using Equation (5.1), the change in internal energy of the fluid can be determined:

$$Q_{12} + W_{12} = U_2 - U_1$$

The heat transfer during compression Q_{12} must be negative according to sign rule B because the heat is transferred out of the fluid. Thus:

$$-10 + 37.5 = U_2 - U_1$$

$$\therefore U_2 - U_1 = +27.5 \text{ kJ}$$

If there was no heat transfer out of the fluid during the change of state and the process became isentropic, the internal energy of the fluid would accordingly increase more and potentially be available for conversion into useful work output in another process.

5.2 FLUID SPECIFIC HEATS

In Chapter 4, the specific heat of a fluid was defined and was used to demonstrate that heat transfers could be evaluated during the change of state of a fluid in a thermal system. It is now possible to be more exact because the number of ways that the state can change in a non-flow process is restricted to five.

The definition of the specific heat was the heat required to raise the temperature of unit mass of substance by one degree. A polytropic non-flow process is the general one to which the formula for specific heat applies. From Equation (4.5):

$$c = \frac{dQ}{m\,dT}$$

Clearly, there can be no specific heat in either a constant temperature ($dT = 0$) or constant entropy ($dQ = 0$) non-flow process. That leaves two other possible specific heats, at constant pressure and constant volume.

In a constant pressure non-flow process, Equation (5.3) shows that the heat transfer is equal to the change in enthalpy of the fluid. Substituting in Equation (4.5) produces the specific heat at constant pressure c_p:

$$c_p = \frac{(dH)_p}{(m\,dT)_p} = \frac{(dh)_p}{(dT)_p} \qquad (5.14)$$

In a constant volume non-flow process, Equation (5.6) shows that the heat transfer is equal to the change in internal energy of the fluid. Substituting in Equation (4.5) produces the specific heat at constant volume c_v:

$$c_v = \frac{(dU)_v}{(m\,dT)_v} = \frac{(du)_v}{(dT)_v} \qquad (5.15)$$

As before, neither equation can be integrated unless the specific heat of the fluid is constant. In general, the specific heat varies with temperature but, if the temperature difference is not too large and the fluid remains in one phase, the specific heat may be assumed to stay approximately constant. However, if the fluid is changing phase, as in water changing phase to steam in a boiler, the specific heat is most definitely not constant and it is not possible to integrate Equations (5.14) and (5.15).

For those situations where the specific heat may be assumed to remain approximately constant, Equation (5.14) integrated gives

$$(H_2 - H_1)_p = (m(h_2 - h_1))_p = (mc_p(T_2 - T_1))_p \qquad (5.16)$$

and Equation (5.15) integrated gives

$$(U_2 - U_1)_v = (m(u_2 - u_1))_v = (mc_v(T_2 - T_1))_v \qquad (5.17)$$

In fact, Equations (5.16) and (5.17) must apply to both non-flow and flow processes because, with m, c_p and c_v constant, they link property values at the end states only, and properties are independent of the process followed. For flow processes, the equations are written:

$$(\dot{m}(h_2 - h_1))_p = (\dot{m}c_p(T_2 - T_1))_p$$

128 THE NON-FLOW ENERGY EQUATION

$$(\dot{m}(u_2 - u_1))_v = (\dot{m}c_v(T_2 - T_1))_v$$

Some typical values which occur regularly are as follows. For air:

$$\{c_p\}_{air} = 1.005 \text{ kJ/kg K}$$

$$\{c_v\}_{air} = 0.718 \text{ kJ/kg K}$$

And for water:

$$\{c_p\}_w = 4.186 \text{ kJ/kg K}$$

5.3 CYCLES CONSISTING OF NON-FLOW PROCESSES

If a fluid is to complete a cycle made up of non-flow processes, at least two processes are required. In practice, due to the mechanical realities of any system, more than two processes are usually necessary. If a reciprocating piston cylinder mechanism is considered as the basis for the cycle, an even number of processes generally need to be defined because the piston has to complete two strokes in order to get back to its original position.

Engine or compressor cycles can be envisaged in theory which encompass any of the above five non-flow processes. The problem is to decide which combination will best satisfy the requirements specified. For example, an engine might be wanted with the most power possible within a given volume capacity, or a compressor needed with a high efficiency. In practice, however, there are only a few cycles which can be made to work successfully in a reciprocating piston cylinder mechanism, or indeed, in any rotary mechanism.

EXAMPLE 5.6

A reciprocating piston cylinder compressor is to be built in which the fluid undergoes the following four non-flow reversible processes operating in a cycle:

1. State 1 to state 2—polytropic compression of the fluid in the cylinder from a pressure of 100 kPa to a pressure of 1 MPa, the volume decreasing from its value at BDC of 0.08 m³ to some intermediate position.

2. State 2 to state 3—constant pressure delivery of the fluid to a storage tank during which the volume further decreases to the TDC value of 0.01 m³.

3. State 3 to state 4—polytropic expansion of the fluid remaining in the clearance volume to a pressure of 100 kPa during which the piston moves from TDC to some intermediate position.

4. State 4 to state 1—constant pressure induction of fluid at 100 kPa from a storage well until the piston reaches BDC and a cycle is completed.

Sketch the cycle on a graph of fluid pressure versus volume and determine the net work input required to drive the compressor per cycle. The fluid obeys the polytropic relationship and the index of both expansion and compression is 1.2.

Determine the net heat transfer from the compressor during one cycle.

Solution

The graph of fluid pressure versus volume for the cycle is shown in Figure 5.9. Care must be exercised with any interpretation of the cycle because the actual mass of the fluid changes from one state to another. For example, between states 2 and 3, some of the fluid is expelled to a storage tank, and between states 4 and 1, some of the fluid is induced from a storage well.

CYCLES CONSISTING OF NON-FLOW PROCESSES

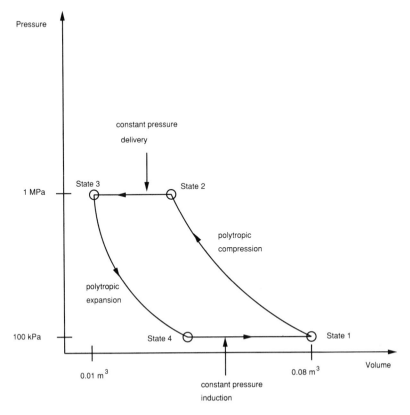

Fig. 5.9 *Example 5.6—a graph of the fluid pressure against volume as the fluid completes a compression cycle made up of four non-flow reversible processes.*

From state 1 to state 2, the non-flow reversible process is polytropic. The work transfer W_{12} during the change of state is given by Equation (5.12) as

$$W_{12} = -\frac{p_2 V_2 - p_1 V_1}{1 - n}$$

Here V_2 is unknown, but using Equation (5.10):

$$p_2 V_2^{1.2} = p_1 V_1^{1.2}$$

$$\therefore \quad V_2 = \frac{0.08 \times 1^{1/1.2}}{10^{1/1.2}} = 0.012 \text{ m}^3$$

$$\therefore \quad W_{12} = -\frac{10 \times 10^5 \times 0.012 - 10^5 \times 0.08}{1 - 1.2} = +2 \times 10^4 \text{ J}$$

From state 2 to state 3, the non-flow reversible process is constant pressure. The work transfer W_{23} during the change of state is given by Equation (5.2) as

$$W_{23} = -p_2(V_3 - V_2)$$

$$\therefore \quad W_{23} = -10 \times 10^5 \times (0.01 - 0.012) = +0.2 \times 10^4 \text{ J}$$

From state 3 to state 4 is another polytropic non-flow reversible process. Using Equation (5.12) again, the work transfer W_{34} during the change of state is

$$W_{34} = -\frac{p_4 V_4 - p_3 V_3}{1 - n}$$

Here V_4 is an unknown but using Equation (5.10):

$$p_4 V_4^{1.2} = p_3 V_3^{1.2}$$

$$\therefore \quad V_4 = \frac{0.01 \times 10^{1/1.2}}{1^{1/1.2}} = 0.068 \text{ m}^3$$

$$\therefore \quad W_{34} = -\frac{10^5 \times 0.068 - 10 \times 10^5 \times 0.01}{1 - 1.2} = -1.6 \times 10^4 \text{ J}$$

From state 4 back to state 1 is another constant pressure non-flow reversible process. From Equation (5.2), the work transfer W_{41} during the change of state is

$$W_{41} = -p_4(V_1 - V_4)$$

$$\therefore \quad W_{41} = -10^5 \times (0.08 - 0.068) = -0.12 \times 10^4 \text{ J}$$

The net work energy transfer in the cycle $\sum(W)_{\text{cyc}}$ is the sum of the work transfers in each of the four non-flow reversible processes that make up the cycle. Thus:

$$\sum(W)_{\text{cyc}} = W_{12} + W_{23} + W_{34} + W_{41}$$

$$\therefore \quad \sum(W)_{\text{cyc}} = +2 \times 10^4 + 0.2 \times 10^4 - 1.6 \times 10^4 - 0.12 \times 10^4 = +4.8 \text{ kJ}$$

The positive answer implies that there is a net work input into the system which should be the case with a compressor.

From the first law of thermodynamics, Equation (4.1) relates the net work and the net heat transfers in a cycle:

$$\sum(Q)_{\text{cyc}} + \sum(W)_{\text{cyc}} = 0$$

$$\therefore \quad \sum(Q)_{\text{cyc}} = -4.8 \text{ kJ}$$

Hence, there is a net heat transfer out of the compressor.

Practical reciprocating air compressors approximately follow the cycle described in the example, but are somewhat difficult to analyse because of the mass of fluid changing throughout the cycle. Some air always remains in the clearance volume while the rest is induced at approximately atmospheric conditions and exhausted at a higher pressure. The cycle goes anticlockwise in recognition of the convention that a compressor requires a net work input.

EXAMPLE 5.7

A reciprocating piston cylinder engine is to be built in which a fluid undergoes the following four non-flow reversible processes operating in a cycle:

1. State 1 to state 2—isentropic compression of the fluid from a temperature of 290 K to a temperature of 830 K.
2. State 2 to state 3—constant pressure heat addition until the temperature of the fluid is 2000 K.
3. State 3 to state 4—isentropic expansion of the fluid until its temperature reaches 990 K.
4. State 4 to state 1—constant volume drop in temperature of the fluid back to its original condition.

CYCLES CONSISTING OF NON-FLOW PROCESSES

The specific heats of the fluid at constant volume and at constant pressure may be assumed constant throughout the cycle and to have values of 0.7 kJ/kg K and 1.1 kJ/kg K respectively. If the mass flow rate of fluid is 0.05 kg/s, determine the net heat transfer per second in the cycle.

Also estimate the overall thermal efficiency of the proposed engine.

Solution

Although the cycle consists of four non-flow processes, the analysis is done for a mass flow rate of fluid of 0.05 kg/s rather than for a mass of fluid of so many kilograms because the cycle is repeated a number of times per second which produces the same end result. It becomes a matter of using time-dependent values in accordance with Section 3.4.1.

The graph of fluid pressure versus volume for the cycle is shown in Figure 5.10. The cycle is in the clockwise direction and should produce a net output of work for a net heat transfer input, as expected for an engine.

State 1 to state 2 is an isentropic non-flow process. From Equation (5.8), the heat transfer $\dot{Q}_{12}$ during the change of state is zero. Thus $\dot{Q}_{12} = 0$.

State 2 to state 3 is a constant pressure non-flow reversible process. From Equation (5.3), the heat transfer $\dot{Q}_{23}$ during the change of state is given by

$$\dot{Q}_{23} = \dot{m}(h_3 - h_2)$$

Because the specific heat remains constant throughout, substitution of Equation (5.16) for the constant pressure process gives

$$\dot{Q}_{23} = \dot{m}c_p(T_3 - T_2) = 0.05 \times 1.1 \times (2000 - 830) = 64.4 \text{ kJ/s}$$

State 3 to state 4 is another isentropic non-flow process. From Equation (5.8), the heat transfer $\dot{Q}_{34}$ during the change of state is again zero. Thus $\dot{Q}_{34} = 0$.

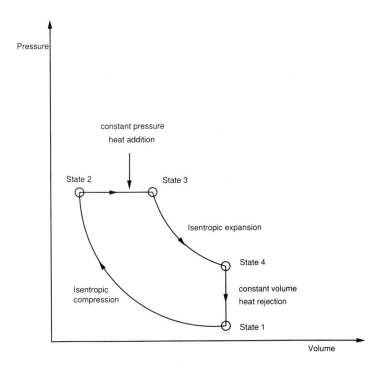

Fig. 5.10 *Example 5.7—a graph of the fluid pressure against volume as the fluid completes an engine cycle made up of four non-flow reversible processes.*

132 THE NON-FLOW ENERGY EQUATION

State 4 to state 1 is a constant volume non-flow reversible process. The heat transfer $\dot{Q}_{41}$ during the change of state is given by Equation (5.6) as

$$\dot{Q}_{41} = \dot{m}(u_1 - u_4)$$

As the specific heat remains constant throughout the cycle, substitution of Equation (5.17) for a constant volume process gives

$$\dot{Q}_{41} = \dot{m} c_v (T_1 - T_4) = 0.05 \times 0.7 \times (290 - 990) = -24.5 \text{ kJ/s}$$

The net heat transfer per second in the cycle $\sum(\dot{Q})_{\text{cyc}}$ is the sum of the heat transfers in each non-flow reversible process. Thus:

$$\sum(\dot{Q})_{\text{cyc}} = \dot{Q}_{12} + \dot{Q}_{23} + \dot{Q}_{34} + \dot{Q}_{41} = 0 + 64.4 + 0 - 24.5 = 39.9 \text{ kJ/s}$$

Hence, there is a net heat transfer into the cycle.

From the first law of thermodynamics, Equation (4.1) gives

$$\sum(\dot{Q})_{\text{cyc}} + \sum(\dot{W})_{\text{cyc}} = 0$$

$$\therefore \quad \sum(\dot{W})_{\text{cyc}} = -\sum(\dot{Q})_{\text{cyc}} = -39.9 \text{ kJ/s}$$

The negative sign indicates that there is a net power output from the cycle and that the system is indeed behaving as an engine. For the overall thermal efficiency η, the useful net energy output is the net power output, and the energy input is the heat input per second in going from state 2 to state 3. Thus, using sign rule C, η is given by

$$\eta = \frac{\sum(\dot{W})_{\text{cyc}}}{\dot{Q}_{23}} = \frac{39.9}{64.4} = 62\%$$

The cycle followed by this engine is the ideal Ackroyd–Stuart cycle after the gentleman who proposed it. It is now more commonly known as the ideal Diesel cycle in recognition of the German engineer who made it work satisfactorily! Practical diesel engines can approach efficiencies of 40% under the right circumstances.

5.4 EFFECT OF THE TYPE OF FLUID

This chapter has been concerned with non-flow processes only, and the NFEE can be seen to be a particular form of the conservation of energy applied to a non-flow process. The next logical step would be to apply the concepts of the conservation of energy to flow processes. However, it is more useful before doing that, to consider firstly whether the type of fluid being used in the thermal system has any bearing upon the energy transfers that take place. In fact, to carry out a complete analysis of all the possible non-flow processes requires more information than has been elicited so far. All the examples in this chapter have been solved without reference to any special characteristics of the fluid and some knowledge of those is necessary in order to proceed further.

Only two fluids will be considered in detail, air behaving as a perfect gas in Chapter 6, and water changing phase to steam in Chapter 7. These fluids, for obvious reasons, are the two most often used in thermal systems. Refrigerants are another common fluid, but an analysis of a system using steam will be very similar to an analysis using a refrigerant so, in practice, there is no need to study it separately.

6
The Fluid as a Perfect Gas

Gases are fluids that exist in the vapour state at NTP. Oxygen, nitrogen, carbon monoxide, carbon dioxide, helium, methane and air (a mixture of gases) are commonly found in thermal systems. A Perfect Gas is one which obeys a given set of formulae. Real gases tend towards becoming perfect gases at low pressures and temperatures. However, it is reasonable to treat real gases as perfect gases under most conditions of pressure and temperature and to apply the perfect gas formulae, because predicted results obtained in making this assumption are approximately the same as those achieved in practice. For the purposes of this book, all gases will be treated as perfect.

The formulae which perfect gases follow can be shown to be true both theoretically and experimentally. They result from the fact that changes in the property values are gradual and can be described fairly easily in a mathematical way. It is not necessary to give a detailed analysis for their derivation, below is a summary of those that are relevant.

6.1 FORMULAE WHICH APPLY TO A PERFECT GAS

6.1.1 The equation of state

The Equation of State for a perfect gas relates the four main properties of the gas, namely the pressure, temperature, mass and volume, together. It can be written for a given mass of gas m, or for a steady mass flow rate $\dot{m}$, as follows:

$$\frac{pV}{mT} = \frac{p\dot{V}}{\dot{m}T} = \frac{pv}{T} = \frac{p}{\rho T} = \text{const.} = R \qquad (6.1)$$

The constant is called the Gas Constant and given the symbol R. It is only constant for a given gas and varies from gas to gas. The Universal Gas Constant $\tilde{R}$ is, as its name implies, a constant for all gases. It has the value of 8.3145 kJ/kmol K and is equal to the gas constant R of a particular gas multiplied by its Molecular Weight (MW). Thus:

Gas constant: R (units J/kg K)

Universal gas constant: $\tilde{R}$ (units J/kmol K)

where $R = \tilde{R}/\text{MW}$.

Some values of R for common gases are given in Table 6.1. Air is included because it is a mixture of gases, mostly oxygen and nitrogen. Its molecular weight and other property values can be determined on a mass proportionate basis, knowing the respective values of

134 THE FLUID AS A PERFECT GAS

Table 6.1 *Values of the gas constant R for common gases.*

Gas	Symbol	Molecular weight	Gas constant (J/kg K)
Oxygen	O_2	32	260
Nitrogen	N_2	28	297
Hydrogen	H_2	2	4 157
Helium	He	4	2 079
Carbon monoxide	CO	28	297
Carbon dioxide	CO_2	44	189
Methane	CH_4	16	520
Sulphur dioxide	SO_2	64	130
Air		29	287

the constituents. Each gas has a symbol to identify it, and the symbols in the table are for the molecule of the gas, the condition in which it exists naturally, rather than the atom. The molecular weight is derived from the periodic table of elements.

The equation of state can be applied to a perfect gas undergoing a process between two identifiable states, and is equally applicable whether it is a flow or a non-flow process, and a reversible or an irreversible process, because it is a relationship between the properties at the end two states and the property values are independent of the process. In this case, between state 1 and state 2 the equation can be written:

$$\frac{p_1 V_1}{m T_1} = \frac{p_1 \dot{V}_1}{\dot{m} T_1} = \frac{p_1 v_1}{T_1} = \frac{p_1}{\rho_1 T_1} = \frac{p_2 V_2}{m T_2} = \frac{p_2 \dot{V}_2}{\dot{m} T_2} = \frac{p_2 v_2}{T_2} = \frac{p_2}{\rho_2 T_2} \qquad (6.2)$$

EXAMPLE 6.1

A container of volume 0.8 m³ contains helium, assumed to be a perfect gas, at a pressure of 500 kPa and a temperature of 300 K. It is heated until the temperature is 400 K (Figure 6.1). What is the mass of the helium, its final pressure and specific volume?

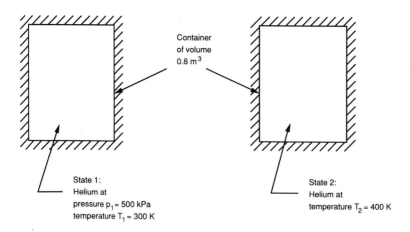

Fig. 6.1 Example 6.1—helium being heated in a constant volume container.

Solution

The gas constant $\{R\}_{he}$ for helium is given in Table 6.1 as

$$\{R\}_{he} = 2079 \text{ J/kg K}$$

Let the helium go from state 1 to state 2 during the constant volume non-flow heat addition process. The mass of helium $\{m\}_{he}$ is given by the equation of state Equation (6.1), at state 1 as follows:

$$\{p_1 V_1\}_{he} = \{mRT_1\}_{he}$$

$$\therefore \{m\}_{he} = \frac{5 \times 10^5 \times 0.8}{2079 \times 300} = 0.64 \text{ kg}$$

The specific volume of the helium at state 1 $\{v_1\}_{he}$ is

$$\{v_1\}_{he} = \frac{\{V\}_{he}}{\{m\}_{he}} = \frac{0.8}{0.64} = 1.25 \text{ m}^3/\text{kg}$$

When the container is heated, the mass of helium remains the same, as does both the total volume and the specific volume of the gas because the container is of fixed size. Thus:

$$\{V_1\}_{he} = \{V_2\}_{he} = 0.8 \text{ m}^3$$

$$\{v_1\}_{he} = \{v_2\}_{he} = 1.25 \text{ m}^3/\text{kg}$$

The final pressure of the gas $\{p_2\}_{he}$ is given by the equation of state, (Equation (6.2)), as

$$\frac{\{p_2 V_2\}_{he}}{\{T_2\}_{he}} = \frac{\{p_1 V_1\}_{he}}{\{T_1\}_{he}}$$

$$\therefore \{p_2\}_{he} = \frac{5 \times 10^5 \times 400}{300} = 667 \text{ kPa}$$

The pressure of the helium increases in direct proportion to the temperature in a constant volume process, as a result of the heat addition making the molecules more active.

6.1.2 Joule's law

In the 1840s, Joule carried out a series of experiments from which he deduced that the internal energy of a gas is directly proportional to its temperature, and this is now known as Joule's law. Thus:

$$U = \Phi(T) \tag{6.3}$$

The experiments conducted involved the free expansion of a number of gases at different pressures and temperatures, much as in Example 6.2. In fact, there was a flaw in the experiments that Joule performed which render the results invalid. However, it can be proved theoretically that Joule's law is correct for a perfect gas and it is accepted for the purposes of this book. With a different experiment and modern instrumentation, it can be shown that real gases only approximate to Joule's law at low pressures.

Recalling Equation (5.17):

$$(U_2 - U_1)_v = (m(u_2 - u_1))_v = (mc_v(T_2 - T_1))_v$$

THE FLUID AS A PERFECT GAS

If Joule's law is valid and the internal energy is a function only of temperature, the specific heat at constant volume c_v, which has already been assumed constant in order to derive Equation (5.17), must also only be a function of temperature. In that case, it is no longer necessary to restrict Equation (5.17) to a constant volume process, it must be applicable to any process undergone by a perfect gas, and the equation can be written:

$$U_2 - U_1 = m(u_2 - u_1) = mc_v(T_2 - T_1) \tag{6.4}$$

The equation is valid for a flow process as much as a non-flow process, in which case it is written:

$$\dot{m}(u_2 - u_1) = \dot{m}c_v(T_2 - T_1)$$

It is also valid for a reversible and irreversible process because, with c_v constant, it is composed of properties which depend only upon the condition of the gas at the end states, not upon the nature of the process itself.

If it is valid for any process, Equation (5.15) for a perfect gas can be written:

$$c_v = \frac{dU}{m\,dT} = \frac{du}{dT} \tag{6.5}$$

There is no need to consider the subscript on the right-hand side of Equation (6.5) for a perfect gas.

The definition of enthalpy H (Equation (5.4)) is

$$H = U + pV$$

For a perfect gas, the equation of state (Equation (6.1)) indicates that the product of the pressure and volume of the gas is proportional to the temperature. As, from Joule's law, the internal energy is also a function of temperature only, it must also be that the enthalpy of the gas is a temperature-dependent property. In other words:

$$H = \Phi(T) \tag{6.6}$$

But Equation (5.16) is

$$(H_2 - H_1)_p = (m(h_2 - h_1))_p = (mc_p(T_2 - T_1))_p$$

For the same reasons that apply to the internal energy, the specific heat at constant pressure must also be a function of temperature and the equation must be applicable to any process undergone by the gas. Therefore, for a perfect gas it can be written:

$$H_2 - H_1 = m(h_2 - h_1) = mc_p(T_2 - T_1) \tag{6.7}$$

$$c_p = \frac{dU}{m\,dT} = \frac{du}{dT} \tag{6.8}$$

Again, as for internal energy, the equations are valid for a reversible or irreversible process and for a flow or non-flow process, in which case Equation (6.7) becomes

$$\dot{m}(h_2 - h_1) = \dot{m}c_p(T_2 - T_1)$$

FORMULAE WHICH APPLY TO A PERFECT GAS

For perfect gases, therefore, substitution for the specific internal energy change during a process can be made in terms of the specific heat at constant volume and the temperature difference, and for specific enthalpy change in terms of the specific heat at constant pressure and the temperature difference. As temperature and specific heat can be measured easily, it makes sense to work in terms of these properties rather than the internal energy and enthalpy.

EXAMPLE 6.2

A rigid insulated container is divided into two compartments, one four times as large as the other, by a membrane (Figure 6.2). The smaller compartment is filled with a perfect gas at a pressure of 800 kPa and the larger compartment is evacuated. When the membrane is broken, the gas expands to fill the entire container. What is the pressure of the gas after the free expansion process?

Solution

Let state 1 be when the gas occupies the smaller upper compartment and state 2 be when the gas occupies the whole container.

As the gas expands out of the small compartment when the membrane is broken, its volume increases to fill the entire container but its pressure decreases because the same number of molecules now occupy the larger volume. If the boundary is drawn inside the container walls, as in Figure 6.2, the fluid undergoes a non-flow process. Considering the NFEE (Equation (5.1)):

$$Q_{12} + W_{12} = U_2 - U_1$$

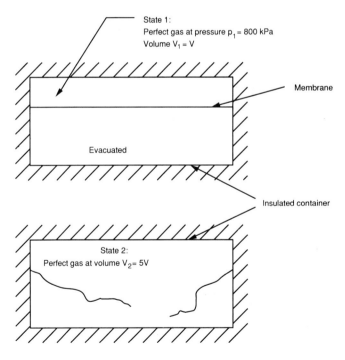

Fig. 6.2 *Example 6.2—a container divided by a membrane in which the upper smaller compartment is filled with a perfect gas and the lower larger compartment is evacuated.*

where $Q_{12} = 0$ (because the container is insulated implying no heat transfer through the walls) and $W_{12} = 0$ (because the fluid expands under the influence of its pressure not because of a work transfer) therefore $U_2 = U_1$. But for a perfect gas, Joule's law (Equation (6.3)) is $U = \Phi(T)$. Hence, for the gas expanding from the smaller compartment to fill the entire container $T_1 = T_2$. As the larger compartment is four times the size of the smaller compartment $V_2 = 5V_1$.

Applying the equation of state (Equation (6.2)) given that the mass of the gas is constant:

$$\frac{p_1 V_1}{m T_1} = \frac{p_2 V_2}{m T_2}$$

$$\therefore \quad p_2 = \frac{8 \times 10^5}{5} = 1.6 \times 10^5 \text{ Pa}$$

The gas pressure decreases in inverse proportion to the volume during the constant temperature process because the same number of molecules occupy a larger volume.

6.1.3 Relationship between specific heats

The definition of specific enthalpy (Equation (5.4)) is

$$h = u + pv$$

The difference between the specific enthalpy at state 2 and state 1 can be written as

$$h_2 - h_1 = (u_2 - u_1) + (p_2 v_2 - p_1 v_1)$$

Substituting from the equation of state (Equation (6.1)) for a perfect gas:

$$h_2 - h_1 = (u_2 - u_1) + R(T_2 - T_1)$$

Substituting from Equations (6.4) and (6.7) for a perfect gas gives

$$c_p(T_2 - T_1) = c_v(T_2 - T_1) + R(T_2 - T_1)$$

$$\therefore \quad c_p = c_v + R \tag{6.9}$$

As the gas constant for a perfect gas is always a positive value, Equation (6.9) implies that the specific heat of a perfect gas at constant pressure is always greater than the specific heat at constant volume. The above equation can also be shown to be true by considering the differential form for perfect gases in which the specific heats vary with temperature in which case, as the gas constant for a particular gas is a fixed value, the difference between the specific heats must remain constant.

The ratio of the specific heat at constant pressure to the specific heat at constant volume of a perfect gas is given a special symbol γ and, if the assumption is made for a perfect gas that the specific heats are constant, it also is constant. Thus:

$$\gamma = \frac{c_p}{c_v} \tag{6.10}$$

The value of γ for air is taken as 1.4.

FORMULAE WHICH APPLY TO A PERFECT GAS

Table 6.2 Example 6.3—values of R and c_p for common gases.

Gas	R (J/kg K)	c_p (J/kg K)	c_v (J/kg K)	γ
O_2	260	918		
N_2	297	1 040		
H_2	4 157	14 310		
He	2 079	5 193		
CO	297	1 040		
CO_2	189	846		
CH_4	520	2 226		
SO_2	130	610		
Air	287	1 005		

Table 6.3 Example 6.3—values of R, c_p, c_v and γ for common gases.

Gas	R (J/kg K)	c_p (J/kg K)	c_v (J/kg K)	γ
O_2	260	918	658	1.395
N_2	297	1 040	743	1.4
H_2	4 157	14 310	10 153	1.409
He	2 079	5 193	3 114	1.67
CO	297	1 040	743	1.4
CO_2	189	846	657	1.288
CH_4	520	2 226	1 706	1.305
SO_2	130	610	480	1.27
Air	287	1 005	718	1.4

EXAMPLE 6.3

The gas constant R and the specific heat at constant pressure c_p at a temperature of 300 K for all the gases listed in Table 6.1 are shown in Table 6.2. Complete the table by finding the specific heat at constant volume and the ratio of the specific heats, again at a temperature of 300 K, assuming all the gases to be perfect.

Solution

The specific heat at constant volume c_v is given by Equation (6.9) and the ratio of specific heats γ by Equation (6.10). Table 6.3 is a completed version of Table 6.2. Of course, if any two of the values of R, c_p, c_v and γ for a perfect gas are known, the other two values can be determined.

6.1.4 Entropy of a perfect gas

For a fluid undergoing a non-flow process, the first law of thermodynamics applied between the two end states (Equation (5.1)) may be written:

$$Q_{12} + W_{12} = U_2 - U_1$$

In differential form, the equation is

$$dQ + dW = dU$$

Substituting for the work and heat transfer by Equations (3.1) and (3.2) respectively, thereby assuming reversible non-flow processes, and for the internal energy of a perfect gas by Equation (6.5), gives

$$mT \, ds - mp \, dv = mc_v \, dT$$

$$\therefore \quad ds = c_v \frac{dT}{T} + p \frac{dv}{T}$$

Substituting from the equation of state (Equation (6.1)) gives

$$ds = c_v \frac{dT}{T} + R \frac{dv}{v}$$

Integrating, assuming that for a perfect gas the specific heat at constant volume and the gas constant are constant, gives

$$s_2 - s_1 = c_v \ln \frac{T_2}{T_1} + R \ln \frac{v_2}{v_1} \tag{6.11}$$

Substituting from the equation of state (Equation (6.2)) and using the relationship between the specific heats of a perfect gas (Equation (6.9)) yields two other similar equations:

$$s_2 - s_1 = c_p \ln \frac{T_2}{T_1} - R \ln \frac{p_2}{p_1} \tag{6.12}$$

$$s_2 - s_1 = c_v \ln \frac{p_2}{p_1} + c_p \ln \frac{v_2}{v_1} \tag{6.13}$$

Equations (6.11), (6.12) and (6.13) are valid for a perfect gas undergoing any process, non-flow or flow, reversible or irreversible, because they merely relate the change in the value of the property entropy to the change in the value of two other properties, given that the specific heats and R are constant, and values of properties at any state are independent of the process. This is so despite the fact that the equations are derived from the first law of thermodynamics which has certain restrictions placed upon it, and that substitution was made for W and Q with formulae that only apply to a non-flow reversible process. The equations are in terms of the specific entropy s, but can be converted to entropy S by multiplying by the mass of gas m. In other words:

$$S_2 - S_1 = m(s_2 - s_1)$$

6.2 DETERMINATION OF PROPERTY VALUES

In Chapter 2, methods were described for determining the values of the mass, volume, pressure and temperature of a fluid. In Chapter 4, methods were suggested for finding the specific heat. It can be seen from the above equations that it is now possible to calculate the values of the internal energy difference (from Equation (6.4)), the enthalpy difference (from Equation (6.7)) and the entropy difference (from Equations (6.11), (6.12) and (6.13)) of a perfect gas as it undergoes a change of state. The fact that it is only the changes in these properties that can be worked out is not a disadvantage because it is not necessary to know the absolute values at any state in order to discover the work and heat transfer during a process. It is possible to stipulate a reference state at which they are zero, for example the state at which the pressure and the temperature reach zero for a perfect gas, but it is not essential for the purposes of examining a thermal system. So, although the existence of the properties internal energy, enthalpy and entropy of a perfect gas have either been assumed, or derived theoretically, they can be used in any analysis because values can be ascribed to them.

EXAMPLE 6.4

Helium, assumed to be a perfect gas, is being used as the fluid in a piston cylinder mechanism. When the piston is at TDC, 0.08 kg of the helium is at a pressure of 3 MPa and a temperature of 800 K (Figure 6.3). If heat is added to the helium while the piston stays at the TDC position, what is the change in the entropy of the helium if its temperature increases to 2000 K?

If, instead, heat is added to the helium while the piston moves towards BDC while maintaining the pressure of the gas constant, what is now the change in the entropy of the helium when its temperature reaches 2000 K (Figure 6.4)?

Finally, if the heat is added in such a way that the gas expands and the piston moves towards BDC but the temperature of the gas remains constant, what is now the change in the entropy of the helium if its volume increases fourfold (Figure 6.5)?

For the helium, take the values of the specific heats from Table 6.2.

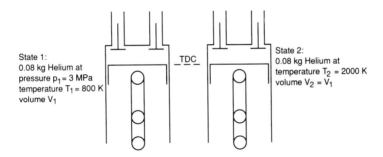

Fig. 6.3 *Example 6.4 — heat addition to helium in a piston cylinder mechanism during a constant volume non-flow process.*

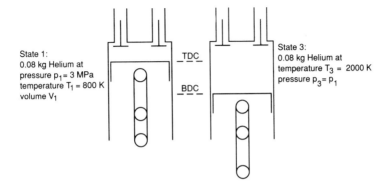

Fig. 6.4 *Example 6.4 — heat addition to helium in a piston cylinder mechanism during a constant pressure non-flow process.*

142 THE FLUID AS A PERFECT GAS

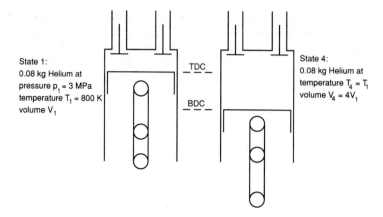

Fig. 6.5 *Example 6.4—heat addition to helium in a piston cylinder mechanism during a constant temperature non-flow process.*

Solution
From Table 6.2 for helium:

$$\{R\}_{he} = 2079 \text{ J/kg K}$$

$$\{c_p\}_{he} = 5193 \text{ J/kg K}$$

$$\{c_v\}_{he} = 3114 \text{ J/kg K}$$

When the heat is added at constant volume, the change in the specific entropy of the gas between state 1 and state 2 is best given by Equation (6.11):

$$\{s_2 - s_1\}_{he} = \left\{ c_v \ln \frac{T_2}{T_1} + R \ln \frac{v_2}{v_1} \right\}_{he}$$

$$\therefore \{s_2 - s_1\}_{he} = 3114 \ln \frac{2000}{800} = 2853.3 \text{ J/kg K}$$

The actual entropy rise is

$$\{S_2 - S_1\}_{he} = \{m (s_2 - s_1)\}_{he} = 0.08 \times 2853.3 = 228.3 \text{ J/K}$$

When the heat is added at constant pressure between state 1 and state 3, the change in the specific entropy of the gas is best given by Equation (6.12):

$$\{s_3 - s_1\}_{he} = \left\{ c_p \ln \frac{T_3}{T_1} - R \ln \frac{p_3}{p_1} \right\}_{he}$$

$$\therefore \{s_3 - s_1\}_{he} = 5193 \ln \frac{2000}{800} = 4758.3 \text{ J/kg K}$$

The actual entropy rise is

$$\{S_3 - S_1\}_{he} = \{m (s_3 - s_1)\}_{he} = 0.08 \times 4758.3 = 380.7 \text{ J/K}$$

When the heat is added at constant temperature between state 1 and state 4, the change in the specific entropy of the gas is best given by Equation (6.11) again:

$$\{s_4 - s_1\}_{he} = \left\{c_v \ln \frac{T_4}{T_1} + R \ln \frac{v_4}{v_1}\right\}_{he}$$

$$\therefore \{s_4 - s_1\}_{he} = 2079 \ln 4 = 2882.1 \text{ J/kg K}$$

The actual entropy rise is

$$\{S_4 - S_1\}_{he} = \{m\ (s_4 - s_1)\}_{he} = 0.08 \times 2882.1 = 230.6 \text{ J/K}$$

In fact, the increase of entropy during a process is an important number and can indicate something about the efficiency of that process, as will be seen in more detail in Chapters 8 and 12.

6.3 NON-FLOW PROCESSES WITH PERFECT GASES

In Chapter 5, five non-flow processes were identified, a constant pressure, a constant volume, a constant temperature, a constant entropy and a polytropic process. If the fluid is a perfect gas, it is now possible to substitute into all of the equations derived in Chapter 5, the equations developed so far in Chapter 6, that is Equations (6.1)–(6.13). This can help considerably with the analysis of a thermal system. In fact, it can now be shown that the first four processes are polytropic processes with different values for the index n when the fluid is a perfect gas, as follows.

6.3.1 Constant pressure non-flow process with a perfect gas

A polytropic non-flow process follows a relationship between the pressure and volume of the gas as in Equation (5.10):

$$pV^n = \text{const}.$$

If $n = 0$, this equation reduces to $p = \text{const}$. Therefore, a constant pressure non-flow process is a polytropic non-flow process with the index $n = 0$ and must obey all the polytropic non-flow relationships of Chapter 5 with that substitution made. This is true for any fluid but, in practice, is particularly applicable to a perfect gas which often does follow the polytropic Equation (5.10).

6.3.2 Constant volume non-flow process with a perfect gas

Taking Equation (5.10) again for a polytropic non-flow process:

$$pV^n = \text{const}.$$

Rearranging gives

$$p^{1/n}V = \text{const}.^{1/n} = \text{const}.$$

If $n = \infty$, this becomes $V = \text{const}$. Therefore, a constant volume non-flow process is also a polytropic non-flow process, but with the index $n = \infty$ and must obey all the polytropic relationships of Chapter 5 with that substitution made. For example, letting $n = \infty$ in Equation (5.12) gives the work output as zero, which is the same result as Equation (5.5). As with the constant pressure non-flow process, it is particularly applicable to a perfect gas which often does follow the polytropic Equation (5.10).

6.3.3 Constant temperature non-flow process with a perfect gas

The equation of state, Equation (6.1), is

$$pV = mRT$$

For a fixed mass of gas m at a constant temperature T, this becomes

$$pV = \text{const.} = pV^1$$

In other words, in comparison to the polytropic Equation (5.10), a constant temperature non-flow process is a polytropic non-flow process but with the index $n = 1$, and must obey all the polytropic relationships of Chapter 5 with that substitution made. However, unlike the constant pressure and constant volume processes, this is only true if the fluid is a perfect gas because it has been shown to be so using the equation of state which applies only to a perfect gas.

Substituting $n = 1$ into Equation (5.12) does not give a sensible answer for the work transfer so some care must be exercised. However, it is possible now to determine the work transfer W in a constant temperature non-flow reversible process when the fluid is a perfect gas, as follows.

The work transfer when the fluid goes from state 1 to state 2 is given by Equation (3.1) as

$$W_{12} = -\int_1^2 p \, dV$$

For a perfect gas, the equation of state (Equation (6.1)) is

$$pV = mRT$$

Substituting for p gives

$$W_{12} = -\int_1^2 \frac{mRT \, dV}{V}$$

For a given mass of gas m with a gas constant R, as the process is isothermal, the product of mRT is constant. Therefore, integrating the above gives

$$W_{12} = -mRT \ln \frac{V_2}{V_1}$$

This expression for the work transfer of a perfect gas in an isothermal non-flow process can be written in terms of other properties of the gas by substituting from the equation of state (Equation (6.2)).

Further, considering the NFEE for a fluid undergoing a non-flow process, Equation (5.1):

$$Q_{12} + W_{12} = (U_2 - U_1)$$

From Joule's law (Equation (6.3)) $U = \Phi(T)$. If the process is isothermal, the temperature remains constant and so must the internal energy. The NFEE becomes

$$Q_{12} = -W_{12} = mRT \ln \frac{V_2}{V_1} \qquad (6.14)$$

In other words, the heat transfer of a perfect gas undergoing an isothermal non-flow process is numerically equal to the work transfer.

6.3.4 Constant entropy non-flow process with a perfect gas

Consider the change of entropy of a perfect gas in going from state 1 to state 2 as given by Equation (6.13):

$$s_2 - s_1 = c_v \ln \frac{p_2}{p_1} + c_p \ln \frac{v_2}{v_1}$$

If the process is isentropic, $s_2 = s_1$ and the equation becomes

$$0 = \ln \frac{p_2}{p_1} + \frac{c_p}{c_v} \ln \frac{v_2}{v_1}$$

Substituting from Equation (6.10) gives

$$0 = \ln \frac{p_2}{p_1} + \gamma \ln \frac{v_2}{v_1}$$

$$\therefore \quad 0 = \ln \frac{p_2 v_2^\gamma}{p_1 v_1^\gamma}$$

Antilogging gives

$$p_2 v_2^\gamma = p_1 v_1^\gamma = \text{const.} \qquad (6.15)$$

Or

$$p_2 V_2^\gamma = p_1 V_1^\gamma = \text{const.} \qquad (6.16)$$

In other words, an isentropic non-flow process is a polytropic non-flow process, but with the index $n = \gamma$ and must obey all the polytropic non-flow relationships of Chapter 5 with that substitution made when the fluid is a perfect gas. In fact, Equations (6.15) and (6.16) are also valid for a flow process, and a reversible or irreversible process, because they are derived from Equation (6.13) to which these conditions apply.

6.3.5 Polytropic non-flow process with a perfect gas

Equation (5.10) for a polytropic non-flow process between state 1 and state 2:

$$p_1 V_1^n = p_2 V_2^n$$

THE FLUID AS A PERFECT GAS

combined with Equation (6.2), the equation of state for a perfect gas:

$$\frac{p_1 V_1}{m T_1} = \frac{p_1 v_1}{T_1} = \frac{p_2 V_2}{m T_2} = \frac{p_2 v_2}{T_2}$$

gives rise to the following set of relationships for a perfect gas:

$$\frac{T_2}{T_1} = \frac{p_2^{(n-1)/n}}{p_1^{(n-1)/n}} = \frac{V_2^{1-n}}{V_1^{1-n}} = \frac{v_2^{1-n}}{v_1^{1-n}} \qquad (6.17)$$

Not only is Equation (6.17) valid for any polytropic non-flow process undergone by a perfect gas between state 1 and state 2, it is also true for any constant pressure non-flow process in which case $n = 0$, any constant volume non-flow process in which case $n = \infty$, any constant temperature non-flow process in which case $n = 1$ and any constant entropy non-flow process in which case $n = \gamma$. In the latter case, it is true for an isentropic flow process because the equation:

$$pV^\gamma = \text{const}.$$

was derived under those circumstances. This is extremely useful when certain flow processes are analysed in Chapter 8.

It is often easier to determine the work and heat transfer in any non-flow process with a perfect gas by using the polytropic relationships, namely:

$$W_{12} = -\frac{m(p_2 v_2 - p_1 v_1)}{1 - n} \qquad (5.12)$$

$$Q_{12} = m(u_2 - u_1) - W_{12}$$

which becomes, for a perfect gas:

$$Q_{12} = mc_v(T_2 - T_1) - W_{12} \qquad (6.18)$$

where $n = 0$ for a constant pressure process, $n = \infty$ for a constant volume process, $n = 1$ for a constant temperature process and $n = \gamma$ for a constant entropy process. Equations (5.12) and (6.18), combined with the special relationships for a perfect gas (Equations (6.1)–(6.17)), are sufficient, with one exception. When the process is isothermal, substitution of $n = 1$ into Equation (5.12) gives an incorrect answer, and it is necessary to resort to Equation (6.14).

A plot of fluid pressure against volume for the four different types of polytropic non-flow process that a perfect gas can undergo is shown in Figure 6.6.

Interest is centred upon the two quadrants where n varies from zero to infinity and where it has a positive value. It is possible to conceive of systems where n has a negative value, for example when the pressure and volume of the gas both increase, but these are not of practical use as yet. The slope of any curve is an important consideration when a reciprocating engine or compressor cycle is under consideration.

EXAMPLE 6.5

A reciprocating piston cylinder engine is to be built based upon the ideal Stirling cycle. The fluid is helium, assumed a perfect gas, for which the specific heat at constant pressure $\{c_p\}_\text{he}$ is 5193 J/kg K. It is intended to use 0.005 kg

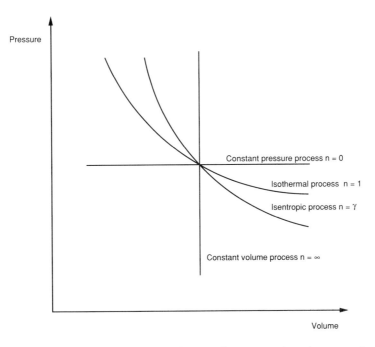

Fig. 6.6 *Graph of pressure versus volume for a perfect gas undergoing a constant pressure, constant volume, constant temperature and constant entropy non-flow process.*

of helium and for it to remain entirely within the engine as it undergoes the following four reversible non-flow processes in completing a cycle, which it does 20 times per second:

1. State 1 to state 2—isothermal compression from a pressure of 100 kPa and temperature of 300 K through a volume ratio of 8.
2. State 2 to state 3—constant volume heat addition of 1760 kJ/kg.
3. State 3 to state 4—isothermal expansion to the maximum volume.
4. State 4 to state 1—constant volume heat rejection back to the initial state.

Sketch the cycle on a graph of pressure versus volume. Determine the maximum cycle temperature and pressure, the power produced by the cycle, the heat rejected in going from state 4 to state 1, and the overall thermal efficiency of the engine.

Solution

The cycle plotted as a graph of helium pressure against volume is shown in Figure 6.7.

For the helium, of molecular weight 4, the gas constant $\{R\}_{he}$ from Table 6.1 is

$$\{R\}_{he} = 2.079 \text{ kJ/kg K}$$

The specific heat at constant volume $\{c_v\}_{he}$ is given by Equation (6.9) as

$$\{c_p - c_v\}_{he} = \{R\}_{he}$$
$$\therefore \quad \{c_v\}_{he} = 5.193 - 2.079 = 3.114 \text{ kJ/kg K}$$

THE FLUID AS A PERFECT GAS

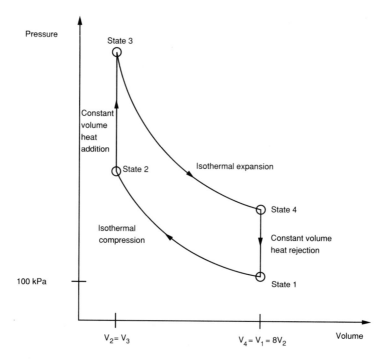

Fig. 6.7 *Example 6.5—graph of the fluid pressure against volume for helium undergoing an ideal stirling cycle made up of four non-flow reversible processes.*

The volume ratio of the gas is

$$\frac{\{V_1\}_{he}}{\{V_2\}_{he}} = \frac{\{v_1\}_{he}}{\{v_2\}_{he}} = \frac{\{V_4\}_{he}}{\{V_3\}_{he}} = \frac{\{v_4\}_{he}}{\{v_3\}_{he}} = 8$$

In going from state 1 to state 2, the non-flow reversible process is isothermal. Hence:

$$\{T_1\}_{he} = \{T_2\}_{he} = 300 \text{ K}$$

Using the equation of state (Equation (6.2))

$$\frac{\{p_2 v_2\}_{he}}{\{T_2\}_{he}} = \frac{\{p_1 v_1\}_{he}}{\{T_1\}_{he}}$$

$$\therefore \quad \{p_2\}_{he} = 8 \times 1 \times 10^5 = 8 \times 10^5 \text{ Pa}$$

As the non-flow reversible process is isothermal, the work transfer W_{12} and the heat transfer Q_{12} are both given by Equation (6.14) as

$$Q_{12} = -W_{12} = \left\{ mRT_1 \ln \frac{V_2}{V_1} \right\}_{he} = 0.005 \times 2079 \times 300 \ln \frac{1}{8} = -6.5 \text{ kJ}$$

Therefore, in going from state 1 to state 2, there is a work input and a heat output.

State 2 to state 3 is a constant volume heat addition non-flow reversible process for which the work transfer W_{23}, from Equation (5.12) with $n = \infty$, is zero. Thus $W_{23} = 0$. The heat transfer Q_{23} is given as being $+1760$ kJ/kg,

NON-FLOW PROCESSES WITH PERFECT GASES

positive because it is a heat addition, and also by Equation (6.18) as

$$Q_{23} = \{mc_v(T_3 - T_2)\}_{he} - W_{23}$$

$$\therefore \quad 1760 \times 10^3 \times 0.005 = 0.005 \times 3114(\{T_3\}_{he} - 300)$$

$$\therefore \quad \{T_3\}_{he} = 865.2 \text{ K}$$

where $\{T_3\}_{he}$ is the maximum temperature reached by the helium in the cycle. During the constant volume non-flow reversible process, $\{V_2\}_{he} = \{V_3\}_{he}$. Applying the equation of state (Equation (6.2)) gives

$$\frac{\{p_3 V_3\}_{he}}{\{T_3\}_{he}} = \frac{\{p_2 V_2\}_{he}}{\{T_2\}_{he}}$$

$$\therefore \quad \{p_3\}_{he} = \frac{8 \times 10^5 \times 865.2}{300} = 23.1 \times 10^5 \text{ Pa}$$

where $\{p_3\}_{he}$ is the maximum pressure reached by the helium in the cycle. State 3 to state 4 is another isothermal non-flow reversible process for which the work transfer W_{34} and the heat transfer Q_{34} are given by Equation (6.14) as

$$Q_{34} = -W_{34} = \{mRT_3 \times \ln \frac{V_4}{V_3}\}_{he} = 0.005 \times 2079 \times 865.2 \, \ln(8) = +18.7 \text{ kJ}$$

Hence, there is a work output and heat addition during this isothermal non-flow reversible process.

Finally, state 4 to state 1 is another constant volume non-flow reversible process for which the work transfer W_{41} is again zero by Equation (5.12) with $n = \infty$, and the heat transfer Q_{23} is given by Equation (6.18) as

$$Q_{41} = \{mc_v(T_1 - T_4)\}_{he} - W_{41}$$

It is possible to determine the unknown temperatures from the equation of state and find Q_{41}, as in process 23, but another approach is as follows. The net work transfer in the cycle $\sum(W)_{cyc}$ is the sum of all the work transfers in each non-flow reversible process. Thus:

$$\sum(W)_{cyc} = W_{12} + W_{23} + W_{34} + W_{41} = +6.5 + 0 - 18.7 + 0 = -12.2 \text{ kJ}$$

There is a net output of work from the cycle which is indeed behaving as an engine. If the cycle is repeated 20 times per second, the net power output of the engine $\sum(\dot{W})_{cyc}$ is

$$\sum(\dot{W})_{cyc} = -12.2 \times 20 = -244 \text{ kW}$$

From the first law of thermodynamics (Equation (4.1)):

$$\sum(Q)_{cyc} = -\sum(W)_{cyc} = +12.2 \text{ kJ}$$

The net heat transfer is the sum of the heat transfers in each of the non-flow reversible processes that make up the cycle. Therefore:

$$\sum(Q)_{cyc} = Q_{12} + Q_{23} + Q_{34} + Q_{41}$$

$$\therefore \quad 12.2 = -6.5 + 1760 \times 0.005 + 18.7 + Q_{41}$$

$$\therefore \quad Q_{41} = -8.8 \text{ kJ}$$

150   THE FLUID AS A PERFECT GAS

In a Stirling cycle, the heat rejected in one constant volume non-flow reversible process is the same as the heat supplied in the other constant volume non-flow reversible process, and the performance of an engine based upon the cycle rests upon their being a perfect heat exchanger in which to transfer the heat. Assuming this to be the case, the heat supply to the engine occurs as the fluid undergoes the isothermal process from state 3 to state 4. Therefore, the overall thermal efficiency η, using sign rule C, is given by

$$\eta = \frac{\sum (W)_{cyc}}{Q_{34}} = \frac{12.2}{18.7} = 65.2\%$$

This is a very high efficiency, but is the ideal case. Practical engines based upon the ideal Stirling cycle have external combustion of the fuel and much interest in them centres upon the use of coal or nuclear heat as the energy source. Their practical efficiency is also comparable to that of a diesel engine. The main disadvantage is the difficulties encountered in designing an economic heat exchanger.

EXAMPLE 6.6

A reciprocating piston cylinder engine is to be built based upon the ideal Diesel cycle. The fluid is air, assumed a perfect gas, for which the gas constant $(R)_{air}$ is 287 J/kg K, and the specific heat at constant volume $(c_v)_{air}$ is 718 J/kg K. A mass of air of 0.005 kg undergoes the following four reversible non-flow processes in completing the cycle 30 times per second:

1. State 1 to state 2—adiabatic compression from an initial pressure of 100 kPa and specific volume of 0.83 m³/kg to a pressure of 5 MPa.
2. State 2 to state 3—constant pressure expansion until the volume is 0.2 times the initial value at state 1.
3. State 3 to state 4—adiabatic expansion until the volume is equal to its initial value at state 1.
4. State 4 to state 1—constant volume drop in pressure back to the initial state.

Sketch the cycle on a graph of fluid pressure versus specific volume. Evaluate the power output of the engine, the net heat transfer and the cycle efficiency.

Solution

The cycle plotted as a graph of air pressure against specific volume is shown in Figure 6.8.
The specific heat at constant pressure of the air $\{c_p\}_{air}$ is given by Equation (6.9) as

$$\{c_p - c_v\}_{air} = \{R\}_{air}$$
$$\therefore \quad \{c_p\}_{air} = 287 + 718 = 1005 \text{ J/kg K}$$

The ratio of the specific heats $\{\gamma\}_{air}$ is given by Equation (6.10) as

$$\{\gamma\}_{air} = \frac{\{c_p\}_{air}}{\{c_v\}_{air}} = \frac{1005}{718} = 1.4$$

State 1 to state 2 is a reversible adiabatic (isentropic) non-flow process. The specific volume of the air at state 2 $\{v_2\}_{air}$ is given by Equation (6.15) as

$$\{p_1 v_1^\gamma\}_{air} = \{p_2 v_2^\gamma\}_{air}$$
$$\therefore \quad \{v_2\}_{air} = \frac{\{p_1^{1/\gamma} v_1\}_{air}}{\{p_2^{1/\gamma}\}_{air}} = \frac{1^{1/1.4} \times 0.83}{50^{1/1.4}} = 0.051 \text{ m}^3/\text{kg}$$

NON-FLOW PROCESSES WITH PERFECT GASES 151

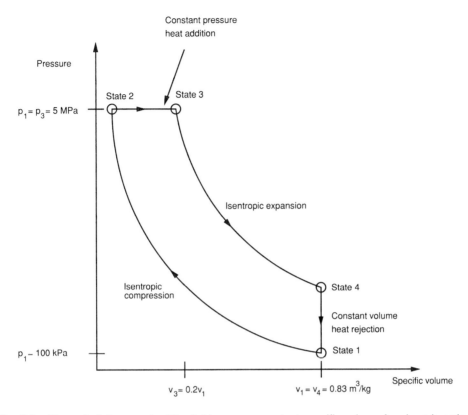

Fig. 6.8 *Example 6.6—graph of the fluid pressure against specific volume for air undergoing an ideal Diesel cycle made up of four non-flow reversible processes.*

The temperature of the air at states 1 and 2, $\{T_1\}_{air}$ and $\{T_2\}_{air}$ respectively, is given by the equation of state (Equation (6.1)) as

$$\{p_1 v_1\}_{air} = \{RT_1\}_{air}$$

$$\therefore \quad \{T_1\}_{air} = \frac{1 \times 10^5 \times 0.83}{287} = 289.2 \text{ K}$$

$$\{p_2 v_2\}_{air} = \{RT_2\}_{air}$$

$$\therefore \quad \{T_2\}_{air} = \frac{50 \times 10^5 \times 0.051}{287} = 888.5 \text{ K}$$

The work transfer in the isentropic non-flow process W_{12} is given by Equation (5.12), with $n = \gamma$, as

$$W_{12} = -\frac{\{m(p_2 v_2 - p_1 v_1)\}_{air}}{\{1 - \gamma\}_{air}}$$

$$\therefore \quad W_{12} = -\frac{0.005 \times 10^5 \times (50 \times 0.051 - 1 \times 0.83)}{1 - 1.4} = +2.2 \text{ kJ}$$

152   THE FLUID AS A PERFECT GAS

The heat transfer in the isentropic non-flow process Q_{12}, by definition, is zero. Thus $Q_{12} = 0$. This can be checked from Equation (6.18)

$$Q_{12} = mc_v(T_2 - T_1) - W_{12} = mc_v(T_2 - T_1) + \frac{m(p_2v_2 - p_1v_1)}{1 - \gamma}$$

Substituting from the equation of state (Equation (6.1)) gives

$$Q_{12} = mc_v(T_2 - T_1) + \frac{mR(T_2 - T_1)}{1 - \gamma} = m(T_2 - T_1)\left(c_v + \frac{R}{1 - \gamma}\right)$$

But from Equations (6.9) and (6.10):

$$c_p - c_v = R$$

$$\therefore \quad c_v\left(\frac{c_p}{c_v} - 1\right) = R$$

$$\therefore \quad c_v(\gamma - 1) = R$$

$$\therefore \quad c_v = \frac{R}{\gamma - 1}$$

Substituting in the equation for Q_{12} gives

$$Q_{12} = m(T_2 - T_1)\left(\frac{R}{\gamma - 1} + \frac{R}{1 - \gamma}\right) = 0$$

State 2 to state 3 is a reversible constant pressure non-flow heat addition until the volume is 0.2 of its value at state 1. Hence:

$$\{p_2\}_{\text{air}} = \{p_3\}_{\text{air}} = 5 \text{ MPa} = 50 \times 10^5 \text{ Pa}$$

$$\{v_3\}_{\text{air}} = 0.2\{v_1\}_{\text{air}} = 0.2 \times 0.83 = 0.166 \text{ m}^3/\text{kg}$$

The temperature of the air at state 3 $\{T_3\}_{\text{air}}$ is given by the equation of state (Equation (6.1)) as

$$\{p_3 v_3\}_{\text{air}} = \{RT_3\}_{\text{air}}$$

$$\therefore \quad \{T_3\}_{\text{air}} = \frac{50 \times 10^5 \times 0.166}{287} = 2892 \text{ K}$$

The work transfer in the constant pressure non-flow reversible process W_{23} is given by Equation (5.12) with $n = 0$ as

$$W_{23} = -\{mp_2(v_3 - v_2)\}_{\text{air}} = -0.005 \times 50 \times 10^5 \times (0.166 - 0.051) = -2.9 \text{ kJ}$$

The heat transfer in the constant pressure non-flow reversible process Q_{23} is given by Equation (6.18) as

$$Q_{23} = \{mc_v(T_3 - T_2)\}_{\text{air}} - W_{23}$$

$$= 0.005 \times 718 \times (2892 - 888.5) + 2.9 \times 10^3 = 10.0 \text{ kJ}$$

State 3 to state 4 is another isentropic non-flow process during which the specific volume of the air expands back to its original value at state 1. Thus:

$$\{v_4\}_{\text{air}} = \{v_1\}_{\text{air}} = 0.83 \text{ m}^3/\text{kg}$$

The pressure of the air at state 4 $\{p_4\}_{air}$ is given by Equation (6.15) as

$$\{p_3 v_3^\gamma\}_{air} = \{p_4 v_4^\gamma\}_{air}$$

$$\therefore \{p_4\}_{air} = \frac{50 \times 10^5 \times 0.166^{1.4}}{0.83^{1.4}} = 5.25 \times 10^5 \text{ Pa}$$

The temperature of the air at state 4 $\{T_4\}_{air}$ is given by the equation of state (Equation (6.1)) as

$$\{p_4 v_4\}_{air} = \{RT_4\}_{air}$$

$$\therefore \{T_4\}_{air} = \frac{5.25 \times 10^5 \times 0.83}{287} = 1518.3 \text{ K}$$

The work transfer in the isentropic non-flow process W_{34} is given by Equation (5.12) with $n = \gamma$ as

$$W_{34} = -\frac{\{m(p_4 v_4 - p_3 v_3)\}_{air}}{\{1 - \gamma\}_{air}}$$

$$\therefore W_{34} = -\frac{0.005 \times 10^5 \times (5.25 \times 0.83 - 50 \times 0.166)}{1 - 1.4} = -4.9 \text{ kJ}$$

The heat transfer in the isentropic non-flow process Q_{34}, by definition, is zero. Thus $Q_{34} = 0$.

State 4 to state 1 is a constant volume non-flow reversible drop in pressure. The work transfer W_{41} from Equation (5.12) with $n = \infty$, is zero, thus $W_{41} = 0$. The heat transfer Q_{41} is given by Equation (6.18) as

$$Q_{41} = \{mc_v(T_1 - T_4)\}_{air} + W_{41} = 0.005 \times 718 \times (289.2 - 1518.3) = -4.4 \text{ kJ}$$

The net work transfer from the cycle $\sum(W)_{cyc}$ is the sum of the work transfer in each non-flow reversible process. Thus:

$$\sum(W)_{cyc} = W_{12} + W_{23} + W_{34} + W_{41} = +2.2 - 2.9 - 4.9 + 0 = -5.6 \text{ kJ}$$

If the cycle is completed 30 times per second, the net power output $\sum(\dot{W})_{cyc}$ is

$$\sum(\dot{W})_{cyc} = \sum(W)_{cyc} \times 30 = -5.6 \times 30 = -168 \text{ kW}$$

As a check, by the first law of thermodynamics (Equation (4.1)) the sum of the net work transfer in the cycle $\sum(W)_{cyc}$ and the net heat transfer $\sum(Q)_{cyc}$ in the cycle should equal zero:

$$\sum(Q)_{cyc} = Q_{12} + Q_{23} + Q_{34} + Q_{41} = 0 + 10.0 + 0 - 4.4 = 5.6 \text{ kJ}$$

$$\therefore \sum(Q)_{cyc} + \sum(W)_{cyc} = 5.6 - 5.6 = 0$$

Therefore the results agree. The overall thermal efficiency of the cycle η is the ratio of the net work transfer to the heat supplied, which is in non-flow reversible process 23. Thus, using sign rule C:

$$\eta = \frac{\sum(W)_{cyc}}{Q_{23}} = \frac{5.6}{10.0} = 56\%$$

This is, of course, high for a diesel engine, but the analysis is based upon ideal reversible processes. Practical diesel engines utilise internal combustion of the fuel and can only be considered to follow the ideal cycle in a general way. The fuel is injected into the cylinder at high pressure and, as the change in volume of the air during the compression

154   THE FLUID AS A PERFECT GAS

process is relatively large, it is ignited by the temperature of the air after compression, which eliminates the need for electrical ignition. Reliability of the engines is therefore improved and efficiencies approaching 40% are possible.

EXAMPLE 6.7

A reciprocating piston cylinder engine is to be built based upon the ideal Otto cycle. The fluid is air, assumed a perfect gas, for which the gas constant $\{R\}_{air}$ is 287 J/kg K, and the specific heat at constant volume $\{c_v\}_{air}$ is 718 J/kg K. The air undergoes the following four reversible non-flow processes.

1. State 1 to state 2 — adiabatic compression from an initial pressure of 100 kPa and specific volume 0.8 m³/kg through a volume ratio, called the compression ratio, of 8.
2. State 2 to state 3 — constant volume heat addition until the temperature of the air is 2000 K.
3. State 3 to state 4 — adiabatic expansion until the volume is equal to its initial value at state 1.
4. State 4 to state 1 — constant volume drop in pressure back to the initial state.

Although the cycle comprises four non-flow processes, the analysis is to be carried out assuming a mass flow rate of air of 0.05 kg/s.

Sketch the cycle on a graph of fluid pressure versus specific volume. Evaluate the power output of the engine, the net heat transfer and the cycle efficiency.

If the fuel to be used is petrol of calorific value 43 000 kJ/kg and cost \$0.05/kWh, determine the mass flow rate of the petrol and the annual running cost of the engine if it is to be run 5000 hours per year and there are no losses due to the combustion.

Examine the effects of adding a turbocharger to the engine. This can be idealised on the fluid pressure against specific volume diagram by allowing the isentropic expansion to proceed until the pressure of the air is the same as at state 1, and completing the cycle by having a constant pressure decrease in volume back to the original state.

Solution

The cycle plotted as a graph of air pressure against specific volume is shown in Figure 6.9.

The specific heat at constant pressure of the air $\{c_p\}_{air}$ is given by Equation (6.9) as

$$\{c_p - c_v\}_{air} = \{R\}_{air}$$

$$\therefore \quad \{c_p\}_{air} = 287 + 718 = 1005 \text{ J/kg K}$$

The ratio of the specific heats $\{\gamma\}_{air}$ is given by Equation (6.10) as

$$\{\gamma\}_{air} = \frac{\{c_p\}_{air}}{\{c_v\}_{air}} = \frac{1005}{718} = 1.4$$

State 1 to state 2 is a reversible adiabatic non-flow process. The specific volume of the air at state 2 $\{v_2\}_{air}$ is

$$\{v_2\}_{air} = \frac{\{v_1\}_{air}}{8} = \frac{0.8}{8} = 0.1 \text{ m}^3/\text{kg}$$

The pressure of the air at state 2 $\{p_2\}_{air}$ is given by Equation (6.15) as

$$\{p_1 v_1^\gamma\}_{air} = \{p_2 v_2^\gamma\}_{air}$$

$$\therefore \quad \{p_2\}_{air} = \frac{\{p_1 v_1^\gamma\}_{air}}{\{v_2^\gamma\}_{air}} = 1 \times 10^5 \times (8)^{1.4} = 18.4 \times 10^5 \text{ Pa}$$

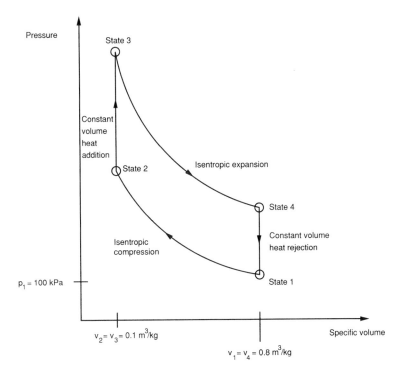

Fig. 6.9 *Example 6.7— graph of the fluid pressure against specific volume for air undergoing an ideal Otto cycle made up of four non-flow reversible processes.*

The temperature of the air at states 1 and 2, $\{T_1\}_{\text{air}}$ and $\{T_2\}_{\text{air}}$ respectively, is given by the equation of state (Equation (6.1)) as

$$\{p_1 v_1\}_{\text{air}} = \{RT_1\}_{\text{air}}$$

$$\therefore \quad \{T_1\}_{\text{air}} = \frac{1 \times 10^5 \times 0.8}{287} = 278.7 \text{ K}$$

$$\{p_2 v_2\}_{\text{air}} = \{RT_2\}_{\text{air}}$$

$$\therefore \quad \{T_2\}_{\text{air}} = \frac{18.4 \times 10^5 \times 0.1}{287} = 640.4 \text{ K}$$

The work transfer in the isentropic non-flow process $\dot{W}_{12}$ is given by Equation (5.12), with $n = \gamma$, as

$$\dot{W}_{12} = -\frac{\{\dot{m}(p_2 v_2 - p_1 v_1)\}_{\text{air}}}{\{1 - \gamma\}_{\text{air}}}$$

$$\therefore \quad \dot{W}_{12} = -\frac{0.05 \times 10^5 \times (18.4 \times 0.1 - 1 \times 0.8)}{1 - 1.4} = +13 \text{ kW}$$

The heat transfer in the isentropic non-flow process $\dot{Q}_{12}$, by definition, is zero. Thus $\dot{Q}_{12} = 0$.

State 2 to state 3 is a reversible constant volume non-flow heat addition until the temperature $\{T_3\}_{\text{air}}$ is 2000 K. Hence:

$$\{v_2\}_{\text{air}} = \{v_3\}_{\text{air}} = 0.1 \text{ m}^3/\text{kg}$$

156 THE FLUID AS A PERFECT GAS

The pressure of the air at state 3 $\{p_3\}_{air}$ is given by the equation of state (Equation (6.1)) as

$$\{p_3 v_3\}_{air} = \{RT_3\}_{air}$$

$$\therefore \{p_3\}_{air} = \frac{287 \times 2000}{0.1} = 57.4 \times 10^5 \text{ Pa}$$

The work transfer in the constant volume non-flow reversible process $\dot{W}_{23}$ is given by Equation (5.12) with $n = \gamma$ as $\dot{W}_{23} = 0$. The heat transfer in the constant volume non-flow reversible process $\dot{Q}_{23}$ is given by Equation (6.18) as

$$\dot{Q}_{23} = \{\dot{m} c_v (T_3 - T_2)\}_{air} - \dot{W}_{23} = 0.05 \times 718 \times (2000 - 640.4) = 48.8 \text{ kW}$$

State 3 to state 4 is another isentropic non-flow process during which the specific volume of the air expands back to its original value at state 1. Thus:

$$\{v_4\}_{air} = \{v_1\}_{air} = 0.8 \text{ m}^3/\text{kg}$$

The pressure of the air at state 4 $\{p_4\}_{air}$ is given by Equation (6.15) as

$$\{p_3 v_3^\gamma\}_{air} = \{p_4 v_4^\gamma\}_{air}$$

$$\therefore \{p_4\}_{air} = \frac{57.4 \times 10^5 \times 0.1^{1.4}}{0.8^{1.4}} = 3.1 \times 10^5 \text{ Pa}$$

The temperature of the air at state 4 $\{T_4\}_{air}$ is given by the equation of state (Equation (6.1)) as

$$\{p_4 v_4\}_{air} = \{RT_4\}_{air}$$

$$\therefore \{T_4\}_{air} = \frac{3.1 \times 10^5 \times 0.8}{287} = 864.1 \text{ K}$$

The work transfer in the isentropic non-flow process $\dot{W}_{34}$ is given by Equation (5.12) with $n = \gamma$ as

$$\dot{W}_{34} = -\frac{\{\dot{m}(p_4 v_4 - p_3 v_3)\}_{air}}{\{1 - \gamma\}_{air}}$$

$$\therefore \dot{W}_{34} = -\frac{0.05 \times 10^5 \times (3.1 \times 0.8 - 57.4 \times 0.1)}{1 - 1.4} = -40.8 \text{ kW}$$

The heat transfer in the isentropic non-flow process $\dot{Q}_{34}$, by definition, is zero. Thus: $\dot{Q}_{34} = 0$.

State 4 to state 1 is a reversible non-flow constant volume drop in pressure. The work transfer $\dot{W}_{41}$ from Equation (5.12) with $n = \infty$, is zero. Thus: $\dot{W}_{41} = 0$. The heat transfer $\dot{Q}_{41}$ is given by Equation (6.18) as

$$\dot{Q}_{41} = \{\dot{m} c_v (T_1 - T_4)\}_{air} - \dot{W}_{41} = 0.05 \times 718 \times (278.7 - 864.1) = -21.0 \text{ kW}$$

The net work transfer from the cycle $\sum(\dot{W})_{cyc}$ is the sum of the work transfer in each reversible non-flow process. Thus:

$$\sum(\dot{W})_{cyc} = \dot{W}_{12} + \dot{W}_{23} + \dot{W}_{34} + \dot{W}_{41}$$

$$\therefore \sum(\dot{W})_{cyc} = +13 + 0 - 40.8 + 0 = -27.8 \text{ kW}$$

NON-FLOW PROCESSES WITH PERFECT GASES

As a check, by the first law of thermodynamics (Equation (4.1)), the sum of the net work transfer in the cycle $\sum(\dot{W})_{cyc}$ and the net heat transfer in the cycle $\sum(\dot{Q})_{cyc}$ should equal zero:

$$\sum(\dot{Q})_{cyc} = \dot{Q}_{12} + \dot{Q}_{23} + \dot{Q}_{34} + \dot{Q}_{41} = 0 + 48.8 + 0 - 21.0 = 27.8 \text{ kW}$$

$$\therefore \sum(\dot{Q})_{cyc} + \sum(\dot{W})_{cyc} = 27.8 - 27.8 = 0$$

Therefore the results agree. The overall thermal efficiency of the cycle η is the ratio of the net work to the heat supplied, which is in non-flow reversible process 23. Thus, using sign rule C:

$$\eta = \frac{\sum(\dot{W})_{cyc}}{\dot{Q}_{23}} = \frac{27.8}{48.8} = 57\%$$

For a heat supply of 48.8 kW from the combustion of petrol with no losses, the mass flow rate of the petrol $\{\dot{m}\}_{fuel}$ is given by Equation (4.7) as

$$\dot{Q}_{23} = \dot{Q}_{comb} = \{\dot{m}\}_{fuel} CV$$

$$\therefore \{\dot{m}\}_{fuel} = \frac{48.8}{43000} = 0.001\,135 \text{ kg/s}$$

The annual running cost of the engine $\$_{ARC}$ is given by Equation (4.9) as

$$\$_{ARC} = EN_{hrs}\$_{kWh} = 48.8 \times 5000 \times 0.05 = \$12\,200 \text{ per year}$$

When there is a turbocharger fitted, the air pressure against specific volume diagram is as shown in Figure 6.10. State 3 to state 5 is still an isentropic non-flow process but now

$$\{p_5\}_{air} = \{p_1\}_{air} = 100 \text{ kPa}$$

The specific volume of the air at state 5 $\{v_5\}_{air}$ is given by Equation (6.15) as

$$\{p_3 v_3^\gamma\}_{air} = \{p_5 v_5^\gamma\}_{air}$$

$$\therefore \{v_5\}_{air} = \frac{57.4^{1/1.4} \times 0.1}{1^{1/1.4}} = 1.8 \text{ m}^3\text{/kg}$$

The temperature of the air at state 5 $\{T_5\}_{air}$ is given by the equation of state (Equation (6.1)) as

$$\{p_5 v_5\}_{air} = \{RT_5\}_{air}$$

$$\therefore \{T_5\}_{air} = \frac{1 \times 10^5 \times 1.8}{287} = 627.2 \text{ K}$$

The work transfer in the isentropic non-flow process $\dot{W}_{35}$ is given by Equation (5.12) with $n = \gamma$ as

$$\dot{W}_{35} = -\frac{\{\dot{m}(p_5 v_5 - p_3 v_3)\}_{air}}{\{1-\gamma\}_{air}}$$

$$\therefore \dot{W}_{35} = -\frac{0.05 \times 10^5 \times (1 \times 1.8 - 57.4 \times 0.1)}{1 - 1.4} = -49.3 \text{ kW}$$

The heat transfer in the isentropic non-flow process $\dot{Q}_{35}$, by definition, is zero. Thus $\dot{Q}_{35} = 0$.

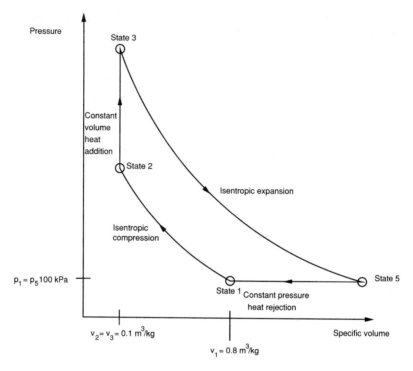

Fig. 6.10 *Example 6.7—graph of the fluid pressure against specific volume for air undergoing a cycle made up of four non-flow reversible processes.*

State 5 to state 1 is a non-flow reversible constant pressure decrease in volume. The work transfer $\dot{W}_{51}$ from Equation (5.12) with $n = 0$, is

$$\dot{W}_{51} = -\{\dot{m}p_5(v_1 - v_5)\}_{\text{air}} = -0.05 \times 10^5 \times (0.8 - 1.8) = +5 \text{ kW}$$

The heat transfer $\dot{Q}_{51}$ is given by Equation (6.18) as

$$\dot{Q}_{51} = \{\dot{m}c_v(T_1 - T_5)\}_{\text{air}} - \dot{W}_{51}$$
$$\therefore \quad \dot{Q}_{51} = 0.05 \times 718 \times (278.7 - 627.2) - 5 \times 10^3 = -17.5 \text{ kW}$$

The net work transfer from the cycle with the turbocharger $\sum(\dot{W}')_{\text{cyc}}$ is the sum of the work transfer in each non-flow reversible process. Thus:

$$\sum(\dot{W}')_{\text{cyc}} = \dot{W}_{12} + \dot{W}_{23} + \dot{W}_{34} + \dot{W}_{51}$$
$$\therefore \quad \sum(\dot{W}')_{\text{cyc}} = +13 + 0 - 49.3 + 5 = -31.3 \text{ kW}$$

As a check, by the first law of thermodynamics (Equation (4.1)) the sum of the net work transfer in the cycle $\sum(\dot{W}')_{\text{cyc}}$ and the net heat transfer in the cycle $\sum(\dot{Q}')_{\text{cyc}}$ should equal zero:

$$\sum(\dot{Q}')_{\text{cyc}} = \dot{Q}_{12} + \dot{Q}_{23} + \dot{Q}_{34} + \dot{Q}_{51} = 0 + 48.8 + 0 - 17.5 = 31.3 \text{ kW}$$

$$\therefore \sum (\dot{Q}')_{\text{cyc}} + \sum (\dot{W}')_{\text{cyc}} = 31.3 - 31.3 = 0$$

Therefore the results agree. The overall thermal efficiency of the cycle η is the ratio of the net work transfer to the heat supplied, which is in non-flow reversible process 23. Thus, using sign rule C:

$$\eta = \frac{\sum (\dot{W}')_{\text{cyc}}}{\dot{Q}_{23}} = \frac{31.3}{48.8} = 64\%$$

As the heat supply is the same, the mass flow rate of the fuel and the annual running cost of the engine remain the same. Hence, the engine fitted with the turbocharger produces more work because the area of the pressure volume diagram is larger, and is more efficient because no extra heat is supplied to produce the increased work. But it will have a greater capital cost.

As with the diesel engine, practical petrol engines utilise internal combustion and can only be considered to follow the ideal Otto cycle approximately. The fuel enters the cylinder with the air through the inlet manifold and, because the change in volume of the mixture during the compression process is relatively small, the fuel then has to be ignited by a sparking plug. Their efficiencies are lower than an equivalent diesel, but are suitable for smaller power outputs exactly because the change in volume during compression is small, resulting in a lower capacity engine.

It can be seen that for an initial air pressure and temperature of 100 kPa and 278.7 K respectively, without the turbocharger, the air pressure and temperature at state 4 are 310 kPa and 864.1 K respectively. In a real engine, the pressure difference to the atmosphere is used to drive the products of combustion out through the exhaust valve. All the products are emitted at 864.1 K, a waste of energy as referred to in Section 1.2.1. Even with the turbocharger fitted, in which case the pressure is reduced to its minimum value of 100 kPa, the temperature of the products of combustion at state 5 is 627.2 K. This will still represent a considerable waste of energy and explains why the efficiency of the engine is inherently low. It is not possible to reduce the temperature at exhaust to something approaching the atmospheric value.

EXAMPLE 6.8

A cylinder, closed by a piston at TDC position, contains 0.082 kg of air, assumed a perfect gas, at a temperature of 0°C and a pressure of 1.03 MPa. The gas undergoes a reversible polytropic non-flow expansion as the piston moves to BDC during which its volume increases fourfold (Figure 6.11). Determine the magnitude of the work and heat transfers during the reversible non-flow process when the index of expansion $n = \gamma$, the ratio of the specific heats, in which case the expansion is isentropic, when $n = 1.3$, 1.2 and 1.1, and when $n = 1$ and the process is isothermal. The specific heat at constant pressure of the air is constant and equal to 1005 J/kg K.

Solution

The gas constant $\{R\}_{\text{air}}$ for air is given in Table 6.1:

$$\{R\}_{\text{air}} = 287 \text{ J/kg K}$$

Equations (6.9) and (6.10) give the relationships between the specific heats as

$$\{\gamma\}_{\text{air}} = \frac{\{c_p\}_{\text{air}}}{\{c_v\}_{\text{air}}}$$

$$\{c_p - c_v\}_{\text{air}} = \{R\}_{\text{air}}$$

$$\therefore \{c_v\}_{\text{air}} = 718 \text{ J/kg K} \quad \text{and} \quad \{\gamma\}_{\text{air}} = 1.4$$

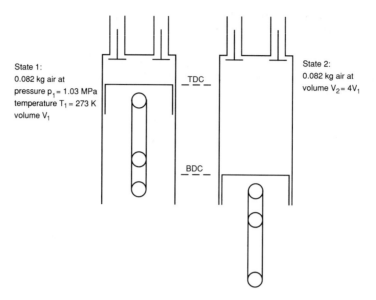

Fig. 6.11 *Example 6.8—a piston cylinder mechanism in which the air undergoes a number of polytropic non-flow reversible processes.*

Let the air go from state 1 to state 2. The final temperature $\{T_2\}_\text{air}$ in each case is given by Equation (6.17):

$$\frac{\{T_2\}_\text{air}}{\{T_1\}_\text{air}} = \frac{\{V_2^{1-n}\}_\text{air}}{\{V_1^{1-n}\}_\text{air}}$$

where

$$\{T_1\}_\text{air} = 0\,°\text{C} = 273\text{ K} \quad \text{and} \quad \frac{\{V_2\}_\text{air}}{\{V_1\}_\text{air}} = 4$$

The work transfer W_{12} in each case is given by Equation (5.12):

$$W_{12} = -\frac{\{m(p_2 v_2 - p_1 v_1)\}_\text{air}}{1-n}$$

When the fluid is a perfect gas, substitution from the equation of state, (Equation (6.1)) gives

$$W_{12} = -\frac{\{mR(T_2 - T_1)\}_\text{air}}{1-n}$$

When $n = 1$, it is necessary to use the isothermal expression for the work transfer (Equation (6.14)) as follows:

$$Q_{12} = -W_{12} = \left\{mRT \ln \frac{V_2}{V_1}\right\}_\text{air}$$

The heat transfer Q_{12} in each case, except where $n = 1$, is given by Equation (6.18) as

$$Q_{12} = \{mc_v(T_2 - T_1)\}_\text{air} - W_{12}$$

Hence it is possible to draw up Table 6.4 for the solution. Consideration of such tables allows the best process to be chosen to satisfy a particular requirement if a reciprocating engine or compressor is being designed.

Table 6.4 *Example 6.8—values of n, $\{T_2\}_{air}$, W_{12} and Q_{12}.*

Process type	Value of n	$\{T_2\}_{air}$ (K)	W_{12} (kJ)	Q_{12} (kJ)
Isentropic	$n = \gamma = 1.4$	156.8	−6.8	0
Polytropic	$n = 1.3$	180.1	−7.3	1.8
Polytropic	$n = 1.2$	206.9	−7.8	3.9
Polytropic	$n = 1.1$	237.7	−8.3	6.2
Isothermal	$n = 1.0$	273.0	−8.9	8.9

6.4 FLOW PROCESSES WITH PERFECT GASES

While there are many applications for perfect gases in non-flow processes such as occur in reciprocating piston cylinder machinery, there are also very many applications for perfect gases in flow processes, and these are considered in detail in Chapter 8.

7 The Fluid as Water/Steam

After air, water/steam is the most common fluid used in a thermal system and it is necessary to consider it as liquid water, steam vapour and as water changing phase to steam or steam changing phase to water. While the relationships between the properties when the fluid is in either the liquid or vapour state are fairly straightforward, this is no longer the case when the fluid is changing phase. It is not possible to describe changes in the property values by means of fairly simple formulae, as with a perfect gas. Instead, recourse must be made to tables or charts to obtain the necessary information.

7.1 TEMPERATURE/VOLUME RELATIONSHIP FOR WATER/STEAM

Consider a container filled with 1 kg of water placed upon a lighted natural gas ring. Initially the water is at approximately atmospheric conditions, say a pressure of 101.325 kPa and a temperature of 293 K. As heat is transferred to the water from the combustion of the natural gas, the water temperature rises and its volume increases very slightly. When the temperature reaches 373.15 K, it remains constant for a while, even though heat is still being added, while it changes phase to steam, and this is accompanied by a large increase in volume. Once all the water has become steam, further addition of heat results in the temperature rising and the steam expanding in volume slightly. The relationship between the temperature and the volume of the water/steam is shown in Figure 7.1.

When the water changes phase to steam, it is doing so under constant temperature and constant pressure conditions, 373.15 K and 101.325 kPa in this case. The temperature is called the Boiling-point and the water is said to Boil or Evaporate. The two properties, pressure and temperature, are not independent of each other during the change of phase. Instead, the temperature is dependent upon the pressure and the boiling-point varies with pressure. The heat that is added during the change of phase is known as the Latent Heat of Vaporisation. It is possible for the change of phase to occur in the other direction, for the steam to become water. In this case, no heat supply is required, rather the latent heat is given out by the fluid to the atmosphere. The phase change is referred to as Condensation.

When the water has reached the boiling-point corresponding to the given pressure, the water is said to be Saturated, and when it has changed phase to become steam, also at the boiling-point, it is called saturated steam, sometimes dry saturated steam. In between, it is a mixture of saturated water and saturated steam in certain proportions, and it is called a Mixture or Wet Vapour.

164   THE FLUID AS WATER/STEAM

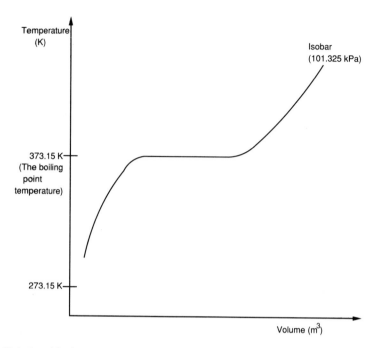

Fig. 7.1 *Relationship between the temperature and volume of water/steam at a constant pressure of 101.325 kPa.*

Beyond the boiling-point, where the temperature is increasing, the fluid is known as Superheated steam, and below the boiling-point, where the temperature is decreasing, the fluid is called Unsaturated water.

When the same heating procedure is carried out on the 1 kg of water in the container but under conditions of higher pressure, its initial volume is slightly less than that at atmospheric pressure, the boiling-point temperature is raised, and the increase in the volume of the water as it changes phase to steam decreases in comparison, as shown in Figure 7.2.

Eventually, when the pressure reaches 22.11 MPa, there is only a slight inflexion point in the curve to indicate the boiling-point, which is at 647.3 K. There is no longer a period when heat is added and the temperature is seen to remain constant while the volume increases. The latent heat of vaporisation is zero. Therefore, the change in the volume of the water/steam during the change of phase is an indication of the amount of heat transfer, into or out of the fluid. The inflexion point is called the Critical Point. Above the critical pressure, there is no definite transition from water to steam and the two phases cannot be identified individually.

When the same heating procedure is carried out on the 1 kg of water in the container but under conditions of lower pressure, its initial volume is slightly greater than at atmospheric pressure, the boiling-point temperature is lowered, and there is an increase in the volume of the water as it changes phase to steam, and in the latent heat of vaporisation, as shown again in Figure 7.2. It is possible to lower the pressure to 611.2 Pa, in which case the boiling-point falls to 273.16 K and ice, water and steam coexist. This is called the Triple Point of Water, the volume of the mixture depending upon the proportions of ice, saturated water and saturated steam.

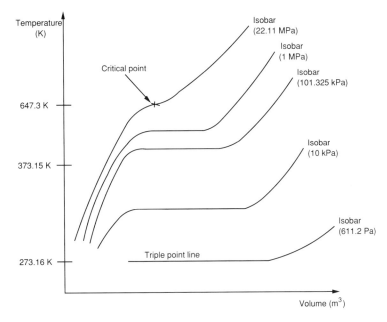

Fig. 7.2 *Relationship between the temperature and volume of water/steam at various pressures.*

7.2 PRESSURE/VOLUME RELATIONSHIP FOR WATER/STEAM

A temperature against volume graph for water/steam is of limited use. Of far more relevance is a pressure against volume graph, as shown in Figure 7.3, because the area under the curve is an indication of the work transfer in any thermal process. Figure 7.3 takes the data of Figures 7.1 and 7.2, but shows isotherms, lines of constant temperature, rather than isobars, lines of constant pressure.

Line AB represents the change in volume of the fluid at the triple point of water, where the ice, water and steam phases all coexist. Point C is the critical point of water at which there is only an inflection to indicate the boiling-point and change of phase from water to steam. Curve ACB is called the Saturation Curve. Curve AC is the saturated water curve and whenever the water conditions are such that it falls on this curve, it must be in the saturated water state. Curve CB is the saturated steam curve and whenever the steam conditions are such that it falls on this curve, it must be in the saturated steam state. Within the area bounded by ACBA, the fluid is a mixture of saturated water and saturated steam in some proportion. In this area the fluid changes phase under conditions of constant pressure and constant temperature, which can only be represented by horizontal lines between curve AC and curve BC.

To the left of curve AC, the fluid is in the unsaturated water state, and to the right of curve CB, the fluid is in the unsaturated steam state, called superheated steam. Above the critical point, water changes phase to steam without a recognisable boiling-point being identified, in other words there is no definite transition from the liquid to the vapour state, and the steam is called Supercritical.

166 THE FLUID AS WATER/STEAM

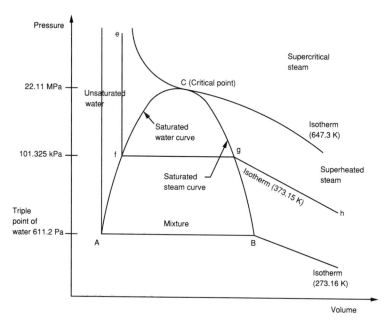

Fig. 7.3 *Relationship between the pressure and volume of water/steam at various temperatures.*

For thermal systems, interest in the properties of water/steam lies in the region of pressure between the critical and triple points. Taking isotherm efgh as an example, when the fluid conditions are such that it lies on line ef, it is unsaturated water. At point f it becomes saturated water. At point g it is saturated steam, and between points f and g it is changing phase from saturated water to saturated steam and is a mixture of the two. On line gh it is superheated steam.

7.3 OTHER PROPERTY RELATIONSHIPS FOR WATER/STEAM

If values for some of the other properties of water/steam of interest in a thermal system, such as the entropy, internal energy and the enthalpy, can be found, they too can be plotted on the graph of pressure versus volume. As with a perfect gas, it is possible to find differences in these properties when the fluid changes state. For internal energy, the NFEE (Equation (5.1)) applies to any non-flow process, so that the internal energy change of water/steam may be determined from a knowledge of the work and heat transfer during a change of state in a non-flow process. The enthalpy change is given by Equation (5.4). For the entropy change, Equation (3.2) applies only to a reversible non-flow process. However, as the entropy is a property and exists at the two end states independent of the process undergone by the water/steam between the states, it is possible to find the change of entropy from a knowledge of other property values through the NFEE as follows:

$$Q_{12} + W_{12} = U_2 - U_1$$

Substituting from Equations (3.1) and (3.2) gives

$$\int_1^2 T\,dS - \int_1^2 p\,dV = \int_1^2 dU$$

$$\therefore\ S_2 - S_1 = \int_1^2 \frac{dU}{T} + \int_1^2 \frac{p\,dV}{T}$$

Although Equations (3.1) and (3.2) apply specifically to reversible non-flow processes, this equation relates only fluid property values which are independent of the process. As the water/steam undergoes any non-flow process, measurements of its pressure, temperature and volume can be made and plotted on a graph of p/T against V. The actual slope of the graph is not too important because the change in entropy depends only upon the end states. The internal energy change is found from the NFEE as discussed above so that a graph of $1/T$ against U can be drawn. The change in the entropy of the water/steam between the end states of the process is given by summing the areas under the curve on each graph.

By international agreement, it has been deemed that at the triple point of water, where the pressure is 611.2 Pa and the temperature is 273.16 K, the internal energy and the entropy of saturated water shall be zero. The enthalpy can be found from Equation (5.4) which is its definition. Hence, individual values for the entropy, internal energy and enthalpy can be determined, although they are, in fact, meaningless. Fortunately, it is only differences in these property values during a change of state that are of consequence in a thermal system. The individual values may be plotted on the graph of pressure against volume of Figure 7.3, but they have not been included because it results in a profusion of lines.

7.3.1 Specific heat

The specific heat at constant pressure and the specific heat at constant volume of liquid water and steam vapour both vary with temperature but, when analysing thermal systems, it is usually within a suitable order of accuracy to assume that they are both constant. This is certainly all right for water, but strictly only true for steam vapour over a reasonably small range of temperature. Therefore, when the water/steam exists in the liquid or vapour states, Equations (5.15) and (5.16) are applicable.

In addition, the difference between the specific heat at constant pressure and the specific heat at constant volume of liquids is very small. This is because the change of volume of liquids with temperature during a constant pressure process is insignificant and the constant pressure process is also effectively a constant volume process too. As a consequence, generally only the specific heat at constant pressure of liquids is quoted and used. There is no need to restrict Equation (5.15) to any type of process and the subscript is not necessary. Therefore, for water in the liquid phase, Equation (6.7) is applicable.

Further, steam vapour, under many circumstances, can be treated as a perfect gas and, as such, will approximately obey the equations derived in Chapter 6. In this case, substitution for the enthalpy and internal energy in terms of the specific heats can be made in accordance with Equations (6.4), (6.5), (6.7) and (6.8), with no restriction on the type of process.

However, the specific heats at constant pressure and constant volume of liquid water are very different from those of steam vapour. There is no question of assuming that the specific heats are constant during the change of phase and all the equations derived assuming constant specific heats are not valid.

168 THE FLUID AS WATER/STEAM

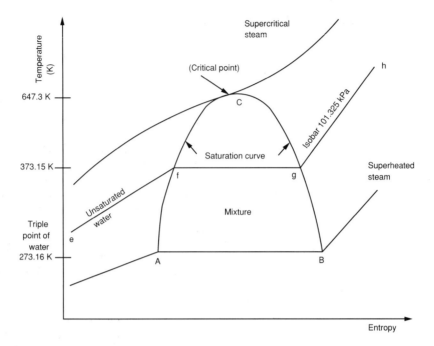

Fig. 7.4 *Relationship between the temperature and entropy of water/steam at various pressures.*

7.4 TEMPERATURE/ENTROPY RELATIONSHIP FOR WATER/STEAM

A plot of the properties temperature versus entropy for the water/steam between the critical and triple points is also useful because it gives an indication of the heat transfer in any process, and this is shown in Figure 7.4 for lines of constant pressure.

As with the pressure against volume graph, it is possible to identify the saturation curve within which the fluid changes phase, and the unsaturated water and superheat regions. The properties volume, internal energy and enthalpy have been left off for clarity.

7.5 DIAGRAMS FOR PERFECT GASES

A gas was defined in Chapter 6 as being a fluid which existed naturally in the vapour state at NTP. In other words, its critical point occurs at a very low temperature and pressure. The characteristic of a gas is that, if it is assumed to be perfect, variations in its property values can be described by a series of simple formulae. In fact, the same will apply to steam when it is in a condition above its critical point, but this rarely happens in practical thermal systems. However, it is often reasonable to apply the perfect gas formula to steam when it is in the superheat condition. The changes in the property values there, as seen in Figures 7.3 and 7.4, are gradual and the equations of Chapter 6 will give approximate answers which are accurate enough for most circumstances. Considering the isotherm efgh in Figure 7.3, however, it is not possible to derive formulae which will describe the changes in the property

7.6 DRYNESS FRACTION

Considering the isotherm efgh in Figure 7.3, at point f the fluid is saturated water, at point g it is saturated steam, and in between it is a mixture of the two as it changes phase at constant pressure and constant temperature. The Dryness Fraction is defined as 'the mass of saturated steam within the total mass of the mixture'. It is given the symbol X and is written:

$$X = \frac{\text{Mass of saturated steam}}{\text{Mass of mixture}} \quad (7.1)$$

Hence, the value of X varies from 0 at point f to 1 at point g.

Consider further the simplified pressure against volume graph of Figure 7.5, where, in fact, the specific volume is plotted. When the fluid is saturated water and lies on curve AC, let the values of all the properties there have a subscript f. Thus, the specific volume of the saturated water at point f is v_f. Similarly, when the fluid is saturated steam and lies on curve CB, let the values of all the properties have a subscript g. Thus, the specific volume of the saturated steam at point g is v_g.

As the dryness fraction X varies from 0 at point f to 1 at point g, so the specific volume v of the mixture varies in the same way between the value v_f at point f and v_g at point g. The proportions must be the same and it is possible to write:

$$X = \frac{v - v_f}{v_g - v_f}$$

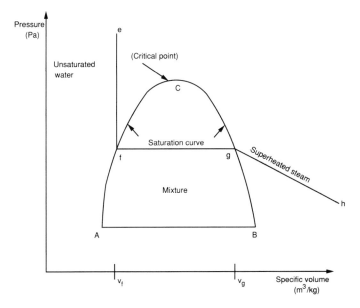

Fig. 7.5 *Relationship between the pressure and specific volume of water/steam for one isotherm.*

$$\therefore \quad v = Xv_g + (1-X)v_f \tag{7.2}$$

Equation (7.2) allows the specific volume of the mixture to be determined from a knowledge of the dryness fraction and the specific volumes of the fluid in the saturated water and saturated steam condition. In practice, the value of v_g at point B is 206.1 m³/kg and the value of v at point C, the critical point, is 0.003 17 m³/kg. Obviously, therefore, values of v_f tend to be quite small in comparison to values of v_g and it is a reasonable assumption under most circumstances to write:

$$v = Xv_g \tag{7.3}$$

In the same way that an equation for the determination of the specific volume of a mixture has been derived, it is possible to write down similar equations for the specific entropy s, the specific internal energy u and the specific enthalpy h of the mixture, as in Equation (7.2) because the simplifying assumption for Equation (7.3) is not valid. Hence:

$$s = Xs_g + (1-X)s_f \tag{7.4}$$
$$u = Xu_g + (1-X)u_f \tag{7.5}$$
$$h = Xh_g + (1-X)h_f \tag{7.6}$$

7.7 STEAM TABLES

It is intended to refer to the booklet *Thermodynamic and Transport Properties of Fluids*, 4th edition, arranged by G.F.C. Rogers and Y.R. Mayhew, published by Basil Blackwell, and commonly referred to as the Steam Tables, whenever the properties of water/steam are required. This is an extremely useful publication, packed full of information, not just on water/steam but on other fluids too. Only a limited part of the tables is considered in Sections 7.7.1–7.7.3. This encompasses the superheat, and saturated water and steam conditions, but there are other tables concerning the properties of water/steam which may be important.

From the steam tables, it is possible to determine the values of the pressure, temperature, specific volume, specific entropy, specific internal energy and specific enthalpy for unsaturated water, saturated water, a mixture of saturated water and saturated steam, saturated steam and superheated steam, for the fluid with pressures between the critical and triple points. Obviously, not every value can be listed and some interpolation is required.

7.7.1 Superheated steam

The superheated steam tables have pressure and temperature as the main axes and the property values start at the saturated steam condition corresponding to any given pressure. The pressure range is up to the critical point where the pressure is 22.11 MPa. Above that pressure is the supercritical region. If a particular value of pressure or temperature required is not listed in the tables, it is necessary to interpolate between the values that are listed, assuming that the variation in any property value is linear.

Only two fluid properties need be known in the superheat tables for all the others to be determined. These do not have to be the temperature and pressure — it is perfectly possible to find the pressure and temperature knowing the specific entropy and specific volume for example. But it is rather tedious to do so. Only examples where either the fluid pressure or temperature is given will be considered here.

EXAMPLE 7.1
What is the specific volume, specific entropy, specific internal energy and specific enthalpy of superheated steam when:

(a) The pressure is 3 MPa and the temperature is 400 °C?
(b) The pressure is 3 MPa and the temperature is 420 °C?
(c) The pressure is 3.3 MPa and the temperature is 400 °C?
(d) The pressure is 3.3 MPa and the temperature is 420 °C?

Solution
(a) A pressure of 3 MPa and a temperature of 400 °C are listed in the superheat tables, so the specific volume $\{v\}_{st}$, specific entropy $\{s\}_{st}$, specific internal energy $\{u\}_{st}$ and specific enthalpy $\{h\}_{st}$ can be read directly as follows:

$$\{v\}_{st} = 0.0993 \text{ m}^3/\text{kg}$$

$$\{s\}_{st} = 6.921 \text{ kJ/kg K}$$

$$\{u\}_{st} = 2933 \text{ kJ/kg}$$

$$\{h\}_{st} = 3231 \text{ kJ/kg}$$

(b) A pressure of 3 MPa is listed in the superheat tables, but only temperatures of 400 and 450 °C. To find the values of $\{v\}_{st}$, $\{s\}_{st}$, $\{u\}_{st}$ and $\{h\}_{st}$ at 420 °C, it is necessary to interpolate between the values given at 400 and 450 °C, assuming a linear relationship. Hence:

$$\frac{420 - 400}{450 - 400} = \frac{\{v\}_{st} - 0.0993}{0.1078 - 0.0993} = \frac{\{s\}_{st} - 6.921}{7.082 - 6.921} = \frac{\{u\}_{st} - 2933}{3020 - 2933} = \frac{\{h\}_{st} - 3231}{3343 - 3231}$$

$$\therefore \{v\}_{st} = 0.1027 \text{ m}^3/\text{kg}$$

$$\{s\}_{st} = 6.985 \text{ kJ/kg K}$$

$$\{u\}_{st} = 2968 \text{ kJ/kg}$$

$$\{h\}_{st} = 3276 \text{ kJ/kg}$$

(c) A temperature of 400 °C is listed in the superheat tables, but only pressures of 3 and 4 MPa. To find the values of $\{v\}_{st}$, $\{s\}_{st}$, $\{u\}_{st}$ and $\{h\}_{st}$ at 3.3 MPa, it is necessary to interpolate between the values given at 3 MPa and 4 MPa, assuming a linear relationship. Hence:

$$\frac{3.3 - 3}{4 - 3} = \frac{\{v\}_{st} - 0.0993}{0.0733 - 0.0993} = \frac{\{s\}_{st} - 6.921}{6.769 - 6.921} = \frac{\{u\}_{st} - 2933}{2921 - 2933} = \frac{\{h\}_{st} - 3231}{3214 - 3231}$$

$$\therefore \{v\}_{st} = 0.0915 \text{ m}^3/\text{kg}$$

$$\{s\}_{st} = 6.875 \text{ kJ/kg K}$$

$$\{u\}_{st} = 2929 \text{ kJ/kg}$$

$$\{h\}_{st} = 3226 \text{ kJ/kg}$$

Table 7.1 Example 7.1—values of $\{v\}_{st}$, $\{s\}_{st}$, $\{u\}_{st}$ and $\{h\}_{st}$ at various pressures.

Pressure	Property	Unit	Temperature		
			400°C	420°C	450°C
3 MPa	$\{v\}_{st}$	m³/kg	0.0993	0.1027	0.1078
	$\{s\}_{st}$	kJ/kg K	6.921	6.985	7.082
	$\{u\}_{st}$	kJ/kg	2933	2968	3020
	$\{h\}_{st}$	kJ/kg	3231	3276	3343
3.3 MPa	$\{v\}_{st}$	m³/kg	0.0915	0.0947	0.0995
	$\{s\}_{st}$	kJ/kg K	6.875	6.940	7.038
	$\{u\}_{st}$	kJ/kg	2929	2964	3017
	$\{h\}_{st}$	kJ/kg	3226	3271	3339
4 MPa	$\{v\}_{st}$	m³/kg	0.0733	0.0760	0.0800
	$\{s\}_{st}$	kJ/kg K	6.769	6.835	6.935
	$\{u\}_{st}$	kJ/kg	2921	2957	3010
	$\{h\}_{st}$	kJ/kg	3214	3260	3330

(d) Neither a pressure of 3.3 MPa nor a temperature of 420°C is listed in the superheat tables, only pressures of 3 and 4 MPa and temperatures of 400 and 450°C. To find the values of $\{v\}_{st}$, $\{s\}_{st}$, $\{u\}_{st}$ and $\{h\}_{st}$ at 3.3 MPa and 420°C, it is first of all necessary to interpolate between 3 and 4 MPa to find out the values of the properties at 3.3 MPa for both 400 and 450°C, and to interpolate at 3.3 MPa between 400 and 450°C to find out the values of the properties at 420°C. Carrying out the interpolation as above assuming linear relationships at all times, leads to Table 7.1. When a double interpolation is required in the superheat tables, it is easier to use a chart, but only tables are considered here.

EXAMPLE 7.2

What is the temperature, specific volume, specific internal energy and specific enthalpy of superheated steam when:

(a) The pressure is 1 MPa and the specific entropy is 6.926 kJ/kg K?
(b) The pressure is 1 MPa and the specific entropy is 6.950 kJ/kg K?
(c) The pressure is 1.2 MPa and the specific entropy is 6.800 kJ/kg K?

Solution

(a) At a pressure of 1 MPa, a specific entropy of 6.926 kJ/kg K is shown in the tables, so the temperature $\{T\}_{st}$, specific volume $\{v\}_{st}$, specific internal energy $\{u\}_{st}$ and specific enthalpy $\{h\}_{st}$ can be read directly as follows:

$$\{T\}_{st} = 250°C$$
$$\{v\}_{st} = 0.2328 \text{ m}^3/\text{kg}$$
$$\{u\}_{st} = 2711 \text{ kJ/kg}$$
$$\{h\}_{st} = 2944 \text{ kJ/kg}$$

(b) At a pressure of 1 MPa, a specific entropy of 6.926 kJ/kg K is shown in the tables under a temperature of 250°C, and a specific entropy of 7.124 kJ/kg K under a temperature of 300°C. The required specific entropy of 6.95 kJ/kg K falls between these two but is not listed. However, the temperature $\{T\}_{st}$ must lie between 250 and 300°C in the same ratio as does the specific entropy. Similarly for the specific volume $\{v\}_{st}$, specific internal energy

Table 7.2 *Example 7.2 — Values of* $\{v\}_{st}$, $\{s\}_{st}$, $\{u\}_{st}$ *and* $\{h\}_{st}$ *at 1.2 MPa.*

Pressure	Property	Unit	Temperature	
			200 °C	250 °C
1.2 MPa	$\{v\}_{st}$	m³/kg	0.1766	0.2005
	$\{s\}_{st}$	kJ/kg K	6.598	6.840
	$\{u\}_{st}$	kJ/kg	2613	2705
	$\{h\}_{st}$	kJ/kg	2816	2936

$\{u\}_{st}$ and specific enthalpy $\{h\}_{st}$. Assuming a linear interpolation, the property values are as follows:

$$\frac{6.95 - 6.926}{7.124 - 6.926} = \frac{\{T\}_{st} - 250}{300 - 250} = \frac{\{v\}_{st} - 0.2328}{0.2580 - 0.2328} = \frac{\{u\}_{st} - 2711}{2794 - 2711} = \frac{\{h\}_{st} - 2944}{3052 - 2944}$$

$$\therefore \{T\}_{st} = 256.1\,°C$$

$$\{v\}_{st} = 0.2359\ m^3/kg$$

$$\{u\}_{st} = 2721\ kJ/kg$$

$$\{h\}_{st} = 2957\ kJ/kg$$

(c) Neither a pressure of 1.2 MPa, nor a specific entropy of 6.8 kJ/kg K is listed in the superheat tables. Firstly the property values at a pressure of 1.2 MPa must be derived by interpolating between the values given at 1 and 1.5 MPa. It is necessary to do this only in the region of specific entropy of value 6.8 kJ/kg K. Assuming a linear interpolation, the results are given in Table 7.2. The known specific entropy value of the steam of 6.8 kJ/kg K lies between the values of 6.598 kJ/kg K at a temperature of 200 °C and 6.84 kJ/kg K at a temperature of 250 °C. The unknown property values must be in the same proportion as the specific entropy, assuming a linear interpolation. Hence:

$$\frac{6.8 - 6.598}{6.84 - 6.598} = \frac{\{T\}_{st} - 200}{250 - 200} = \frac{\{v\}_{st} - 0.1766}{0.2005 - 0.1766} = \frac{\{u\}_{st} - 2613}{2705 - 2613} = \frac{\{h\}_{st} - 2816}{2936 - 2816}$$

$$\therefore \{T\}_{st} = 241.7\,°C$$

$$\{v\}_{st} = 0.1965\ m^3/kg$$

$$\{u\}_{st} = 2690\ kJ/kg$$

$$\{h\}_{st} = 2916\ kJ/kg$$

7.7.2 Saturated water and steam

There are two sets of tables for saturated water and steam. The first has temperature, the boiling-point, as the main axis, ranging from the triple point value of 0.01–100 °C, and is mainly used for air-conditioning problems. The second set has pressure as the main axis, ranging from the triple point pressure of 611.2 Pa to the critical point pressure of 22.11 MPa. When the fluid is a mixture of saturated water and steam, the pressure and temperature are no longer independent of each other. Therefore, corresponding to any given temperature, there is a particular pressure p_{fg}, and to any given pressure, there is a particular temperature, the boiling-point T_{fg}. For each pressure and temperature, values of v_g, u_f, u_g, h_f, h_g, s_f and s_g are given. These are the saturated water and saturated steam values, and to find the property values of any mixture, it is necessary to know the dryness fraction. Also shown are columns for h_{fg} and s_{fg}. These are simply the difference between the saturated specific enthalpy and specific entropy values, in other words

$$h_{fg} = (h_g - h_f) \quad \text{and} \quad s_{fg} = (s_g - s_f)$$

Not every temperature or pressure in the range mentioned above is given in the tables, but it is usually within a suitable order of accuracy to assume the nearest listed value. As with the superheat tables, only examples whereby either the pressure or the temperature of the mixture is known will be considered.

EXAMPLE 7.3

Find the unknown steam property values of pressure, temperature, specific volume, specific internal energy, specific enthalpy, specific entropy and dryness fraction, when:

(a) The pressure is 1.2 MPa and the steam is in the saturated water condition;
(b) The pressure is 1.2 MPa and the steam is in the saturated steam condition;
(c) The pressure is 1.2 MPa and the steam is a mixture with a dryness fraction of 0.8;
(d) The pressure is 1.24 MPa and the steam is a mixture with a dryness fraction of 0.8;
(e) The pressure is 1.2 MPa and the steam has a specific enthalpy of 2000 kJ/kg;
(f) The pressure is 1.2 MPa and the steam has a specific volume of 0.15 m³/kg;
(g) The temperature is 70 °C and the steam is in the saturated steam condition;
(h) The temperature is 130 °C and the steam is in the saturated water condition;
(i) The temperature is 160 °C and the steam is a mixture with a dryness fraction of 0.5.

Solution

(a) Saturated water takes the subscript f values. At a pressure of 1.2 MPa therefore:

$$\{T_{fg}\}_{st} = 188\,°C \text{ (the boiling-point)}$$
$$\{v\}_{st} = v_f \text{ (not listed as very small)}$$
$$\{u\}_{st} = u_f = 797 \text{ kJ/kg}$$
$$\{h\}_{st} = h_f = 798 \text{ kJ/kg}$$
$$\{s\}_{st} = s_f = 2.216 \text{ kJ/kg K}$$
$$X = 0 \text{ (for saturated water)}$$

(b) Saturated steam takes the subscript g values. At a pressure of 1.2 MPa therefore:

$$\{T_{fg}\}_{st} = 188\,°C \text{ (the boiling-point)}$$
$$\{v\}_{st} = v_g = 0.1632 \text{ m}^3/\text{kg}$$
$$\{u\}_{st} = u_g = 2588 \text{ kJ/kg}$$
$$\{h\}_{st} = h_g = 2784 \text{ kJ/kg}$$
$$\{s\}_{st} = s_g = 6.523 \text{ kJ/kg K}$$
$$X = 1 \text{ (for saturated steam)}$$

(c) At a dryness fraction X of 0.8, it is necessary to use Equations (7.3)–(7.6). At a pressure of 1.2 MPa therefore:

$\{T_{fg}\}_{st} = 188\,°C$ (the boiling-point, constant as the steam changes from saturated water to saturated steam)

$\{v\}_{st} = Xv_g = 0.8 \times 0.1632 = 0.1306 \text{ m}^3/\text{kg}$

$\{u\}_{st} = Xu_g + (1 - X)u_f = 0.8 \times 2588 + (1 - 0.8) \times 797 = 2230 \text{ kJ/kg}$

$\{h\}_{st} = Xh_g + (1 - X)h_f = 0.8 \times 2784 + (1 - 0.8) \times 798 = 2387 \text{ kJ/kg}$

$\{s\}_{st} = Xs_g + (1 - X)s_f = 0.8 \times 6.523 + (1 - 0.8) \times 2.216 = 5.662 \text{ kJ/kg K}$

(d) A pressure of 1.24 MPa is not given, but property values at pressures of 1.2 MPa and 1.3 MPa are. It is possible to interpolate between the values at 1.2 and 1.3 MPa to find $\{T_{fg}\}_{st}$, v_g, u_f, u_g, h_f, h_g, s_f and s_g at a pressure of 1.24 MPa. However, it is usually within a sufficient order of accuracy to take those values at the nearest listed pressure, 1.2 MPa in this case. At a dryness fraction of 0.8, therefore, the answers are the same as for part (c).

(e) In this case, the dryness fraction is not known. However, the steam must be a mixture because, at a pressure of 1.2 MPa, its specific enthalpy lies between the saturated water value of $h_f = 797$ kJ/kg and the saturated steam value of 2784 kJ/kg. The known specific enthalpy can be used to find the dryness fraction from Equation (7.6) as follows:

$$\{h\}_{st} = Xh_g + (1 - X)h_f$$

At 1.2 MPa:

$$2000 = X \times 2784 + (1 - X) \times 798 \quad \therefore \quad X = 0.605$$

The other property values can now be found from Equations (7.3)–(7.5):

$\{T_{fg}\}_{st} = 188\,°C$ (as before)

$\{v\}_{st} = Xv_g = 0.605 \times 0.1632 = 0.0987 \text{ m}^3/\text{kg}$

$\{u\}_{st} = Xu_g + (1 - X)u_f = 0.605 \times 2588 + (1 - 0.605) \times 797 = 1881 \text{ kJ/kg}$

$\{s\}_{st} = Xs_g + (1 - X)s_f = 0.605 \times 6.523 + (1 - 0.605) \times 2.216 = 4.822 \text{ kJ/kg K}$

(f) As in part (e) the dryness fraction is not known. The steam could be a mixture or a liquid because, at a pressure of 1.2 MPa, its specific volume is less than the saturated steam value of 0.1632 m³/kg, but the specific volume of saturated water is not given. However, the specific volume of saturated water at the critical point is 0.003 17 m³/kg and, as the known volume is 0.15 m³/kg, it must be a mixture. The specific volume can be used to find the dryness fraction from Equation (7.3) as follows:

$$\{v\}_{st} = Xv_g$$

At 1.2 MPa:

$$0.15 = X \times 0.1632 \quad \therefore \quad X = 0.919$$

The other property values can now be found from Equations (7.4) to (7.6):

$\{T_{fg}\}_{st} = 188\,°C$ (as before)

$\{u\}_{st} = Xu_g + (1 - X)u_f = 0.919 \times 2588 + (1 - 0.919) \times 797 = 2443 \text{ kJ/kg}$

$\{h\}_{st} = Xh_g + (1 - X)h_f = 0.919 \times 2784 + (1 - 0.919) \times 798 = 2623 \text{ kJ/kg}$

$\{s\}_{st} = Xs_g + (1 - X)s_f = 0.919 \times 6.523 + (1 - 0.919) \times 2.216 = 6.174 \text{ kJ/kg K}$

176   THE FLUID AS WATER/STEAM

(g) Saturated steam takes the subscript g values. At a temperature of 70 °C, from the first set of saturated water and steam tables therefore:

$$\{p_{fg}\}_{st} = 31.16 \text{ kPa (the boiling-point pressure)}$$
$$\{v\}_{st} = v_g \text{ (not given)}$$
$$\{u\}_{st} = u_g \text{ (not given)}$$
$$\{h\}_{st} = h_g = 2626.3 \text{ kJ/kg}$$
$$\{s\}_{st} = s_g = 7.755 \text{ kJ/kg K}$$
$$X = 1 \text{ (for saturated steam)}$$

Approximate values for v_g and u_g can be obtained from the second set of saturated water and steam tables. A temperature of 70 °C is not given but the boiling-point corresponding to a pressure of 32 kPa is 70.6 °C. Assuming that this is near enough:

$$\{v\}_{st} = v_g = 4.921 \text{ m}^3/\text{kg}$$
$$\{u\}_{st} = u_g = 2470 \text{ kJ/kg}$$

(h) Saturated water takes the subscript f values. The first set of saturated water and steam tables only goes up to a temperature of 100 °C, but a temperature of 130 °C is given in the second set of tables, at a pressure of 270 kPa. The other property values, therefore, are as follows:

$$\{p_{fg}\}_{st} = 270 \text{ kPa (the boiling-point pressure)}$$
$$\{v\}_{st} = v_f \text{ (not listed as very small)}$$
$$\{u\}_{st} = u_f = 546 \text{ kJ/kg}$$
$$\{h\}_{st} = h_f = 546 \text{ kJ/kg}$$
$$\{s\}_{st} = s_f = 1.634 \text{ kJ/kg K}$$
$$X = 0 \text{ (for saturated water)}$$

(i) A temperature of 160 °C is not given in the second set of saturated water and steam tables, but property values at temperatures of 158.8 °C (corresponding to a pressure of 600 kPa) and 165 °C (corresponding to a pressure of 700 kPa) are. Assuming that the values at 158.8 °C are sufficiently accurate, at a dryness fraction X of 0.5, the other property values can be found from Equations (7.3)–(7.6), as follows:

$$\{p_{fg}\}_{st} = 600 \text{ kPa (the boiling-point pressure)}$$
$$\{v\}_{st} = Xv_g = 0.5 \times 0.3156 = 0.1578 \text{ m}^3/\text{kg}$$
$$\{u\}_{st} = Xu_g + (1-X)u_f = 0.5 \times 2568 + (1-0.5) \times 669 = 1619 \text{ kJ/kg}$$
$$\{h\}_{st} = Xh_g + (1-X)h_f = 0.5 \times 2757 + (1-0.5) \times 670 = 1714 \text{ kJ/kg}$$
$$\{s\}_{st} = Xs_g + (1-X)s_f = 0.5 \times 6.761 + (1-0.5) \times 1.931 = 4.346 \text{ kJ/kg K}$$

7.7.3 Unsaturated water

Property values of water in the unsaturated state vary very little with pressure or temperature from their value in the saturated state. For example, the volume of water contained in a tank under a pressure of 100 kPa is little different to its volume under a pressure of 10 MPa at the same temperature. Similarly, the volume of water under a constant pressure of 100 kPa hardly changes if the temperature is 20 or 99 °C. The same applies to the properties entropy, internal energy and enthalpy and, for this reason, it may be assumed that, within a suitable order of accuracy, when unsaturated water is being used in a thermal system its property values are those as if it were saturated at the same condition and the subscript f values apply. Therefore, there is no need to include a further set of tables for unsaturated water. Sometimes a table of the difference to the saturated value is given for a range of conditions.

If necessary, it is easy to determine the specific internal energy or specific enthalpy for unsaturated water when the specific heats at constant volume and constant pressure are assumed constant. For example, the specific internal energy of unsaturated water u at a temperature T can be determined from the saturated value u_f, corresponding to a boiling-point temperature T_{fg}, as follows:

$$u = u_f - (u_f - u)$$

From Equation (5.16), for a fluid with a constant specific heat:

$$(u_f - u)_v = (c_v(T_{fg} - T))_v$$

However, as stated in Section 7.3.1, it is not necessary to restrict the equation to a constant volume process for a liquid.

$$\therefore \quad u = u_f - c_v(T_{fg} - T) \tag{7.7}$$

Similarly, if the specific heat of the water at constant pressure c_p is constant, the specific enthalpy of unsaturated water h, using Equation (5.15), is

$$h = h_f - c_p(T_{fg} - T) \tag{7.8}$$

7.8 HOW TO DETERMINE THE CONDITION OF WATER/STEAM

It is not always apparent in a thermal system what the condition of the water/steam is. It can only be determined by measuring or knowing two independent properties, which means not the pressure and temperature when the fluid is a mixture. The values listed in the steam tables can be used to discover if it is superheated steam, saturated steam or water, a mixture, or unsaturated water.

EXAMPLE 7.4
State the condition of water/steam given the following property values:

(a) $\{p\}_{st} = 1.5$ MPa, $X = 0$
(b) $\{p\}_{st} = 1.5$ MPa, $X = 1$
(c) $\{p\}_{st} = 1.5$ MPa, $X = 0.8$
(d) $\{p\}_{st} = 1.5$ MPa, $\{v\}_{st} = 0.1317$ m³/kg
(e) $\{p\}_{st} = 1.5$ MPa, $\{s\}_{st} = 2.315$ kJ/kg K

THE FLUID AS WATER/STEAM

Table 7.3 *Example 7.4.*

Condition	Reason
(a) Saturated water	The dryness fraction is zero
(b) Saturated steam	The dryness fraction is one
(c) Mixture	The dryness fraction is between zero and one
(d) Saturated steam	$\{v\}_{st} = v_g$
(e) Saturated water	$\{s\}_{st} = s_f$
(f) Mixture	$\{u\}_{st}$ is between $u_f = 843$ kJ/kg and $u_g = 2595$ kJ/kg
(g) Unsaturated water	$\{h\}_{st} < h_f$ ($h_f = 845$ kJ/kg)
(h) Unsaturated water	$\{T\}_{st} < 198.3\,°C$ (the boiling-point)
(i) Saturated water, saturated steam, or mixture	$\{T\}_{st} = 198.3\,°C$, the boiling-point, which is constant during the change of phase. $\{p_{fg}\}_{st}$ and $\{T_{fg}\}_{st}$ no longer independent
(j) Superheated steam	$\{T\}_{st} > 198.3\,°C$ (the boiling-point)
(k) Superheated steam	$\{h\}_{st} > h_g$ ($h_g = 2792$ kJ/kg)
(l) Saturated steam	$\{s\}_{st} = s_g$
(m) Superheated steam	$\{v\}_{st} > v_g$ ($v_g = 1.673$ m³/kg)
(n) Unsaturated water	$\{h\}_{st} < h_f$ ($h_f = 419.1$ kJ/kg)
(o) Superheated steam	$\{s\}_{st}$ Given in superheat tables at 300 °C
(p) Superheated steam	$\{h\}_{st} = 3025$ kJ/kg at 2 MPa and $\{h\}_{st} = 2995$ kJ/kg at 3 MPa in superheat tables, or at $\{T\}_{st} = 299.2\,°C$ in saturated water and steam tables $\{h\}_{st} > h_g$ ($h_g = 2751$ kJ/kg)

(f) $\{p\}_{st} = 1.5$ MPa, $\{u\}_{st} = 2000$ kJ/kg
(g) $\{p\}_{st} = 1.5$ MPa, $\{h\}_{st} = 700$ kJ/kg
(h) $\{p\}_{st} = 1.5$ MPa, $\{T\}_{st} = 160\,°C$
(i) $\{p\}_{st} = 1.5$ MPa, $\{T\}_{st} = 198.3\,°C$
(j) $\{p\}_{st} = 1.5$ MPa, $\{T\}_{st} = 250\,°C$
(k) $\{p\}_{st} = 1.5$ MPa, $\{h\}_{st} = 3000$ kJ/kg
(l) $\{T\}_{st} = 100\,°C$, $\{s\}_{st} = 7.355$ kJ/kg K
(m) $\{T\}_{st} = 100\,°C$, $\{v\}_{st} = 2.0$ m³/kg
(n) $\{T\}_{st} = 100\,°C$, $\{h\}_{st} = 400$ kJ/kg
(o) $\{T\}_{st} = 300\,°C$, $\{s\}_{st} = 6.541$ KJ/kg K
(p) $\{T\}_{st} = 300\,°C$, $\{h\}_{st} = 3000$ kJ/kg

Solution
The results are shown in Table 7.3.

7.9 WATER/STEAM IN NON-FLOW PROCESSES

Water/steam may be used as the fluid in a non-flow process. In terms of a piston cylinder mechanism, there is little practical application nowadays, although there is nothing to say that reciprocating steam engines will not once again find favour. However, there are a number of circumstances in the operation of thermal systems when the water/steam is considered to undergo a non-flow constant pressure, constant volume, constant temperature, constant entropy or polytropic process. All the equations derived in Chapter 5 are applicable depending upon the type of process followed.

EXAMPLE 7.5

1.2 kg of steam is initially at a pressure of 1 MPa and a temperature of 250°C. Determine its final condition for each of the following reversible non-flow processes:

(a) The rejection of heat to the surroundings of 300 kJ at constant pressure. What is the magnitude and direction of the work transfer in this case?

(b) The rejection of heat at constant volume until the pressure is 200 kPa. What is the magnitude of the heat rejection in this case?

(c) The rejection of heat to the surroundings of 300 kJ at constant temperature. What is the magnitude and direction of the work transfer in this case?

(d) A work output at constant entropy until the pressure is 200 kPa. What is the magnitude of the work transfer in this case?

(e) A polytropic expansion according to the law $pv^{1.2} = \text{const.}$ until the pressure is 200 kPa. What is the magnitude of the work and heat transfer in this case?

Solution

At a pressure of 1 MPa, the boiling-point is 179.9°C. Therefore, at a temperature of 250°C, the steam must be superheated. From the superheat tables, the initial property values at state 1 are as follows:

$$\{p_1\}_{st} = 1 \text{ MPa}$$
$$\{T_1\}_{st} = 250°C$$
$$\{m_1\}_{st} = \{m_2\}_{st} = 1.2 \text{ kg}$$
$$\{v_1\}_{st} = 0.2328 \text{ m}^3/\text{kg}$$
$$\{s_1\}_{st} = 6.926 \text{ kJ/kg K}$$
$$\{u_1\}_{st} = 2711 \text{ kJ/kg}$$
$$\{h_1\}_{st} = 2944 \text{ kJ/kg}$$

(a) For a non-flow reversible constant pressure rejection of heat Q_{12} from state 1 to state 2 of 300 kJ, the heat transfer is proportional to the change in enthalpy of the fluid and Equation (5.3) is applicable:

$$Q_{12} = \{H_2 - H_1\}_{st} = \{m(h_2 - h_1)\}_{st}$$

According to sign rule B, Q_{12} is negative because it is a heat rejection. Thus:

$$-300 = 1.2(\{h_2\}_{st} - 2944)$$
$$\therefore \{h_2\}_{st} = 2694 \text{ kJ/kg}$$

For a constant pressure process:

$$\{p_1\}_{st} = \{p_2\}_{st} = \{p\}_{st} = 1 \text{ MPa}$$

At a pressure of 1 MPa, the specific enthalpy of saturated steam h_g is 2778 kJ/kg. Therefore, as $\{h_2\}_{st} < h_g$, the fluid at state 2 is a mixture. The known specific enthalpy can be used to find the dryness fraction X_2 in Equation (7.6). From the saturated water and steam tables at 1 MPa:

$$\{h_2\}_{st} = X_2 h_g + (1 - X_2) h_f$$
$$\therefore \quad 2694 = X_2 \times 2778 + (1 - X_2) \times 763$$
$$\therefore \quad X_2 = 0.958$$

The specific volume at state 2 $\{v_2\}_{st}$ can now be found from Equation (7.3):

$$\{v_2\}_{st} = X_2 v_g$$
$$\therefore \quad \{v_2\}_{st} = 0.958 \times 0.1944 = 0.1862 \text{ m}^3/\text{kg}$$

The work transfer W_{12} can be found from Equation (5.2) for a constant pressure non-flow reversible process:

$$W_{12} = -\{mp(v_2 - v_1)\}_{st} = -1.2 \times 10 \times 10^5 \times (0.1862 - 0.2328) = +55.9 \text{ kJ}$$

As W_{12} is positive, it is a work input. The process can be represented by a horizontal line on a pressure versus volume diagram. Heat rejection at constant pressure is associated with a work input, a decreasing volume and with the superheated steam tending towards becoming a mixture, and vice versa for constant pressure heat supply.

(b) For a reversible non-flow constant volume rejection of heat from state 1 to state 3 until the pressure is 200 kPa:

$$\{v_1\}_{st} = \{v_3\}_{st} = 0.2328 \text{ m}^3/\text{kg}$$

At a pressure of 200 kPa, the specific volume of saturated steam is 0.8856 m³/kg. As $\{v_3\}_{st} < v_g$ and is also greater than the specific volume at the critical point, the fluid at state 3 is a mixture. The known specific volume can be used to find the dryness fraction X_3 in Equation (7.3). From the saturated water and steam tables at 200 kPa:

$$\{v_3\}_{st} = X_3 v_g \quad \therefore \quad X_3 = \frac{0.2328}{0.8856} = 0.263$$

The specific internal energy at state 3 $\{u_3\}_{st}$ can now be found from Equation (7.5):

$$\{u_3\}_{st} = X_3 u_g + (1 - X_3) u_f$$
$$\therefore \quad \{u_3\}_{st} = 0.263 \times 2530 + (1 - 0.263) \times 505 = 1038 \text{ kJ/kg}$$

The heat transfer Q_{13} can be found from Equation (5.6) for a constant volume non-flow reversible process:

$$Q_{13} = \{m(u_3 - u_1)\}_{st} = 1.2(1038 - 2711) = -2007.6 \text{ kJ}$$

As Q_{13} is negative, it is a heat output. There is no work transfer in a constant volume non-flow reversible process, which can be represented on a pressure versus volume diagram by a vertical line. Heat rejection at constant volume is associated with a decrease in the pressure and with the superheated steam tending towards becoming a mixture, and vice versa for constant volume heat supply.

(c) For a constant temperature process between state 1 and state 4:

$$\{T_1\}_{st} = \{T_4\}_{st} = \{T\}_{st} = 250 \,°\text{C} = 523 \text{ K}$$

For a reversible non-flow constant temperature rejection of heat Q_{14} of 300 kJ, Equation (5.7) is applicable:

$$Q_{14} = \{mT(s_4 - s_1)\}_{st}$$

According to sign rule B, Q_{14} is negative because it is a heat rejection. Thus:

$$-300 = 1.2 \times 523(\{s_4\}_{st} - 6.926)$$

$$\therefore \quad \{s_4\}_{st} = 6.448 \text{ kJ/kg K}$$

At a temperature of 250.3 °C, corresponding to a pressure of 4 MPa, the specific entropy of saturated steam s_g is 6.070 kJ/kg K. Therefore, as $\{s_4\}_{st} > s_g$, the fluid at state 4 is superheated. Interpolating from the superheat tables at 250 °C between a pressure of 2 MPa where the specific entropy is 6.547 kJ/kg K and 3 MPa where the specific entropy is 6.289 kJ/kg K, it is possible to determine the specific internal energy $\{u_4\}_{st}$ of the fluid at state 4:

$$\frac{6.448 - 6.547}{6.289 - 6.547} = \frac{\{u_4\}_{st} - 2681}{2646 - 2681}$$

$$\therefore \quad \{u_4\}_{st} = 2668 \text{ kJ/kg}$$

The work transfer in the non-flow reversible constant temperature process W_{14} can be found from Equation (5.1), the NFEE:

$$Q_{14} + W_{14} = \{m(u_4 - u_1)\}_{st}$$

$$\therefore \quad -300 + W_{14} = 1.2(2668 - 2711)$$

$$\therefore \quad W_{14} = +248.4 \text{ kJ}$$

As W_{14} is positive, it is a work input. The process can be represented by a horizontal line on a temperature versus entropy diagram. Heat rejection at constant temperature is associated with a work input, a decreasing entropy, and with the superheated steam tending towards becoming a mixture, and vice versa for constant temperature heat supply.

(d) For a constant entropy work transfer between state 1 and state 5:

$$\{s_1\}_{st} = \{s_5\}_{st} = 6.926 \text{ kJ/kg K}$$

At a pressure of 200 kPa, the specific entropy of saturated steam is 7.127 kJ/kg K. As $\{s_5\}_{st} < s_g$, the fluid at state 5 is a mixture. The known specific entropy can be used to find the dryness fraction X_5 in Equation (7.4). From the saturated water and steam tables at 200 kPa:

$$\{s_5\}_{st} = X_5 s_g + (1 - X_5) s_f$$

$$\therefore \quad 6.926 = X_5 \times 7.127 + (1 - X_5) \times 1.53$$

$$\therefore \quad X_5 = 0.964$$

The specific internal energy at state 5 $\{u_5\}_{st}$ can be found from Equation (7.5):

$$\{u_5\}_{st} = X_5 u_g + (1 - X_5) u_f$$

$$\therefore \quad \{u_5\}_{st} = 0.964 \times 2530 + (1 - 0.964) \times 505 = 2457 \text{ kJ/kg}$$

The work transfer W_{15} can be found from Equation (5.9) for a constant entropy non-flow reversible process:

$$W_{15} = -\{m(u_1 - u_5)\}_{st} = -1.2 \times (2711 - 2457) = -304.8 \text{ kJ}$$

As W_{15} is negative, it is a work output. There is no heat transfer in a constant entropy non-flow reversible process, which can be represented on a temperature versus entropy diagram by a vertical line. Work output at constant entropy is associated with a decrease in the temperature, and with the superheated steam tending towards becoming a mixture, and vice versa for constant entropy work input.

(e) For a polytropic expansion from state 1 to state 6 where the pressure is 200 kPa according to the law $pv^{1.2}$ = constant, the specific volume at state 6 $\{v_6\}_{st}$ is given as

$$\{p_1 v_1^{1.2}\}_{st} = \{p_6 v_6^{1.2}\}_{st}$$

$$\therefore \quad \{v_6\}_{st} = \frac{10^{1/1.2} \times 0.2328}{2^{1/1.2}} = 0.8901 \text{ m}^3/\text{kg}$$

At a pressure of 200 kPa, the specific volume of saturated steam v_g is 0.8856 m³/kg. Therefore, as $\{v_6\}_{st} > v_g$, the fluid at state 6 is superheated. Interpolating from the superheat tables at 200 kPa between the boiling-point where the specific volume is 0.8856 m³/kg and a temperature of 150 °C where the specific volume is 0.9602 m³/kg, it is possible to determine the specific internal energy $\{u_6\}_{st}$ of the fluid at state 6:

$$\frac{0.8901 - 0.8856}{0.9602 - 0.8856} = \frac{\{u_6\}_{st} - 2530}{2578 - 2530}$$

$$\therefore \quad \{u_6\}_{st} = 2533 \text{ kJ/kg}$$

The work transfer W_{16} can be found from Equation (5.12) for a polytropic non-flow reversible process:

$$W_{16} = -\frac{\{m(p_6 v_6 - p_1 v_1)\}_{st}}{1 - n}$$

$$\therefore \quad W_{16} = -\frac{1.2 \times (2 \times 10^5 \times 0.8901 - 10 \times 10^5 \times 0.2328)}{1 - 1.2} = -328.7 \text{ kJ}$$

As W_{16} is negative, it is a work output.

The heat transfer in the process Q_{16} can be found from Equation (5.1), the NFEE:

$$Q_{16} + W_{16} = \{m(u_6 - u_1)\}_{st}$$

$$\therefore \quad Q_{16} - 328.7 = 1.2 \times (2533 - 2711)$$

$$\therefore \quad Q_{16} = 115.1 \text{ kJ}$$

As Q_{16} is positive, it is a heat input. A polytropic process is only identifiable when the steam is in the superheat region because it is only there that the properties vary in a manner which can be represented by a formula, not when the steam changes phase. On a pressure versus volume diagram, the process is a curve whose shape is dictated by the index of expansion or compression, n in Equations (5.10) and (5.11). A heat input is associated with a work output, a decrease in steam pressure and an increase in steam volume, and vice versa.

7.10 WATER/STEAM IN FLOW PROCESSES

Most applications in thermal systems when the fluid is water/steam involve flow processes. These are dealt with in detail in Chapter 8.

8 Steady Flow Processes

The distinction between a non-flow process and a flow process was made in Chapter 3, but it is now possible to be more specific.

In a non-flow process, the fluid is contained within a boundary, be it physical or imaginary, and it does not cross the boundary at any time, even though the boundary itself may change its shape. Energy may be transferred to or from the working fluid in the form of work and/or heat, and the internal energy of the fluid may change. However, neither the work and heat transfer, nor any of the fluid properties, are time-dependent values (Figure 8.1).

In a flow process, the fluid moves continuously into and out of the boundary. Usually all the fluid enters at one position at the boundary and leaves at another, but this is not always the case. While the fluid is within the system boundary, heat and work may be transferred in the normal manner. But the energy transfers, and some of the fluid properties, are now related to time (Figure 8.2).

A flow process can be steady or unsteady, but only steady flow processes will be considered. To be steady, a flow process must obey the following conditions (Figure 8.3):

1. The mass flow rate of fluid must remain constant throughout the process. This means that the mass flow rate entering the system must be the same as that leaving the system, and conservation of mass flow rate of fluid is maintained. It is possible for the fluid to enter and leave the system at different positions across the boundary. But for the

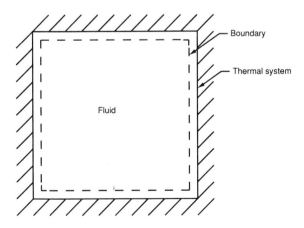

Fig. 8.1 *In non-flow processes, the fluid remains within the boundary.*

184 STEADY FLOW PROCESSES

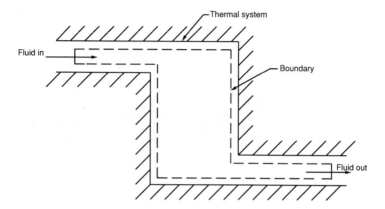

Fig. 8.2 In flow processes, the fluid crosses the boundary.

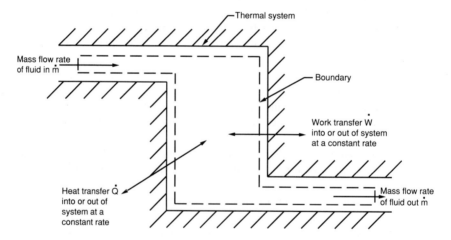

Fig. 8.3 In a steady flow process, the mass flow rate of fluid is constant, the energy transfers take place at a constant rate, and the properties do not change their value at any point with time.

process to be considered steady, all the mass flow rate of fluid, wherever it enters the system boundary, must be equal to all the mass flow rate of fluid leaving the system boundary, again wherever that occurs.

2. Energy transfers by heat and work across the system boundary between the fluid and the surroundings must be at a constant rate. If a rotary turbine in a hydroelectric scheme is producing 100 MW of work output, this remains fixed all the time that the fluid is flowing at a constant rate through the turbine.

3. The fluid properties at each point in the process do not vary with time. Each property may vary from point to point, but at any given point, they maintain a particular value which does not change with time while the fluid is flowing at a constant rate. In previous chapters, the assumption occasionally has been made that the fluid properties are uniform, for example at a given cross-section in a pipe. For the flow process to be

considered steady does not imply that the properties anywhere must also be uniform. The properties may vary over a cross-section, but at any point they may not change their value with time.

All the processes associated with reciprocating piston cylinder machinery discussed so far have been categorised as non-flow. However, most components in a thermal system can be considered not just as flow processes but as steady flow processes, and so the analysis of steady flow processes is both extremely important and very useful. It rests primarily upon two equations, the Continuity Equation, and the Steady Flow Energy Equation (SFEE).

8.1 THE CONTINUITY EQUATION

For steady flow processes, the mass flow rate of fluid entering the system at state 1, $\dot{m}_1$, must equal the mass flow rate of fluid leaving the system at state 2, $\dot{m}_2$. This is effectively an expression of the conservation of the mass flow rate of the fluid and can be written:

$$\dot{m}_1 = \dot{m}_2 = \dot{m} = \text{const}. \tag{8.1}$$

The mass flow rate is the volume flow rate $\dot{V}$ multiplied by the density of the fluid ρ. Hence, Equation (8.1) becomes

$$\dot{m}_1 = \dot{V}_1 \rho_1 = \dot{m}_2 = \dot{V}_2 \rho_2 = \dot{m} = \text{const}. \tag{8.2}$$

If the fluid is incompressible the density remains the same, and not only is the mass flow rate of fluid constant but the volume flow rate is also constant.

The volume flow rate can be further simplified as the product of the velocity of the fluid C and the cross-sectional area of flow A_{xs} (Figure 8.4). Hence, Equation (8.2) becomes:

$$\dot{m}_1 = \rho_1 C_1 (A_{xs})_1 = \dot{m}_2 = \rho_2 C_2 (A_{xs})_2 = \dot{m} \tag{8.3}$$

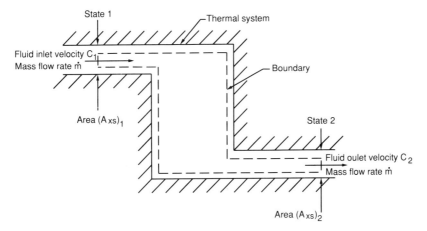

Fig. 8.4 *For the conservation of mass flow rate, the mass flow rate into a system equals the mass flow rate out of the system.*

186 STEADY FLOW PROCESSES

In terms of the specific volume, the equation can be written:

$$\dot{m}_1 = \frac{C_1(A_{xs})_1}{v_1} = \dot{m}_2 = \frac{C_2(A_{xs})_2}{v_2} = \dot{m} \qquad (8.4)$$

EXAMPLE 8.1

Steam, at a pressure of 200 kPa, a temperature of 150°C and a velocity of 30 m/s, flows through a main pipe of diameter 250 mm and into a T-junction. The pipe from both outlets of the T-junction is only 75 mm in diameter. The steam passes through one of the outlet pipes to a cleaning process which it enters with a pressure of 150 kPa and a temperature of 125°C. In the other outlet pipe, the steam passes through a heat exchanger which it leaves as a saturated liquid at a pressure of 120 kPa (Figure 8.5). If the mass flow rate of saturated water leaving the heat exchanger is 10 times the mass flow rate of steam entering the cleaning process, what is the velocity of the water leaving the heat exchanger and the velocity of the steam entering the cleaning process assuming steady flow conditions?

Solution

At 200 kPa and 150°C, the steam in the 250 mm diameter main pipe is superheated. From the superheat tables, its specific volume $\{v\}_{st}$ is

$$\{v\}_{st} = 0.9602 \text{ m}^3/\text{kg}$$

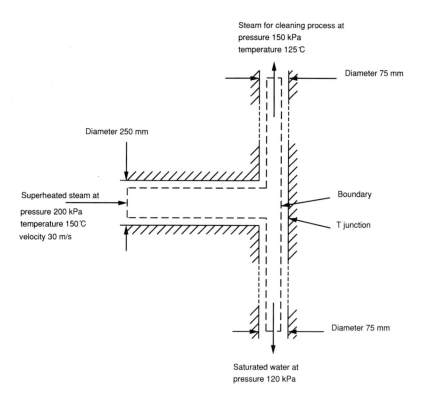

Fig. 8.5 *Example 8.1—superheated steam entering a T-junction, saturated water leaving a heat exchanger and superheated steam entering a cleaning process.*

The cross-sectional area of the main pipe of diameter 250 mm A_{xs} is

$$A_{xs} = 0.25\pi \times 0.25^2 = 0.049 \text{ m}^2$$

The mass flow rate of steam in the main pipe $\{\dot{m}\}_{st}$ can be determined from the continuity equation (Equation (8.4)) as

$$\{\dot{m}\}_{st} = \frac{\{C\}_{st} A_{xs}}{\{v\}_{st}} = \frac{30 \times 0.049}{0.9602} = 1.53 \text{ kg/s}$$

If the mass flow rate of saturated water $\{\dot{m}\}_w$ leaving the heat exchanger is 10 times the mass flow rate of steam entering the cleaning process $\{\dot{m}'\}_{st}$, by the conservation of mass:

$$\{\dot{m}\}_{st} = \{\dot{m}\}_w + \{\dot{m}'\}_{st} \quad \therefore \quad 1.53 = 11\{\dot{m}'\}_{st} = 1.1\{\dot{m}\}_w$$

$$\therefore \quad \{\dot{m}'\}_{st} = 0.139 \text{ kg/s} \quad \text{and} \quad \{\dot{m}\}_w = 1.39 \text{ kg/s}$$

The specific volume of the saturated water at exit from the heat exchanger $\{v\}_w$ is given in the saturated water and steam tables at 120 kPa approximately as

$$\{v\}_w = v_f = 0.1048 \times 10^{-2} \text{ m}^3/\text{kg}$$

The cross-sectional area of the pipe in which the saturated water flows A''_{xs} is

$$A''_{xs} = 0.25\pi \times 0.075^2 = 4.42 \times 10^{-3} \text{ m}^2$$

The velocity of the saturated water at exit from the heat exchanger $\{C\}_w$ is given by the continuity equation (Equation (8.4)) as

$$\{C\}_w = \frac{\{\dot{m} v\}_w}{A''_{xs}} = \frac{1.39 \times 0.001\,048}{0.004\,42} = 0.33 \text{ m/s}$$

At 150 kPa and 125°C, the steam entering the cleaning process is superheated. From the superheat tables, its specific volume $\{v'\}_{st}$, by interpolation, is

$$\{v'\}_{st} = 1.204 \text{ m}^3/\text{kg}$$

The cross-sectional area of the pipe in which the superheated steam flows A'_{xs} is

$$A'_{xs} = 0.25\pi \times 0.075^2 = 4.42 \times 10^{-3} \text{ m}^2$$

The velocity of the superheated steam entering the cleaning process $\{C'\}_{st}$ is given by the continuity equation (Equation (8.4)) as

$$\{C'\}_{st} = \frac{\{\dot{m}' v'\}_{st}}{A'_{xs}} = \frac{0.139 \times 1.204}{0.004\,42} = 37.9 \text{ m/s}$$

In pipes, steam velocities of this magnitude are high but not uncommon, whereas water velocities are usually much lower. In the solution, it does not matter that in the heat exchanger the steam changes phase, the mass flow rate of steam at entry must equal the mass flow rate of saturated water at exit.

EXAMPLE 8.2

Air at a pressure of 225 kPa and a temperature of 280 K flows through a main rectangular air-conditioning duct of width 600 mm and depth 600 mm and into a junction box. The air leaves the junction box in three separate secondary ducts, each of width 300 mm and depth 300 mm, which distribute the air to three different rooms in a

188 STEADY FLOW PROCESSES

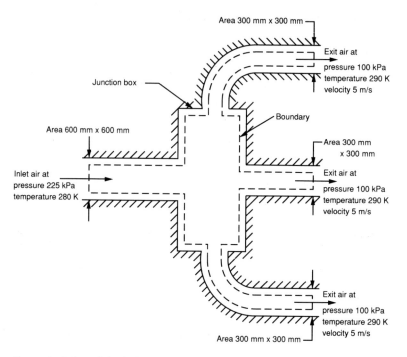

Fig. 8.6 *Example 8.2—air in the main duct of an air conditioning system being divided in a junction box into three secondary ducts.*

building. At exit from each of the three secondary ducts, the air has a velocity of 5 m/s, is at a pressure of 100 kPa and has a temperature of 290 K (Figure 8.6). Determine the mass flow rate of air leaving each secondary duct, the mass flow rate of air in the main duct and the velocity of the air in the main duct.

Solution
The cross-sectional area of each secondary duct A'_{xs} is

$$A'_{xs} = 0.3 \times 0.3 = 0.09 \text{ m}^2$$

If the air leaving each secondary duct has a pressure p' and temperature T', the density of the air leaving each secondary duct ρ', assuming it is a perfect gas, is given by the equation of state (Equation (6.1)) as

$$\{\rho'\}_{air} = \frac{\{p'\}_{air}}{\{RT'\}_{air}} = \frac{1 \times 10^5}{287 \times 290} = 1.2 \text{ kg/m}^3$$

The mass flow rate of air leaving each secondary duct $\{\dot{m}'\}_{air}$ is given by the continuity equation (Equation (8.3)) as

$$\{\dot{m}'\}_{air} = \{\rho' C'\}_{air} A'_{xs} = 1.2 \times 5 \times 0.09 = 0.54 \text{ kg/s}$$

By the conservation of mass, the mass flow rate of air in the main duct $\{\dot{m}\}_{air}$ must be the sum of the mass flow rates of air leaving each secondary duct. As the same mass flow of air leaves each secondary duct, $\{\dot{m}\}_{air}$ is given by

$$\{\dot{m}\}_{air} = 3\{\dot{m}'\}_{air} = 3 \times 0.54 = 1.62 \text{ kg/s}$$

As air is a compressible fluid, its density in the main duct will be different to that at exit from each secondary duct. If the air in the main duct has a pressure $\{p\}_{air}$ and temperature $\{T\}_{air}$, the density of the air in the main duct $\{\rho\}_{air}$ is given by the equation of state (Equation (6.1)) as

$$\{\rho\}_{air} = \frac{\{p\}_{air}}{\{RT\}_{air}} = \frac{2.25 \times 10^5}{287 \times 280} = 2.8 \text{ kg/m}^3$$

The cross-sectional area of the main duct A_{xs} is

$$A_{xs} = 0.6 \times 0.6 = 0.36 \text{ m}^2$$

The velocity of the air in the main duct $\{C\}_{air}$ is given by the continuity equation (Equation (8.3)) as

$$\{C\}_{air} = \frac{\{\dot{m}\}_{air}}{A_{xs}\{\rho\}_{air}} = \frac{1.62}{0.36 \times 2.8} = 1.6 \text{ m/s}$$

It does not matter that the air is compressed in the main duct and at atmospheric pressure when it exits from each secondary duct, the continuity equation for constant mass flow rate at entry and exit is still applicable.

8.2 THE CONSERVATION OF ENERGY

The NFEE (Equation (5.1)) related the work and heat transfer in a non-flow process with the change in the internal energy of the fluid. The SFEE is the same concept but applied to a steady flow process. In this case, the rates of energy transfer must be considered, in other words, the rate of work transfer, the rate of heat transfer and the rate of change of internal energy. In addition, there are some other energies which become relevant as a result of the fluid flowing into and out of the system boundary, such as the kinetic energy and the potential energy of the fluid.

The SFEE is based upon the principle of the conservation of energy, namely that the sum of all the energies transferred into a system at state 1 must be equal to the sum of all the energies transferred out of the system at state 2 (Figure 8.7). It is derived by writing down

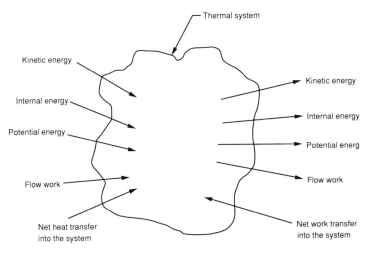

Fig. 8.7 *The conservation of energy applied to a generalised thermal system.*

STEADY FLOW PROCESSES

all the energies entering and leaving the system, and equating them. The relevant energies are given as follows.

8.2.1 Heat transfer term

Heat transfer into a system is positive and heat transfer out of a system is negative, according to the sign convention adopted. As the sign convention is already being applied, it is simplest to represent all the heat transfer into or out of the system by one term Q_{12} which, in effect, becomes the net energy transfer by heat. It is considered as an energy input to the system, the positive value.

8.2.2 Work transfer term

Work transfer into the system is positive and work transfer out of the system is negative, according to the sign convention adopted. But, as with heat transfer, it is simplest to represent the net work transfer by one term $\dot{W}_{12}$, considered as an energy input to the system, the positive value.

8.2.3 Flow work

One part of the work transfer is identified separately and this is particular to fluids flowing into and out of a thermal system. Consider the fluid entering a cross-section of the system in Figure 8.8.

Some work must be done to push the volume of fluid per second $\dot{V}_1$ into the system against the inlet pressure p_1. The work done is called the Flow Work and was introduced in Section 2.3. Equation (2.3) is written:

$$p_0 = p + 0.5\rho C^2$$

Rearranging the equation for a mass flow rate of fluid $\dot{m}$ gives

$$\frac{\dot{m} p_0}{\rho} = \frac{\dot{m} p}{\rho} + 0.5 \dot{m} C^2$$

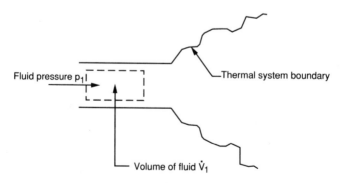

Fig. 8.8 The fluid entering a thermal system requires an energy input called the flow work.

THE STEADY FLOW ENERGY EQUATION (SFEE)

The last term in the equation describes the kinetic energy of the fluid as in Section 8.2.4. The second term, therefore, must also be an energy and must describe the ability of the fluid to achieve some form of activity by virtue of its pressure. The activity may be the ability to do work and, as this is what is required in order to push the volume of fluid per second $\dot{V}_1$ into the system against the inlet pressure p_1, the flow work must take the same form as the second term in the above equation. Hence, the flow work at inlet is

$$(\text{Flow work})_1 = \frac{\dot{m} p_1}{\rho_1} = \dot{m} p_1 v_1 = p_1 \dot{V}_1$$

Similarly there will be a flow work at outlet, given by

$$(\text{Flow work})_2 = \frac{\dot{m} p_2}{\rho_2} = \dot{m} p_2 v_2 = p_2 \dot{V}_2$$

8.2.4 Kinetic energy

This is the energy inherent in the fluid by nature of its velocity and is given by $0.5 \dot{m} C_1^2$ at inlet and $0.5 \dot{m} C_2^2$ at outlet.

8.2.5 Potential energy

This is the energy inherent in the fluid by nature of its height above a given datum and is given by $\dot{m} g z_1$ at inlet and $\dot{m} g z_2$ at outlet.

8.2.6 Internal energy

This is the energy inherent in the fluid by the nature of the molecular activity, as derived in Chapter 5. It is given by $\dot{m} u_1$ at inlet and $\dot{m} u_2$ at outlet.

8.2.7 Other energies

In some applications, energies due to surface tension, electricity, gravity or magnetism may be relevant. They will not be considered further here.

8.3 THE STEADY FLOW ENERGY EQUATION (SFEE)

The SFEE is generated by summing all the energy inputs to the system and equating them with the sum of all the energy outputs, thereby satisfying the conservation of energy. Consider the general thermal system of Figure 8.9.

$$\sum (\text{Energies})_1 = \sum (\text{Energies})_2$$

$$\dot{Q}_{12} + \dot{W}_{12} + \dot{m} \left(\frac{p_1}{\rho_1} + u_1 + \frac{C_1^2}{2} + g z_1 \right) = \dot{m} \left(\frac{p_2}{\rho_2} + u_2 + \frac{C_2^2}{2} + g z_2 \right) \quad (8.5)$$

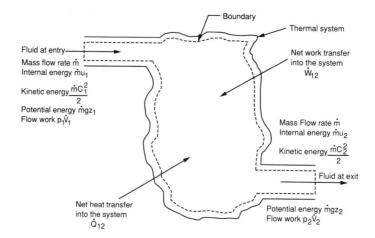

Fig. 8.9 *Energy inputs to and energy outputs from a generalised thermal system.*

But the definition of the specific enthalpy h is

$$h = u + \frac{p}{\rho} = u + pv$$

Substituting in Equation (8.5) gives

$$\dot{Q}_{12} + \dot{W}_{12} + \dot{m}(h_1 + 0.5C_1^2 + gz_1) = \dot{m}(h_2 + 0.5C_2^2 + gz_2) \tag{8.6}$$

This is the SFEE and is probably the most useful equation in thermofluids. It may be applied between the inlet and outlet of a component in a thermal system or at the inlet and outlet of any number of components through which the fluid passes under steady flow conditions and which make up a thermal system. A knowledge of exactly how the system works is not necessary, all that is essential is that the fluid properties in the equation can be assigned values. As such, therefore, it is a very general but usable equation.

It can be applied to the components which make up the majority of thermal systems in which the flow of the fluid is steady, such as boilers, heat exchangers, turbines, compressors, fans, pumps, nozzles, diffusers, mixing chambers and valves. It can also be applied to the pipe- and ductwork which is used to connect the components together in the system. The same equation is relevant whether the fluid is a perfect gas, a liquid, a vapour or changing phase. When the fluid remains in one phase, it is often possible to make the substitution for the specific enthalpy in terms of the product of the specific heat at constant pressure and the temperature difference, as in Equation (5.15).

Fortunately, it is not always necessary to use all the terms in the SFEE when it is being applied. Some of the terms may not be relevant, and others may be of an order of magnitude which make them insignificant. For every application, it is necessary to consider each of the terms in turn and to make a decision as to whether to substitute values in for them, or whether to simply ignore them. For example, there may be no heat transfer into or out of the system boundary under consideration, in which case $\dot{Q}_{12} = 0$. Often the kinetic energy and potential energy terms are of a much smaller order of magnitude than the specific enthalpy term, and can be neglected in comparison. However, one term which is always left in the equation is the specific enthalpy. This indicates how important a fluid property it is because it provides a link between the other types of energy in the SFEE.

The SFEE between the inlet (state 1) and the outlet (state 2) of a thermal system is most commonly written:

$$\dot{Q}_{12} + \dot{W}_{12} = \dot{m}(h_2 - h_1) + 0.5\dot{m}(C_2^2 - C_1^2) + \dot{m}g(z_2 - z_1) \qquad (8.7)$$

Considering the terms, this becomes a net balance of the following energies:

(Heat transfer) + (Work transfer) = (Enthalpy change) + (Kinetic energy change) + (Potential energy change)

and can be written:

$$\dot{Q}_{12} + \dot{W}_{12} = \Delta ENTHALPY + \Delta KE + \Delta PE \qquad (8.8)$$

EXAMPLE 8.3

Air enters an engine at a temperature of 290 K, a velocity of 10 m/s, and with a steady mass flow rate of 0.9 kg/s. It leaves with the same mass flow rate and at the same vertical height, but at a temperature of 700 K and with a velocity of 30 m/s. The engine receives 500 kW net of heat input (Figure 8.10). If the specific heat at constant pressure of the air may be assumed constant and equal to 1.01 kJ/kg K, what is the net power output and the overall thermal efficiency of the engine?

Solution

Considering the SFEE as in Equation (8.7):

$$\dot{Q}_{12} = +500 \times 10^3 \text{ W (+ve by the sign convention because it is a heat input)}$$

$$\{\dot{m}(h_2 - h_1)\}_{air} = \{\dot{m}c_p(T_2 - T_1)\}_{air} \text{ (because the fluid is a perfect gas with a constant specific heat and Equation (6.7) is valid)}$$

$$\{\dot{m}c_p(T_2 - T_1)\}_{air} = 0.9 \times 1.01 \times 10^3 \times (700 - 290) = 372\,690 \text{ W}$$

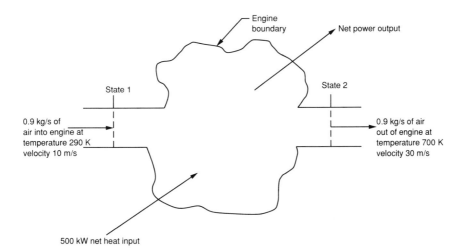

Fig. 8.10 *Example 8.3—the SFEE applied to an air engine which has a heat input and produces a work output.*

$\{0.5\dot{m}(C_2^2 - C_1^2)\}_{\text{air}} = 0.5 \times 0.9 \times (30^2 - 10^2) = 360 \text{ W}$

$z_1 = z_2$ (because the air enters and leaves the engine at the same vertical height)

Substituting in Equation (8.7) gives

$$500 \times 10^3 + \dot{W}_{12} = 372\,690 + 360$$

$$\therefore \quad \dot{W}_{12} = -126.95 \text{ kW}$$

The negative sign indicates that it is a net power output from the engine. The overall thermal efficiency η is the ratio of the net power output to the net heat input. Hence, using sign rule C:

$$\eta = \frac{\dot{W}_{12}}{\dot{Q}_{12}} = \frac{126.95}{500} = 25.4\%$$

It is noticeable that the kinetic energy term has a negligible effect upon the answer for the power output compared to the heat transfer or enthalpy terms. Indeed, it would be quite acceptable to ignore it. For an inlet velocity of 10 m/s, it would require an outlet velocity of over 100 m/s for the kinetic energy term to become significant and this sort of velocity is most unlikely in a practical engine!

EXAMPLE 8.4

Steam enters an engine with a pressure of 500 kPa and a temperature of 200 °C. It passes through the engine at the steady mass flow rate of 9 kg/s, and leaves at a pressure of 150 kPa. The steam enters and leaves the engine at the same velocity but, in doing so, goes through a vertical rise in height of 5 m (Figure 8.11). If the net heat transfer to the engine is 750 kW and the engine produces 250 kW of power output, what is the specific enthalpy and condition of the steam at the exit of the engine?

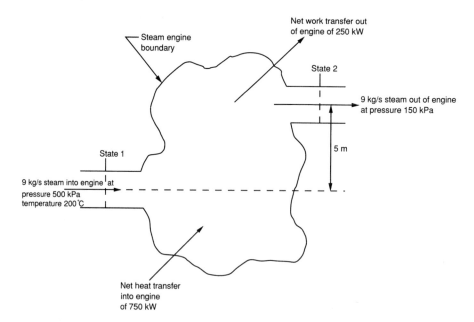

Fig. 8.11 *Example 8.4—the SFEE applied to a steam engine which has a heat input and produces a work output.*

Solution

At entry to the engine, at 500 kPa and 200 °C, the steam is superheated. From the superheat tables, the specific enthalpy of the steam at inlet $\{h_1\}_{st}$ is

$$\{h_1\}_{st} = 2857 \text{ kJ/kg}$$

Considering the SFEE as in Equation (8.7):

$$\dot{Q}_{12} = +750 \times 10^3 \text{ W (+ve by the sign convention because it is a heat input)}$$
$$\dot{W}_{12} = -250 \times 10^3 \text{ W (−ve by the sign convention because it is a power output)}$$
$$\{C_2^2\}_{st} = \{C_1^2\}_{st} \text{ (because the steam enters and leaves the engine with the same velocity)}$$
$$z_2 - z_1 = 5 \text{ m (because the steam leaves at a height of 5 m above the point at which it enters)}$$
$$\{\dot{m}\}_{st} g(z_2 - z_1) = 9 \times 9.81 \times 5 = 441.45 \text{ W}$$

Substituting in Equation (8.7) gives

$$750 \times 10^3 - 250 \times 10^3 = 9(\{h_2\}_{st} - 2857 \times 10^3) + 441.45$$
$$\therefore \{h_2\}_{st} = 2912.5 \text{ kJ/kg}$$

With this value of $\{h_2\}_{st}$ at the exit pressure of 150 kPa, the steam leaving the engine is superheated. Interpolating from the superheat tables gives the steam leaving temperature $\{T_2\}_{st}$ as

$$\{T_2\}_{st} = 219.75 \,°C$$

The potential energy term has a negligible effect upon the answer for the specific enthalpy of the steam at exit, and it would be justifiable to ignore it. This is often the case with the SFEE, some terms are of an order of magnitude which make them insignificant compared to others and, therefore, can be left out.

8.4 THE SFEE APPLIED TO A BOILER/HEAT EXCHANGER

In a Boiler, there is a heat source and a heat sink, with heat transfer taking place between the two. In a fossil-fuel boiler, the heat generation is as a result of the combustion of a fuel such as oil, coal, natural gas, wood or charcoal. In other types of boiler, the heat may be generated from an energy source like the sun. In most boilers, the heat sink is a fluid such as water, which may change phase to become steam to be used elsewhere in a thermal system. A schematic of a typical boiler is shown in Figure 8.12.

The SFEE can be applied to the fluid which is receiving the heat in the boiler. Considering each of the terms in Equation (8.8) in turn:

$\dot{Q}_{12}$ — the primary purpose of the boiler is to transfer heat, so this term is clearly important.

$\dot{W}_{12}$ — a pump is required to move the fluid through the boiler, but this can be considered a separate component, so the work transfer term is zero.

$\Delta ENTHALPY$ — this term is always retained.

ΔKE — in a typical boiler, the fluid enters at approximately the same velocity as it leaves and the change in kinetic energy is small, especially in comparison to the enthalpy term. It is a reasonable assumption, therefore, to neglect the kinetic energy term.

196 STEADY FLOW PROCESSES

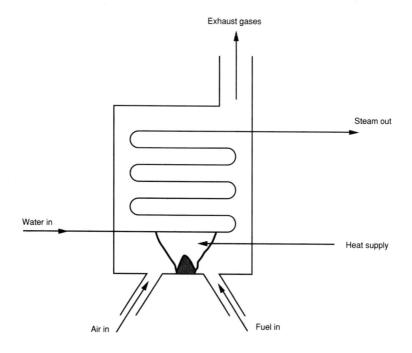

Fig. 8.12 *A schematic of a typical boiler which converts water into steam as a consequence of a heat input from the combustion of a fuel.*

ΔPE — most boilers are not high enough to make any change in potential energy of the fluid significant, and it is a reasonable assumption to neglect the potential energy term.

The SFEE for a boiler is reduced to

$$\dot{Q}_{12} = \dot{m}(h_2 - h_1) \qquad (8.9)$$

When the boiler is a fossil-fuel type, the heat released $\dot{Q}_{\text{comb}}$ by the combustion of a mass flow of fuel $\{\dot{m}\}_{\text{fuel}}$ of calorific value CV is given by Equation (4.6) as

$$\dot{Q}_{\text{comb}} = \{\dot{m}\}_{\text{fuel}} CV$$

Not all the heat released in combustion will end up in the fluid, some will go out through the exhaust of the boiler and some will be transferred out of the walls of the boiler to the atmosphere. Thus there is an energy transfer efficiency η_e given by the ratio of the heat absorbed by the fluid to the heat released in the combustion of the fuel. A typical boiler energy transfer efficiency is 70%, though modern condensing boilers can approach 85%.

A Combustion Chamber can be treated in the same manner as a boiler, the main difference being that the heat sink is the air used for the combustion. Equation (8.9) is valid.

In a Heat Exchanger, heat is transferred from a hot fluid to a colder fluid. There are many different types of heat exchanger on the market depending upon whether the fluids are gases or liquids, and it is not possible here to consider all the available designs. Typically, heat

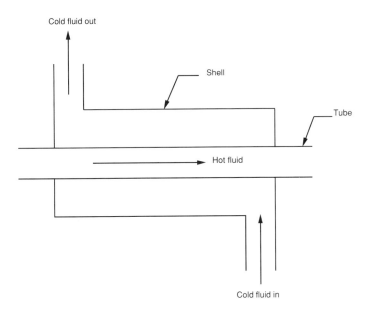

Fig. 8.13 *The SFEE applied to a counterflow heat exchanger.*

exchangers are drawn to show a tube in which the hotter of the two fluids flows, surrounded by a shell in which the colder fluid flows. If the two fluids are both going in the same direction, it is known as a parallel flow heat exchanger, and if they are passing in opposite directions, it is known as a counterflow heat exchanger. As far as the SFEE equation is concerned, it makes no difference in which direction each fluid is moving. A counterflow heat exchanger is shown in Figure 8.13.

The SFEE can be applied to either fluid, but using the same arguments as for the boiler, it will reduce to Equation (8.9). Of course, there will be an energy transfer efficiency between the two fluids because not all the heat leaving the hot fluid will end up in the cold fluid, some will be lost to the atmosphere. A typical heat exchanger calculation was demonstrated in Example 4.2. Heat exchanger energy transfer efficiencies depend upon many factors, but a figure of around 80% would be considered reasonable.

A Condenser is simply a heat exchanger in which one of the fluids is condensing from the vapour to the liquid phase, and an Evaporator is a heat exchanger in which one of the fluids is evaporating from the liquid to the vapour phase. The SFEE of Equation (8.9) is still valid, although it must be recalled that when a fluid changes phase, it does so at constant pressure and constant temperature.

8.4.1 When the fluid is a perfect gas

When a fluid in a heat exchange-type flow process is a perfect gas with a constant specific heat, it is possible to substitute for the specific enthalpy in terms of the specific heat at constant pressure and the temperature difference, in accordance with Equation (6.7). Equation (8.9) becomes

$$\dot{Q}_{12} = \dot{m} c_p (T_2 - T_1) \qquad (8.10)$$

EXAMPLE 8.5

A boiler is required to produce 3000 kg/h of steam at a pressure of 2 MPa and a temperature of 350 °C. Given that the kinetic energy and potential energy terms are negligible, calculate the required rate of heat input assuming that the feed-water is saturated at 2 MPa (Figure 8.14).

What is the rate of heat input to the boiler if the feed water is saturated at a temperature of 70 °C?

Given that the specific heat of water at constant pressure is constant and equal to 4.2 kJ/kg K, what is the rate of heat input to the boiler if the feed-water is at a pressure of 2 MPa and a temperature of 70 °C?

For this last case, if the boiler operated with an energy transfer efficiency of 70%, what would be the energy content of the fuel?

If the fuel was coal costing $0.04/kWh and the boiler ran for 5000 hours per year, what would be the annual running cost of the boiler?

Solution

At exit of the boiler (state 4) where the steam pressure is 2 MPa and temperature 350 °C, the steam is superheated. From the superheat tables, the specific enthalpy of the steam at outlet $\{h_4\}_{st}$ is

$$\{h_4\}_{st} = 3138 \text{ kJ/kg}$$

If the feed-water at inlet is saturated at 20 bar (state 1), the specific enthalpy $\{h_1\}_w$, from the saturated water and steam tables, is

$$\{h_1\}_w = h_f = 909 \text{ kJ/kg}$$

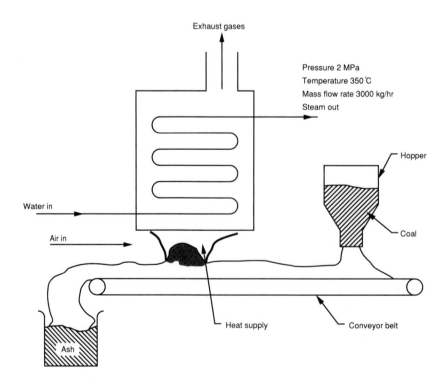

Fig. 8.14 *Example 8.5—a coal boiler which converts water into superheated steam.*

Substituting in Equation (8.9) gives the heat input to the boiler $\dot{Q}_{14}$ as

$$\dot{Q}_{14} = \{\dot{m}\}_{st}(\{h_4\}_{st} - \{h_1\}_w)$$

$$\therefore \dot{Q}_{14} = \frac{3000 \times (3138 - 909)}{3600} = 1858 \text{ kW}$$

If the feed-water at inlet is saturated at 70 °C (state 2), the specific enthalpy $\{h_2\}_w$, from the saturated water and steam tables, is

$$\{h_2\}_w = 293 \text{ kJ/kg}$$

Substituting in Equation (8.9) gives the heat input to the boiler $\dot{Q}_{24}$ as

$$\dot{Q}_{24} = \{\dot{m}\}_{st}(\{h_4\}_{st} - \{h_2\}_w)$$

$$\therefore \dot{Q}_{24} = \frac{3000 \times (3138 - 293)}{3600} = 2371 \text{ kW}$$

If the feed-water at inlet is at 2 MPa and 70 °C (state 3), it is in the unsaturated liquid condition. The specific enthalpy $\{h_3\}_w$, from the saturated water and steam tables, is given by Equation (7.8) as

$$\{h_3\}_w = \{h_f - c_p(T_{fg} - T_1)\}_w = 909 - 4.2 \times (212.4 - 70) = 310.9 \text{ kJ/kg}$$

Substituting in Equation (8.9) gives the heat input to the boiler $\dot{Q}_{34}$ as

$$\dot{Q}_{34} = \{\dot{m}\}_{st}(\{h_4\}_{st} - \{h_3\}_w)$$

$$\therefore \dot{Q}_{34} = \frac{3000 \times (3138 - 310.9)}{3600} = 2356 \text{ kW}$$

Note that $\dot{Q}_{34}$ and $\dot{Q}_{24}$ are almost the same. The condition of the water at entry to the boiler in state 3 is the most representative of the situation in practice, but it can now be seen that taking the saturated value of the specific enthalpy of the inlet water at the appropriate temperature is a quicker way of arriving at roughly the same answer for the heat transfer.

At an energy transfer efficiency η_e of 70%, the heat supplied by the combustion of the coal $\dot{Q}_{comb}$ is

$$\dot{Q}_{comb} = \frac{\dot{Q}_{34}}{\eta_e} = \frac{2356}{0.7} = 3367 \text{ kW}$$

The annual running cost $\$_{ARC}$ is given by Equation (4.9) as

$$\$_{ARC} = E N_{hrs} \$_{kWh} = 3367 \times 5000 \times 0.04 = \$673\,400 \text{ per year}$$

The annual running cost is high because the efficiency of the boiler is only 70%. Use could be made of the 30% of the energy input from the combustion of the fuel that goes out through the exhaust if the air for the combustion is preheated in a heat exchanger, as in Example 4.4.

The same SFEE applies to the boiler whatever the condition of the water at entry and despite the fact that the water changes phase. The heat transfer is given by a relationship between the fluid properties at entry and exit of the boiler, in fact the specific enthalpy in this case.

EXAMPLE 8.6

In a particular heat exchanger, steam at 150 kPa pressure is condensing in the outer shell. Water flows through the tubes at the rate of 1.2 kg/s and increases in temperature from 45 to 90 °C. The water pressure remains approximately constant. The heat exchanger is well lagged with insulation and the energy transfer efficiency is 90% (Figure 8.15). What is the mass flow rate of steam required to maintain these water conditions? The specific heat of water at constant pressure is constant and equal to 4.2 kJ/kg K.

If the steam is produced in a natural-gas-fired boiler of efficiency 75% and natural gas costs $0.04/kWh, what is the annual running cost of the boiler given that the heat exchanger runs for 5000 hours per year?

After a year in operation, the insulation on the heat exchanger is damaged and the energy transfer efficiency falls to 70%. What is the mass flow rate of steam required to maintain the water conditions now, the percentage increase in the mass flow rate of steam and the increase in the annual running cost of the boiler?

Solution

For the water going from state 3 to state 4, the SFEE, Equation (8.9), gives the heat gained $\dot{Q}_{34}$ as

$$\dot{Q}_{34} = \{\dot{m}(h_4 - h_3)\}_w$$

From Section 7.3.1, the enthalpy difference of the water can be substituted for by the product of the specific heat at constant pressure and the temperature difference, in accordance with Equation (6.7), as follows:

$$\dot{Q}_{34} = \{\dot{m}c_p(T_4 - T_3)\}_w = 1.2 \times 4.2 \times (90 - 45) = 226.8 \text{ kW}$$

Using sign rule C, at an energy transfer efficiency of 90%, the heat transferred from the steam going from state 1 to state 2 $\dot{Q}_{12}$ is

$$\eta_e = \frac{\dot{Q}_{34}}{\dot{Q}_{12}} \quad \therefore \quad \dot{Q}_{12} = \frac{226.8}{0.9} = 252 \text{ kW}$$

By sign rule B, this must be negative if it used in a further equation because the heat is transferred from the steam to the water.

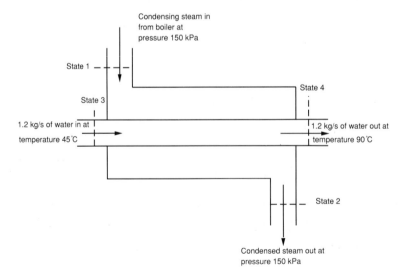

Fig. 8.15 *Example 8.6—a heat exchanger in which water is heated by condensing steam produced in a natural gas fired boiler.*

If the steam is condensing in the shell of the heat exchanger, as it goes from state 1 to state 2, the change in specific enthalpy $\{h_2 - h_1\}_{st}$ is given in the saturated water and steam tables at a pressure of 150 kPa, as

$$\{h_2 - h_1\}_{st} = h_f - h_g = (467 - 2693) = -2226 \text{ kJ/kg}$$

The mass flow rate of steam $\{\dot{m}\}_{st}$ can be found from the SFEE for the steam in the shell of the heat exchanger (Equation (8.9)) as follows (noting that $\dot{Q}_{12}$ is negative):

$$\dot{Q}_{12} = \{\dot{m}(h_2 - h_1)\}_{st} \quad \therefore \quad \{\dot{m}\}_{st} = \frac{-252}{-2226} = 0.113 \text{ kg/s}$$

If the steam is produced in a boiler of efficiency 75%, the energy in the natural gas fuel $\dot{Q}_{comb}$, using sign rule C, is

$$\dot{Q}_{comb} = \frac{\dot{Q}_{12}}{\eta_e} = \frac{252}{0.75} = 336 \text{ kW}$$

At a natural gas cost of $0.04/kWh and running for 5000 hours per year, the annual running cost of the boiler $\$_{ARC}$ is given by Equation (4.9) as

$$\$_{ARC} = EN_{hrs}\$_{kWh} = 336 \times 5000 \times 0.04 = \$67\,200 \text{ per year}$$

If the heat exchanger energy transfer efficiency decreases to 70%, the new mass flow rate of steam required $\{\dot{m}'\}_{st}$ is

$$\{\dot{m}'\}_{st} = \frac{\{\dot{m}\}_{st} \times 0.9}{0.7} = \frac{0.113 \times 0.9}{0.7} = 0.145 \text{ kg/s}$$

The increased mass flow rate of steam $\{\dot{m}''\}_{st}$ is

$$\{\dot{m}''\}_{st} = \{\dot{m}'\}_{st} - \{\dot{m}\}_{st} = 0.145 - 0.113 = 0.032 \text{ kg/s}$$

The percentage increase in the mass flow rate of steam $\{\%\dot{m}\}_{st}$ is

$$\{\%\dot{m}\}_{st} = \frac{\{\dot{m}''\}_{st}}{\{\dot{m}\}_{st}} = \frac{0.032}{0.113} = 28.3\%$$

The increase in the annual running cost of the boiler $\$'_{ARC}$, therefore, is

$$\$'_{ARC} = 0.283 \times 67200 = \$19\,017.6 \text{ per year}$$

In practice, it may not be possible to increase the mass flow rate of condensing steam without increasing the area available for heat transfer. More likely is that the temperature and pressure conditions of the two fluids will have to change.

8.5 THE SFEE APPLIED TO A NOZZLE/DIFFUSER

The objective of a Nozzle is to increase the kinetic energy of a fluid at the expense of a drop in its pressure. Typically, therefore, a fluid enters a nozzle with a relatively high pressure and low velocity, and leaves at a lower pressure but with an increased velocity. To achieve this energy conversion, it is necessary for the cross-sectional area of the flow to decrease.

202 STEADY FLOW PROCESSES

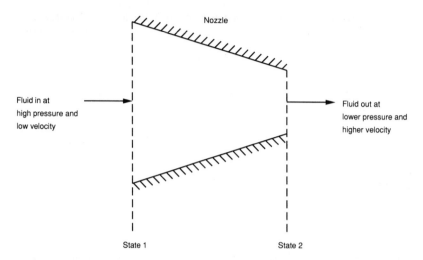

Fig. 8.16 *A schematic of a nozzle in which high-pressure fluid at entry is converted into high kinetic energy fluid at exit.*

Nozzles come in many different shapes, but Figure 8.16 shows a representative schematic of a nozzle.

To understand how a nozzle works, it is necessary to consider Equation (2.3), which relates the fluid static and dynamic pressures, and the continuity equation (Equation (8.3)). The two equations are written:

$$p + 0.5\rho C^2 = p_0 \quad (2.3)$$

$$\dot{m}_1 = \rho_1 C_1 (A_{xs})_1 = \dot{m}_2 = \rho_2 C_2 (A_{xs})_2 \quad (8.3)$$

If the fluid is assumed to be incompressible in the first instance and the flow steady, as the area of the nozzle decreases, the continuity equation predicts that the velocity of the fluid will increase. But in Equation (2.3), if it is assumed initially that the total pressure p_0 remains approximately constant, as the velocity of the fluid increases the static pressure of the fluid must decrease. Hence, nozzles convert high-pressure fluid into high kinetic energy fluid. Even if the fluid is compressible and the total pressure changes, the same effects are valid.

The SFEE can be applied to the fluid flowing through the nozzle. Considering each of the terms in Equation (8.8) in turn:

$\dot{Q}_{12}$ — the amount of heat transfer associated with the fluid passing through the nozzle is difficult to quantify. In addition, it is unlikely to be substantial because most nozzles are not very big, and the time available for the heat transfer to take place is small. In the first instance, therefore, the heat transfer will be assumed to be negligible.

$\dot{W}_{12}$ — a pump is required to move the fluid through the nozzle, but this can be considered a separate component, so the work transfer term is zero.

$\Delta ENTHALPY$ — this term is always retained.

ΔKE — clearly the objective of the nozzle is to increase the kinetic energy of the fluid, so this term is important.

ΔPE — if the fluid flows in a horizontal direction only, this term is zero, but in all nozzles the effects of potential energy may be assumed insignificant.

The SFEE for a nozzle is reduced to

$$0 = (h_2 - h_1) + 0.5(C_2^2 - C_1^2) \tag{8.11}$$

Equation (8.11), relating four property values of the fluid, can be used readily to find one of the property values provided the other three are known. This is possible if appropriate measurements are made. Theoretically it would be advantageous to be able to predict the outlet conditions from a knowledge of the inlet conditions. But as the fluid passes over the surface of the nozzle, the friction forces there oppose its motion. Some of the high pressure of the fluid at entry has to be used to counteract the friction, which means that there is less pressure available for conversion into kinetic energy. The friction forces cause the internal energy of the fluid to rise. The relationship between the internal energy and the pressure of the fluid in the enthalpy term in Equation (8.11) is affected.

The problem can be overcome if a nozzle efficiency is defined. The expansion of the fluid as it passes through the nozzle has already been assumed adiabatic. If it is also assumed to be reversible, the flow process becomes isentropic. Of course, it can never be reversible because friction will oppose the motion again when the fluid flows in the opposite direction, but it is a useful assumption to make because the entropy of the fluid will remain constant. The nozzle efficiency, therefore, can be defined in terms of the ratio of the kinetic energy that the fluid actually achieves to the kinetic energy that it would achieve if it were an isentropic process. If the actual value of the fluid velocity at exit is $(C_2)_{act}$ and the isentropic value is $(C_2)_{isen}$, the nozzle efficiency η_N, for a mass flow rate of fluid $\dot{m}$, becomes

$$\eta_N = \frac{0.5\dot{m}(C_2)_{act}^2}{0.5\dot{m}(C_2)_{isen}^2} = \frac{(C_2)_{act}^2}{(C_2)_{isen}^2} \tag{8.12}$$

The nozzle efficiency is sometimes called the isentropic efficiency because the actual kinetic energy of the fluid at outlet is being compared to the value that would be achieved if the process was isentropic. Values for the nozzle efficiency can be obtained by experiment because $(C_2)_{act}$ can be measured and $(C_2)_{isen}$ determined from Equation (8.11). The nozzle efficiency usually remains approximately constant over a large range of fluid entry conditions, but this information must be gleaned from the nozzle manufacturer. A typical value for a nozzle is 95%, indicating that the pressure required to overcome the friction forces is only small.

The outlet conditions of the fluid from the nozzle can be predicted from a knowledge of the inlet conditions, assuming the flow process to be isentropic, and correcting the answer obtained by utilising the nozzle efficiency. In other words, the more the process tends towards being isentropic, the greater will be the efficiency of the nozzle. This gives an indication of the importance of the fluid property entropy whereby, in many cases, no change in entropy represents an ideal state of affairs.

It is sometimes confusing to realise that the pressure of the fluid is decreasing while it is being squeezed through a smaller cross-sectional area. This is in contrast to a piston cylinder mechanism in which the pressure decreases as the piston falls and the volume of fluid increases. However, the fluid in the piston cylinder device is undergoing a non-flow process, whereas the fluid in the nozzle is undergoing a flow process for which there is an additional kinetic energy term that plays a part in the overall energy transfer.

The objective of a Diffuser is to increase the pressure of a fluid at the expense of a drop in its kinetic energy. Typically a fluid enters a diffuser with a relatively low pressure and high velocity, and leaves at a lower velocity but with an increased pressure. To achieve this energy conversion, it is necessary for the cross-sectional area of the flow to increase. Diffusers, therefore, work in exactly the opposite way to nozzles. Figure 8.17 shows a representative schematic of a diffuser.

The SFEE for a diffuser will reduce to Equation (8.11), exactly the same as for a nozzle and for the same reasons advanced above. In addition, it is necessary to define a diffuser efficiency but, unlike the nozzle, as the primary purpose of a diffuser is to increase the pressure of the fluid, it is defined in terms of the pressure rise. The diffuser efficiency η_{diff} is the ratio of the actual pressure rise of the fluid $(p_2 - p_1)_{\text{act}}$ to the pressure rise that would be achieved if the process was isentropic $(p_2 - p_1)_{\text{isen}}$, giving:

$$\eta_{\text{diff}} = \frac{(p_2 - p_1)_{\text{act}}}{(p_2 - p_1)_{\text{isen}}} \tag{8.13}$$

Again, it is sometimes referred to as the isentropic efficiency as the actual pressure rise is compared to the isentropic value. Values for η_{diff} can be obtained by experiment but they are very dependent upon the design. If the fluid is able to follow the physical boundary of the diffuser, values of up to 85% are achievable. But if the diffuser is shaped such that it requires sudden or significant changes in the direction of the fluid, the fluid will not be able to follow the diffuser contours. This leads to voids and friction which considerably reduces the efficiency. The same effect is apparent in a fluid-measuring device such as a venturi meter, and a fuller description of the possible consequences is given in Section 9.8.1.

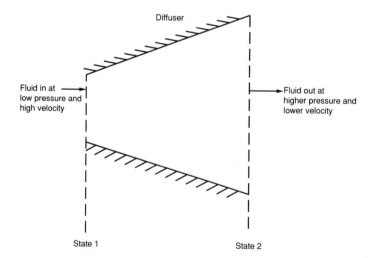

Fig. 8.17 A schematic of a diffuser in which high kinetic energy fluid at entry is converted into high-pressure fluid at exit.

THE SFEE APPLIED TO A NOZZLE/DIFFUSER

As with the nozzle, it is possible to predict the outlet conditions of the fluid from the diffuser from a knowledge of the inlet conditions by assuming the flow process to be isentropic and correcting the answer obtained by the diffuser efficiency.

8.5.1 When the fluid is a perfect gas

When the fluid in a nozzle or diffuser is a perfect gas with a constant specific heat, it is possible to substitute for the specific enthalpy in terms of the specific heat at constant pressure and the temperature difference, in accordance with Equation (6.7). Equation (8.11) becomes

$$0 = c_p(T_2 - T_1) + 0.5(C_2^2 - C_1^2) \quad (8.14)$$

In addition, as the perfect gas is assumed to undergo a reversible adiabatic process for the purposes of the efficiency term, it is possible to apply the equations relating the temperature, pressure and volume of the gas, as in Equation (6.17) with $n = \gamma$, namely:

$$\frac{T_2}{T_1} = \frac{p_2^{(\gamma-1)/\gamma}}{p_1^{(\gamma-1)/\gamma}} = \frac{v_2^{1-\gamma}}{v_1^{1-\gamma}}$$

This equation was shown to be valid for both an isentropic non-flow and isentropic flow process, and it is extremely useful because it allows the outlet conditions to be predicted from a knowledge of the inlet conditions, without recourse having to be made to experiment.

EXAMPLE 8.7

A fluid enters a nozzle of efficiency 95% with a pressure of 150 kPa, a temperature of 150 °C and a velocity of 10 m/s. It is discharged to atmosphere where the pressure is 100 kPa (Figure 8.18). What is the outlet velocity if the fluid is air, assumed to be a perfect gas, the change in specific entropy of the air and the area ratio of the nozzle? For the air, assume that the specific heat at constant pressure is 1.005 kJ/kg K, the ratio of specific heats is 1.4 and the gas constant is 0.287 kJ/kg K.

What is the outlet velocity, change in specific entropy and area ratio of the nozzle if the fluid is steam?

Solution

Assuming that the process is isentropic in the first instance, Equation (6.17) gives the isentropic nozzle outlet temperature of the air $(T_2)_{isen}$ as

$$\frac{\{(T_2)_{isen}\}_{air}}{\{T_1\}_{air}} = \frac{\{p_2^{(\gamma-1)/\gamma}\}_{air}}{\{p_1^{(\gamma-1)/\gamma}\}_{air}}$$

$$\therefore \quad \{(T_2)_{isen}\}_{air} = \frac{423 \times 1^{(1.4-1)/1.4}}{1.5^{(1.4-1)/1.4}} = 376.7 \text{ K}$$

For a perfect gas in a nozzle, the temperature falls with pressure, in accordance with the equation of state, as the velocity increases. The SFEE for a perfect gas in a nozzle is Equation (8.14):

$$0 = \{c_p(T_2 - T_1) + 0.5(C_2^2 - C_1^2)\}_{air}$$

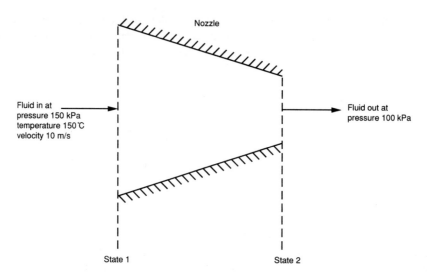

Fig. 8.18 *Example 8.7—a nozzle in which high-pressure fluid at entry is converted into high kinetic energy fluid at exit.*

The isentropic value of the air exit velocity $\{(C_2)_{\text{isen}}\}_{\text{air}}$ can be found if $\{(T_2)_{\text{isen}}\}_{\text{air}}$ is used, as follows:

$$0 = 1.005 \times 10^3 \times (376.7 - 423) + 0.5(\{(C_2)_{\text{isen}}\}_{\text{air}}^2 - 10^2)$$

$$\therefore \{(C_2)_{\text{isen}}\}_{\text{air}} = 305.2 \text{ m/s}$$

The actual value of the air velocity at exit from the nozzle can be found by using the nozzle efficiency (Equation (8.12)):

$$\eta_N = \frac{\{(C_2)_{\text{act}}^2\}_{\text{air}}}{\{(C_2)_{\text{isen}}^2\}_{\text{air}}}$$

$$\therefore \{(C_2)_{\text{act}}\}_{\text{air}} = (0.95)^{0.5} \times 305.2 = 297.5 \text{ m/s}$$

Knowing the actual velocity of the air at exit from the nozzle allows the actual temperature of the air at exit to be found from the SFEE as follows:

$$0 = \{c_p(T_2 - T_1) + 0.5(C_2^2 - C_1^2)\}_{\text{air}}$$

$$\therefore 0 = 1005(\{(T_2)_{\text{act}}\}_{\text{air}} - 423) + 0.5 \times (297.5^2 - 10^2)$$

$$\therefore \{(T_2)_{\text{act}}\}_{\text{air}} = 379 \text{ K}$$

The change in specific entropy of the air can be determined from Equation (6.12), which is applicable to any process. Hence:

$$\{s_2 - s_1\}_{\text{air}} = \left\{c_p \ln \frac{T_2}{T_1} - R \ln \frac{p_2}{p_1}\right\}_{\text{air}}$$

$$\therefore \{s_2 - s_1\}_{\text{air}} = 1.005 \ln \frac{379}{423} - 0.287 \ln \frac{1}{1.5} = 0.006 \text{ kJ/kg K}$$

The actual specific entropy of the air increases as expected, and the greater the increase the more inefficient is the nozzle. If the process was plotted on a graph of fluid temperature against specific entropy, the isentropic change of state would be a vertical line downwards with temperature decrease, and the actual change of state would be a slight curve downwards indicating the specific entropy increase.

The area ratio of the nozzle can be determined from the continuity equation (Equation (8.3)) as follows:

$$\{\rho_1 C_1\}_{\text{air}} (A_{\text{xs}})_1 = \{\rho_2 C_2\}_{\text{air}} (A_{\text{xs}})_2$$

Substituting for the density from the equation of state (Equation (6.2)) gives

$$\frac{\{p_1 C_1\}_{\text{air}} (A_{\text{xs}})_1}{\{RT_1\}_{\text{air}}} = \frac{\{p_2 C_2\}_{\text{air}} (A_{\text{xs}})_2}{\{RT_2\}_{\text{air}}}$$

$$\therefore \quad \frac{150 \times 10 (A_{\text{xs}})_1}{287 \times 423} = \frac{100 \times 297.5 (A_{\text{xs}})_2}{287 \times 379}$$

$$\therefore \quad \frac{(A_{\text{xs}})_1}{(A_{\text{xs}})_2} = 22.1$$

If the fluid is steam, at 150 kPa and 150 °C the steam is superheated. Using the superheat tables, the specific enthalpy $\{h_1\}_{\text{st}}$, specific entropy $\{s_1\}_{\text{st}}$ and specific volume $\{v_1\}_{\text{st}}$ of the steam at entry to the nozzle are

$$\{h_1\}_{\text{st}} = 2773 \text{ kJ/kg}, \quad \{s_1\}_{\text{st}} = 7.420 \text{ kJ/kg K} \quad \text{and} \quad \{v_1\}_{\text{st}} = 1.286 \text{ m}^3\text{/kg}$$

If the process in the nozzle is initially assumed to be isentropic:

$$\{s_2\}_{\text{st}} = \{s_1\}_{\text{st}} = 7.420 \text{ kJ/kg K}$$

At a pressure of 100 kPa and specific entropy of 7.42 kJ/kg K, the steam at exit of the nozzle is still superheated. Interpolating from the superheat tables gives the isentropic value of the specific enthalpy of the steam $\{(h_2)_{\text{isen}}\}_{\text{st}}$ at exit of the nozzle as

$$\{(h_2)_{\text{isen}}\}_{\text{st}} = 2700 \text{ kJ/kg}$$

Substituting in the SFEE for a nozzle, Equation (8.11) gives the isentropic value of the velocity of the steam at exit from the nozzle $\{(C_2)_{\text{isen}}\}_{\text{st}}$ as follows:

$$0 = \{((h_2)_{\text{isen}} - h_1) + 0.5(((C_2)_{\text{isen}})^2 - C_1^2)\}_{\text{st}}$$

$$\therefore \quad 0 = 10^3 \times (2700 - 2773) + 0.5(\{(C_2)_{\text{isen}}\}_{\text{st}}^2 - 10^2)$$

$$\therefore \quad \{(C_2)_{\text{isen}}\}_{\text{st}} = 382.2 \text{ m/s}$$

The actual value of the steam velocity at exit from the nozzle can be found by using the nozzle efficiency (Equation (8.12)):

$$\eta_N = \frac{\{(C_2)_{\text{act}}^2\}_{\text{st}}}{\{(C_2)_{\text{isen}}^2\}_{\text{st}}} \quad \therefore \quad \{(C_2)_{\text{act}}\}_{\text{st}} = (0.95)^{0.5} \times 382.2 = 372.5 \text{ m/s}$$

The actual specific enthalpy of the steam can now be found from the SFEE (Equation (8.11)) as follows:

$$0 = \{((h_2)_{\text{act}} - h_1) + 0.5(((C_2)_{\text{act}})^2 - C_1^2)\}_{\text{st}}$$

$$\therefore \quad 0 = 10^3((h_2)_{\text{act}} - 2773) + 0.5(372.5^2 - 10^2)$$

$$\therefore \quad (h_2)_{\text{act}} = 2704 \text{ kJ/kg}$$

At a pressure of 100 kPa and specific enthalpy of 2704 kJ/kg, the steam at exit of the nozzle is still superheated. Interpolating from the superheat tables gives the actual value of the specific entropy $\{(s_2)_{act}\}_{st}$ and temperature $\{(T_2)_{act}\}_{st}$ of the steam at exit of the nozzle as

$$\{(s_2)_{act}\}_{st} = 7.43 \text{ kJ/kg K} \quad \text{and} \quad \{(T_2)_{act}\}_{st} = 113.9\,°\text{C}$$

The actual specific entropy rise of the steam is

$$\{(s_2)_{act} - (s_1)_{act}\}_{st} = 7.43 - 7.42 = 0.01 \text{ kJ/kg K}$$

As for the air, the actual specific entropy of the steam increases as expected, and the greater the increase the more inefficient is the nozzle. If the process was plotted on a graph of fluid temperature against specific entropy, the isentropic change of state would be a vertical line downwards with temperature decrease, and the actual change of state would be a slight curve downwards indicating the specific entropy increase. The fluid would tend towards becoming a mixture from being superheated although, in this case, it remains entirely in the superheat region.

The area ratio of the nozzle can be determined from the continuity equation (Equation (8.3)) as follows:

$$\frac{\{C_1\}_{st}(A_{xs})_1}{\{v_1\}_{st}} = \frac{\{C_2\}_{st}(A_{xs})_2}{\{v_2\}_{st}}$$

At the exit pressure of 100 kPa and exit enthalpy of 2704 kJ/kg, the exit specific volume of the steam, by interpolation, is given in the superheat tables as

$$\{v_2\}_{st} = 1.763 \text{ m}^3/\text{kg}$$

$$\therefore \quad \frac{10(A_{xs})_1}{1.286} = \frac{372.5(A_{xs})_2}{1.763}$$

$$\therefore \quad \frac{(A_{xs})_1}{(A_{xs})_2} = 27.2$$

In practice, there is a limiting mass flow rate of fluid that can be passed through a nozzle which occurs when the velocity of the fluid reaches the local speed of sound at exit.

EXAMPLE 8.8

Steam, initially at a pressure of 300 kPa, dryness fraction 0.9, and inlet velocity 1000 m/s, passes through a diffuser (Figure 8.19). Calculate the outlet velocity and pressure if the steam temperature at outlet is 300 °C and the isentropic efficiency of the diffuser is 80%. Also determine the entropy increase of the steam and the area ratio required to achieve these exit conditions.

Solution

At a pressure of 300 kPa and with a dryness fraction of 0.9, the steam at inlet to the diffuser is a mixture with a temperature of 133.5 °C. From the saturated water and steam tables, the inlet specific enthalpy, specific entropy and specific volume of the steam $\{h_1\}_{st}$, $\{s_1\}_{st}$ and $\{v_1\}_{st}$ respectively are:

$$\{h_1\}_{st} = X_1 h_g + (1 - X_1) h_f$$

$$\therefore \quad \{h_1\}_{st} = 0.9 \times 2725 + (1 - 0.9) \times 561 = 2508.6 \text{ kJ/kg}$$

$$\{s_1\}_{st} = X_1 s_g + (1 - X_1) s_f$$

$$\therefore \quad \{s_1\}_{st} = 0.9 \times 6.993 + (1 - 0.9) \times 1.672 = 6.461 \text{ kJ/kg K}$$

$$\{v_1\}_{st} = X_1 v_g$$

$$\therefore \quad \{v_1\}_{st} = 0.9 \times 0.6057 = 0.5451 \text{ m}^3/\text{kg}$$

THE SFEE APPLIED TO A NOZZLE/DIFFUSER

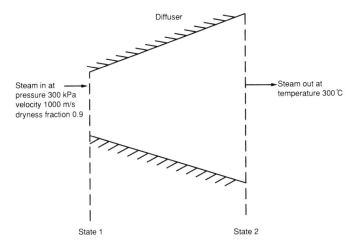

Fig. 8.19 Example 8.8—a diffuser in which high kinetic energy steam at entry is converted into high-pressure steam at exit.

Assuming, in the first instance, that the process in the diffuser is isentropic, the specific entropy of the steam at outlet of the diffuser $\{s_2\}_{st}$ is

$$\{s_2\}_{st} = \{s_1\}_{st} = 6.461 \text{ kJ/kg K}$$

At a temperature of 300 °C and specific entropy of 6.461 kJ/kg K, the steam at outlet of the diffuser is superheated. By interpolation from the superheat tables, the isentropic value of the steam pressure $\{(p_2)_{isen}\}_{st}$ at outlet of the diffuser is

$$\{(p_2)_{isen}\}_{st} = 3.45 \text{ MPa}$$

Using the efficiency of the diffuser (Equation (8.13)) allows the actual pressure of the steam at outlet of the diffuser $\{(p_2)_{act}\}_{st}$ to be found as follows:

$$\eta_{diff} = \frac{\{(p_2 - p_1)_{act}\}_{st}}{\{(p_2 - p_1)_{isen}\}_{st}}$$

$$\therefore \quad 0.8 = \frac{\{(p_2)_{act}\}_{st} - 3}{34.5 - 3}$$

$$\therefore \quad \{(p_2)_{act}\}_{st} = 2.82 \text{ MPa}$$

At a pressure of 2.82 MPa and a temperature of 300 °C, the steam is superheated. By interpolation from the superheat tables, the actual specific enthalpy of the steam at outlet from the diffuser $\{(h_2)_{act}\}_{st}$ is

$$\{(h_2)_{act}\}_{st} = 3000 \text{ kJ/kg}$$

Substituting in the SFEE for a diffuser (Equation (8.11)) gives the actual velocity of the steam at outlet of the diffuser $\{(C_2)_{act}\}_{st}$ as

$$0 = \{(h_2 - h_1)_{act} + 0.5(C_2^2 - C_1^2)_{act}\}_{st}$$

$$\therefore \quad 0 = (3000 - 2508.6) \times 10^3 + 0.5(\{(C_2)_{act}\}_{st}^2 - 1000^2)$$

$$\therefore \quad \{(C_2)_{act}\}_{st} = 131.1 \text{ m/s}$$

At a pressure of 2.82 MPa and a temperature of 300 °C, the actual specific entropy of the steam at outlet from the diffuser $\{(s_2)_{act}\}_{st}$, by interpolation from the superheat tables, is

$$\{(s_2)_{act}\}_{st} = 6.582 \text{ kJ/kg K}$$

The actual specific entropy rise of the steam is

$$\{(s_2)_{act} - (s_1)_{act}\}_{st} = 6.582 - 6.461 = 0.121 \text{ kJ/kg K}$$

The actual specific entropy of the steam increases as expected, and the greater the increase the more inefficient is the diffuser. If the process was plotted on a graph of fluid temperature against specific entropy, the isentropic change of state would be a vertical line upwards with temperature increase, and the actual change of state would be a slight curve upwards indicating the specific entropy increase. The fluid would tend towards becoming superheated from being a mixture, as indeed it does in this case.

The area ratio of the diffuser can be determined from the continuity equation (Equation (8.3)) as follows:

$$\frac{\{C_1\}_{st}(A_{xs})_1}{\{v_1\}_{st}} = \frac{\{C_2\}_{st}(A_{xs})_2}{\{v_2\}_{st}}$$

At the exit pressure of 2.82 MPa and exit temperature of 300 °C, the exit specific volume of the steam is given in the superheat tables as

$$\{v_2\}_{st} = 0.089 \text{ m}^3/\text{kg}$$

$$\therefore \quad \frac{1000(A_{xs})_1}{0.5451} = \frac{131.2(A_{xs})_2}{0.089}$$

$$\therefore \quad \frac{(A_{xs})_1}{(A_{xs})_2} = 0.8$$

A very large decrease in velocity of the steam is achieved with a fairly small area ratio.

8.6 THE SFEE APPLIED TO A ROTARY TURBINE/COMPRESSOR

A Radial Flow Rotary Turbine is shown in Figure 8.20, and a Radial Flow Rotary Compressor is similar except that the fluid travels in the opposite direction. This is a Centrifugal Flow or Radial Flow type because the fluid moves radially in the blading. It is also possible to have an Axial Flow type, and an Axial Flow Compressor is shown in Figure 8.21.

The principle of operation of a centrifugal turbine is as follows. The fluid enters with a low velocity and high pressure. It is converted in the blades of the rotor, which are nozzle-shaped, into a high-velocity low-pressure fluid. The resultant high kinetic energy of the fluid is utilised to drive a shaft around and to produce a power output. The fluid leaves with low pressure and low velocity. In an axial flow device, the same ideas apply except that the rotor blades are aerofoil-shaped and are designed to produce a lower fluid pressure on the top surface of the aerofoil than on the bottom, which results in a force in the direction of rotation of the shaft. Fluid flow over an aerofoil is considered in more detail in Sections 9.5 and 9.6. The stationary guide vanes are used to redirect the fluid between each set of rotor blades.

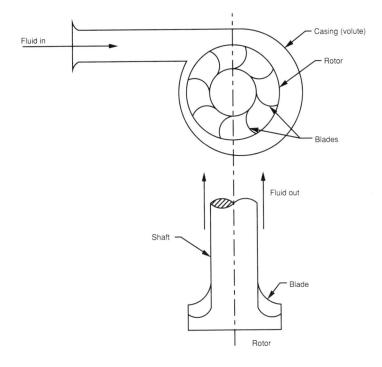

Fig. 8.20 *A radial flow rotary turbine.*

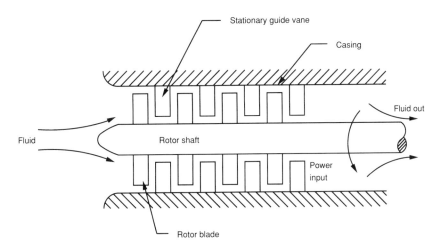

Fig. 8.21 *An axial flow compressor.*

When operating as a compressor, the fluid goes in the opposite direction. It enters with a low pressure and low velocity. The rotor blades are made to rotate by an external power source and, as the fluid flows through them, it receives a considerable amount of kinetic energy. This is converted into pressure by the diffusing shape of the blades and the fluid leaves with a low velocity but high pressure.

The SFEE can be applied to the fluid flowing through a turbine or compressor. Considering each of the terms in Equation (8.8) in turn:

$\dot{Q}_{12}$ — the amount of heat transfer associated with the fluid passing through a compressor or turbine is difficult to quantify. In the first instance, the heat transfer will be assumed to be negligible and the process adiabatic. This is certainly not the case, but the problem can be overcome by applying an efficiency as with the nozzle and diffuser.

$\dot{W}_{12}$ — the objective of a turbine is to produce power and, in a compressor, to receive power. This term, therefore, is important.

$\Delta ENTHALPY$ — this term is always retained.

ΔKE — for both a turbine and compressor, a design objective is that the fluid enters and leaves with low velocity, otherwise insufficient use is being made of the kinetic energy available and this term should be negligible.

ΔPE — any change in height that the fluid undergoes as it passes through a turbine or compressor will only result in a potential energy term that is insignificant compared to the other terms in the SFEE and it will be ignored.

The SFEE for a turbine or compressor, is reduced to

$$\dot{W}_{12} = \dot{m}(h_2 - h_1) \qquad (8.15)$$

Just as with the nozzle and diffuser, use of this equation to predict the work transfer from a knowledge of the inlet conditions is limited unless an efficiency term is defined, based upon the assumption of an isentropic process.

For a turbine, the efficiency η_T is defined as the ratio of the actual power output of the turbine $(\dot{W}_{12})_{act}$ to the power output that would have been achieved if the process was isentropic $(\dot{W}_{12})_{isen}$. Utilising Equation (8.15) gives

$$\eta_T = \frac{(\dot{W}_{12})_{act}}{(\dot{W}_{12})_{isen}} = \frac{(h_2 - h_1)_{act}}{(h_2 - h_1)_{isen}} \qquad (8.16)$$

For a compressor in which the work is put into the system by an external power source, the efficiency η_C is defined the other way around. Thus:

$$\eta_C = \frac{(\dot{W}_{12})_{isen}}{(\dot{W}_{12})_{act}} = \frac{(h_2 - h_1)_{isen}}{(h_2 - h_1)_{act}} \qquad (8.17)$$

These efficiencies, again, are sometimes referred to as the isentropic efficiencies. Turbine efficiencies of over 90% have been achieved while those for a compressor are more like 80–85%. This is because a compressor consists of diffusing sections whereas turbines consist of nozzle sections. The more the process tends towards being isentropic, the greater will be the efficiency of the turbine or compressor, with the isentropic process representing the ideal state of affairs.

A Pump is the term given to a compressor which is increasing the pressure of a liquid, and a Fan is the term given to a simple form of axial flow compressor when it is increasing the pressure of a gas. In fact, a fan is an inefficient compressor because the pressure rise is

8.6.1 When the fluid is a perfect gas

When the fluid in a rotary turbine or compressor is a perfect gas with a constant specific heat, it is possible to substitute for the specific enthalpy in terms of the specific heat at constant pressure and the temperature difference, in accordance with Equation (6.7). Equation (8.15) becomes

$$\dot{W}_{12} = \dot{m} c_p (T_2 - T_1) \tag{8.18}$$

Also, as for the nozzle and diffuser, and for the same reasons, it is possible to use Equation (6.17) which links the temperature, pressure and volume of the fluid.

EXAMPLE 8.9

A steam turbine operates with an inlet pressure of 800 kPa and inlet temperature of 225 °C. The mass flow rate of the steam through the turbine is 18 kg/s and the pressure of the steam at outlet from the turbine is 90 kPa. What is the power output of the turbine if the expansion of the steam is assumed adiabatic and reversible?

In fact, an analysis of the steam leaving the turbine indicates that the exhaust contains water which is discharged at the rate of 0.9 kg/s. What is the actual power output of the turbine, the turbine efficiency, and the change in the specific entropy of the steam during the process (Figure 8.22)?

Solution

At a pressure of 800 kPa and temperature of 225 °C, the steam at inlet is superheated. Interpolating from the superheat tables gives the steam inlet specific enthalpy and inlet specific entropy, $\{h_1\}_{st}$ and $\{s_1\}_{st}$ respectively, as

$$\{h_1\}_{st} = 2896 \text{ kJ/kg} \quad \text{and} \quad \{s_1\}_{st} = 6.9285 \text{ kJ/kg K}$$

Assuming that the expansion process in the turbine is isentropic (reversible and adiabatic), the steam outlet specific entropy $\{s_2\}_{st}$ is

$$\{s_2\}_{st} = \{s_1\}_{st} = 6.9285 \text{ kJ/kg K}$$

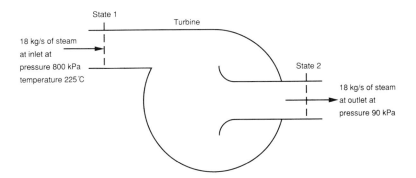

Fig. 8.22 Example 8.9—a steam turbine producing a power output from high-pressure superheated steam at inlet.

At a pressure of 90 kPa and specific entropy of 6.9285 kJ/kg K, the steam at outlet is a mixture. From the saturated water and steam tables, the isentropic dryness fraction and isentropic specific enthalpy of the steam at outlet, $(X_2)_{isen}$ and $\{(h_2)_{isen}\}_{st}$ respectively, are

$$\{s_2\}_{st} = (X_2)_{isen} s_g + (1 - (X_2)_{isen}) s_f$$

$$\therefore \quad 6.9285 = (X_2)_{isen} \times 7.394 + (1 - (X_2)_{isen}) \times 1.27$$

$$\therefore \quad (X_2)_{isen} = 0.924$$

$$\{(h_2)_{isen}\}_{st} = (X_2)_{isen} h_g + (1 - (X_2)_{isen}) h_f$$

$$\therefore \quad \{(h_2)_{isen}\}_{st} = 0.924 \times 2671 + (1 - 0.924) \times 405 = 2499 \text{ kJ/kg}$$

Substituting in the SFEE for a turbine Equation (8.15) gives the isentropic power output $(\dot{W}_{12})_{isen}$ for a mass flow rate of steam $\{\dot{m}\}_{st}$ as

$$(\dot{W}_{12})_{isen} = \{\dot{m}((h_2)_{isen} - h_1)\}_{st}$$

$$\therefore \quad (\dot{W}_{12})_{isen} = 18 \times (2499 - 2896) = -7146 \text{ kW}$$

It is negative because it is a power output, being a turbine. When there is 0.9 kg/s of water in the 18 kg/s of steam in the exhaust from the turbine, the actual dryness fraction of the steam $(X_2)_{act}$, from Equation (7.1), is

$$(X_2)_{act} = \frac{18 - 0.9}{18} = 0.95$$

Hence, the actual specific enthalpy $\{(h_2)_{act}\}_{st}$ and actual specific entropy $\{(s_2)_{act}\}_{st}$ of the steam at outlet of the turbine, where the pressure is 90 kPa and the temperature 96.7 °C, are

$$\{(h_2)_{act}\}_{st} = (X_2)_{act} h_g + (1 - (X_2)_{act}) h_f$$

$$\therefore \quad \{(h_2)_{act}\}_{st} = 0.95 \times 2671 + (1 - 0.95) \times 405 = 2557.7 \text{ kJ/kg}$$

$$\{(s_2)_{act}\}_{st} = (X_2)_{act} s_g + (1 - (X_2)_{act}) s_f$$

$$\therefore \quad \{(s_2)_{act}\}_{st} = 0.95 \times 7.394 + (1 - 0.95) \times 1.27 = 7.088 \text{ kJ/kg K}$$

Substituting in the SFEE for a turbine, (Equation (8.15)) gives the actual power output $(\dot{W}_{12})_{act}$ as

$$(\dot{W}_{12})_{act} = \{\dot{m}((h_2)_{act} - h_1)\}_{st}$$

$$\therefore \quad (\dot{W}_{12})_{act} = 18 \times (2557.7 - 2896) = -6089.4 \text{ kW}$$

The isentropic efficiency of the turbine η_T is given by Equation (8.16) as

$$\eta_T = \frac{(\dot{W}_{12})_{act}}{(\dot{W}_{12})_{isen}} = \frac{6089.4}{7146} = 85.2\%$$

The actual specific entropy rise of the steam is

$$\{(s_2)_{act} - (s_1)_{isen}\}_{st} = 7.088 - 6.9285 = 0.1595 \text{ kJ/kg K}$$

The actual specific entropy of the steam increases as expected, and the greater the increase the more inefficient is the turbine. If the process was plotted on a graph of fluid temperature against specific entropy, the isentropic change of state would be a vertical line downwards with temperature decrease, and the actual change of state would be a slight curve downwards indicating the specific entropy increase. The fluid would tend towards becoming a mixture from being superheated, as indeed it does in this case. The process is typical of what happens to a turbine in steam plant, as in Example 4.6.

EXAMPLE 8.10

Air enters a rotary compressor at a pressure of 100 kPa and a temperature of 99.6 °C, and leaves at a pressure of 1 MPa and temperature of 600 °C. The mass flow rate of the air is constant at 0.8 kg/s (Figure 8.23). What is the actual power required to drive the compressor for the above inlet and outlet conditions? Assume that the air behaves as a perfect gas, that its specific heat at constant pressure is constant and equal to 1.005 kJ/kg K and that its ratio of specific heats is 1.4.

What would be the theoretical power required to drive the compressor if the compression was isentropic and, hence, the outlet temperature was less for the same pressure ratio?

What is the isentropic efficiency of the compressor and the change of specific entropy of the air during the process?

What would be the isentropic efficiency of the compressor and the change of specific entropy if the fluid was steam under the same conditions and it was saturated at entry?

Solution

When the fluid is air, assumed a perfect gas, the SFEE for a compressor, Equation (8.18), gives the actual power input $(\dot{W}_{12})_{act}$ as

$$(\dot{W}_{12})_{act} = \{\dot{m}c_p(T_2 - T_1)_{act}\}_{air}$$

$$\therefore (\dot{W}_{12})_{act} = 0.8 \times 1.005 \times (600 - 99.6) = 402.3 \text{ kW}$$

It is positive as it is a work input, being a compressor. If the compression is assumed isentropic, the isentropic value of the air temperature at exit from the compressor $\{(T_2)_{isen}\}_{air}$ is given by Equation (6.17) as

$$\frac{\{(T_2)_{isen}\}_{air}}{\{T_1\}_{air}} = \frac{\{p_2^{(\gamma-1)/\gamma}\}_{air}}{\{p_1^{(\gamma-1)/\gamma}\}_{air}}$$

$$\therefore \{(T_2)_{isen}\}_{air} = \frac{372.6 \times 10^{(1.4-1)/1.4}}{1^{(1.4-1)/1.4}} = 719.4 \text{ K} = 446.4 \,°C$$

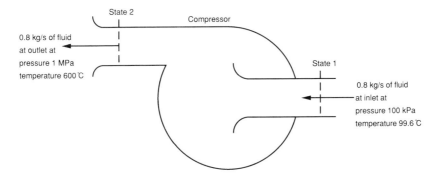

Fig. 8.23 *Example 8.10—a fluid being compressed in a rotary compressor.*

216 STEADY FLOW PROCESSES

The SFEE for a compressor (Equation (8.18)) gives the isentropic power input $(\dot{W}_{12})_{\text{isen}}$ as

$$(\dot{W}_{12})_{\text{isen}} = \{\dot{m}c_p(T_2 - T_1)_{\text{isen}}\}_{\text{air}}$$

$$\therefore (\dot{W}_{12})_{\text{isen}} = 0.8 \times 1.005 \times (446.4 - 99.6) = 278.8 \text{ kW}$$

The isentropic compressor efficiency η_C is given by Equation (8.17) as

$$\eta_C = \frac{(\dot{W}_{12})_{\text{isen}}}{(\dot{W}_{12})_{\text{act}}} = \frac{278.8}{402.3} = 69.3\%$$

The change in specific entropy of the air can be determined from Equation (6.12), which is applicable to any process. Hence:

$$\{s_2 - s_1\}_{\text{air}} = \left\{c_p \ln \frac{T_2}{T_1} - R \ln \frac{p_2}{p_1}\right\}_{\text{air}}$$

$$\therefore \{s_2 - s_1\}_{\text{air}} = 1.005 \ln \frac{873}{372.6} - 0.287 \ln \frac{10}{1} = 0.195 \text{ kJ/kg K}$$

The actual specific entropy of the air increases as expected, and the greater the increase the more inefficient is the compressor. If the process was plotted on a graph of fluid temperature against specific entropy, the isentropic change of state would be a vertical line upwards with temperature increase, and the actual change of state would be a slight curve upwards indicating the specific entropy increase. This is what happens in the compressor of the closed cycle gas turbine, as in Example 4.9.

When the fluid is steam, assumed to be saturated at a pressure of 100 kPa and temperature of 99.6 °C at entry, from the saturated water and steam tables the inlet specific enthalpy and specific entropy, $\{h_1\}_{\text{st}}$ and $\{s_1\}_{\text{st}}$ respectively, are

$$\{h_1\}_{\text{st}} = h_g = 2675 \text{ kJ/kg}$$

$$\{s_1\}_{\text{st}} = s_g = 7.359 \text{ kJ/kg K}$$

At exit from the compressor, if the actual pressure and temperature are 1 MPa and 600 °C respectively, the steam is superheated. From the superheat tables, the actual specific enthalpy $\{(h_2)_{\text{act}}\}_{\text{st}}$ and actual specific entropy $\{(s_2)_{\text{act}}\}_{\text{st}}$ of the steam at exit are

$$\{(h_2)_{\text{act}}\}_{\text{st}} = 3698 \text{ kJ/kg}$$

$$\{(s_2)_{\text{act}}\}_{\text{st}} = 8.028 \text{ kJ/kg K}$$

The SFEE for a compressor (Equation (8.15)) gives the actual power input to the compressor as

$$(\dot{W}_{12})_{\text{act}} = \{\dot{m}(h_2 - h_1)_{\text{act}}\}_{\text{st}}$$

$$\therefore (\dot{W}_{12})_{\text{act}} = 0.8(3698 - 2675) = 818.4 \text{ kW}$$

If the compression process is assumed isentropic, at a pressure of 1 MPa and specific entropy of 7.359 kJ/kg K the steam at exit from the compressor is superheated. Interpolating from the superheat tables gives the isentropic value of the steam specific enthalpy at outlet from the compressor $\{(h_2)_{\text{isen}}\}_{\text{st}}$ as

$$\{(h_2)_{\text{isen}}\}_{\text{st}} = 3196 \text{ kJ/kg}$$

The SFEE for a compressor (Equation (8.15)) gives the isentropic power input to the compressor as

$$(\dot{W}_{12})_{\text{isen}} = \{\dot{m}(h_2 - h_1)_{\text{isen}}\}_{\text{st}} = 0.8 \times (3196 - 2675) = 416.8 \text{ kW}$$

The isentropic compressor efficiency η_C is given by Equation (8.17) as

$$\eta_C = \frac{(\dot{W}_{12})_{\text{isen}}}{(\dot{W}_{12})_{\text{act}}} = \frac{416.8}{818.4} = 50.9\%$$

The actual specific entropy rise of the steam is

$$\{(s_2)_{\text{act}} - (s_1)_{\text{isen}}\}_{\text{st}} = 8.028 - 7.359 = 0.669 \text{ kJ/kg K}$$

The actual specific entropy of the steam increases as expected, and the greater the increase the more inefficient is the compressor. If the process was plotted on a graph of fluid temperature against specific entropy, the isentropic change of state would be a vertical line upwards with temperature increase, and the actual change of state would be a slight curve upwards indicating the specific entropy increase. The fluid would tend towards becoming superheated from the mixture state.

8.7 THE SFEE APPLIED TO AN EXPANSION VALVE

An Expansion Valve is a device which lowers the pressure of a fluid by forcing it through a restriction. A partially opened tap in a pipeline acts like an expansion valve, as does any restriction to the flow. When the fluid forces its way through the smaller passage, its pressure is reduced. A schematic of an expansion valve is shown in Figure 8.24.

The SFEE can be applied to the fluid flowing through the valve. Considering each of the terms in Equation (8.8) in turn:

$\dot{Q}_{12}$ — the amount of heat transfer associated with the fluid passing through the valve is difficult to quantify, but it is unlikely to be substantial because most valves are not very big, and the time available for the heat transfer to take place is small. In the first instance, therefore, the heat transfer will be assumed to be negligible.

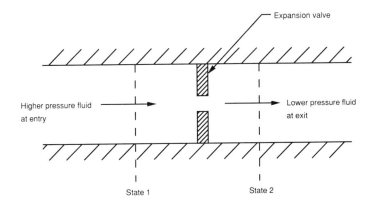

Fig. 8.24 *A schematic of an expansion valve in which the pressure of a fluid is decreased.*

STEADY FLOW PROCESSES

$\dot{W}_{12}$ — a pump is required to move the fluid through the valve, but this can be considered a separate component, so the work transfer term is zero.

ΔENTHALPY — this term is always retained.

ΔKE — generally, the velocity of the fluid before and after the valve is similar, and this term may be considered insignificant.

ΔPE — if the fluid flows in a horizontal direction only, this term is zero, but in all valves the effects of potential energy may be assumed negligible.

The SFEE for an expansion valve is reduced to

$$0 = h_2 - h_1 \qquad (8.19)$$

In other words, the specific enthalpy of the fluid as it passes through the valve remains constant. This is approximately true, and any errors arising from the assumption are likely to be small compared to errors arising as a consequence of other assumptions made when analysing a thermal system.

An expansion valve is a decidedly irreversible process in that it is physically impossible to force the fluid in the opposite direction. It is, therefore, not relevant to consider whether the valve can be isentropic and to define some sort of efficiency.

8.7.1 When the fluid is a perfect gas

When the fluid expanding through an expansion valve is a perfect gas with a constant specific heat, it is possible to substitute for the specific enthalpy in terms of the specific heat at constant pressure and the temperature difference, in accordance with Equation (6.7). In this case, Equation (8.16) reduces to

$$0 = T_2 - T_1 \qquad (8.20)$$

EXAMPLE 8.11

Steam, at a pressure of 500 kPa and dryness fraction 0.98, suffers a pressure drop of 350 kPa as it passes through an expansion valve (Figure 8.25). What is the change in temperature and specific entropy of the steam?

Solution

With a pressure of 500 kPa and dryness fraction of 0.98, the steam before the expansion valve is a mixture. From the saturated water and steam tables, the specific enthalpy, specific entropy and temperature, $\{h_1\}_{st}$, $\{s_1\}_{st}$ and $\{T_1\}_{st}$ respectively, of the steam are

$$\{h_1\}_{st} = X_1 h_g + (1 - X_1) h_f = 0.98 \times 2749 + (1 - 0.98) \times 640 = 2707 \text{ kJ/kg}$$

$$\{s_1\}_{st} = X_1 s_g + (1 - X_1) s_f = 0.98 \times 6.822 + (1 - 0.98) \times 1.86 = 6.723 \text{ kJ/kg K}$$

$$\{T_1\}_{st} = 151.8\,°C$$

The SFEE for the valve (Equation (8.19)) gives the specific enthalpy of the steam after the valve $\{h_2\}_{st}$ as

$$\{h_2\}_{st} = \{h_1\}_{st} = 2707 \text{ kJ/kg}$$

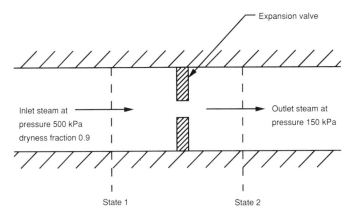

Fig. 8.25 *An expansion valve in which the pressure of steam is reduced.*

At a pressure of 150 kPa and specific enthalpy 2707 kJ/kg, the steam after the expansion valve is superheated. Interpolating from the superheat tables gives the specific entropy and temperature of the steam, $\{s_2\}_{st}$ and $\{T_2\}_{st}$ respectively, as

$$\{s_2\}_{st} = 7.257 \text{ kJ/kg K}$$

$$\{T_2\}_{st} = 118.2\,°C$$

The change in specific entropy and temperature of the steam as it passes through the expansion valve is

$$\{s_2 - s_1\}_{st} = 7.257 - 6.723 = 0.534 \text{ kJ/kg K}$$

$$\{T_2 - T_1\}_{st} = 118.2 - 151.8 = -33.6\,°C$$

Hence, in an expansion valve, the entropy of the steam increases, it is an irreversible process, and the temperature decreases.

8.8 THE SFEE APPLIED TO A MIXING PROCESS

In a Mixing Process, two or more streams of fluid, all at the same pressure, are mixed together to form just one stream. The SFEE can be applied to the mixing process, but it is necessary to consider each individual stream of fluid at entry separately. Taking the terms in Equation (8.8) in turn:

$\dot{Q}_{12}$— the amount of heat transfer associated with the fluids passing through the mixing process is difficult to quantify, but it is unlikely to be substantial because the time available for the heat transfer to take place is small. Also the mixing box should be lagged. In the first instance, therefore, the heat transfer will be assumed to be negligible.

$\dot{W}_{12}$— a pump is required to move the fluids through the mixing process, but this can be considered a separate component, so the work transfer term is zero.

$\Delta ENTHALPY$—this term is always retained.

220 STEADY FLOW PROCESSES

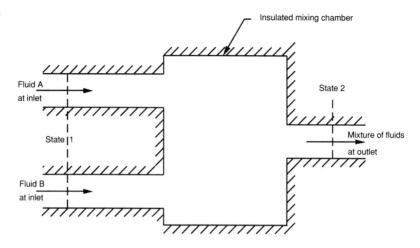

Fig. 8.26 *A mixing process in which two fluids are adiabatically mixed.*

ΔKE — generally, the velocities of the fluids before and after the mixing process are similar, and this term may be considered insignificant.

ΔPE — if the fluid flows in a horizontal direction only, this term is zero, but in all mixing processes the effects of potential energy may be assumed negligible.

The SFEE for a mixing process is reduced to

$$(\text{Enthalpy of fluid mixture})_2 = \sum (\text{Enthalpy of each fluid})_1$$

The enthalpy is a mass-dependent term and it is necessary to consider the product of the mass and specific enthalpy of each fluid at entry (Figure 8.26). For example, if two fluids A and B are mixing together, Equation (8.8) becomes

$$\dot{m}_2 h_2 = (\dot{m}_1 h_1)_A + (\dot{m}_1 h_1)_B \tag{8.21}$$

From the conservation of mass (Equation (8.1)) gives

$$\dot{m}_2 = (\dot{m}_1)_A + (\dot{m}_1)_B$$

As with the expansion valve, a mixing process is highly irreversible and it is not relevant to consider an efficiency term based upon isentropic conditions.

8.8.1 When the fluid is a perfect gas

When the fluids in the mixing process are perfect gases with constant specific heats, it is possible to substitute for the specific enthalpy in terms of the specific heat at constant pressure and the temperature difference, in accordance with Equation (6.7). Equation (8.18) becomes

$$\dot{m}_2 (c_p)_2 T_2 = (\dot{m}_1 (c_p)_1 T_1)_A + (\dot{m}_1 (c_p)_1 T_1)_B \tag{8.22}$$

EXAMPLE 8.12

In an adiabatic mixing chamber, 4 kg/s of steam A of dryness fraction 0.9 and pressure 500 kPa are mixed with 10 kg/s of steam B at the same pressure but with a temperature of 300 °C (Figure 8.27). What is the final temperature of the mixture?

Solution

For steam A, at a pressure of 500 kPa and dryness fraction of 0.9, it is a mixture. From the saturated water and steam tables, its specific enthalpy $\{h_A\}_{st}$ is

$$\{h_A\}_{st} = (Xh_g + (1-X)h_f)_A = 0.9 \times 2749 + (1-0.9) \times 640 = 2538 \text{ kJ/kg}$$

For steam B, at a pressure of 500 kPa and temperature of 300 °C, it is superheated. From the superheat tables, its specific enthalpy $\{h_B\}_{st}$ is

$$\{h_B\}_{st} = 3065 \text{ kJ/kg}$$

From the conservation of mass (Equation (8.1)) the mass flow rate of the mixture $\{\dot{m}_2\}_{st}$ is

$$\{\dot{m}_2\}_{st} = \{(\dot{m}_1)_A + (\dot{m}_1)_B\}_{st} = 4 + 10 = 14 \text{ kg/s}$$

Using the SFEE for a mixing process (Equation (8.21)) gives the specific enthalpy of the mixture $\{h_2\}_{st}$ as

$$\{\dot{m}_2 h_2\}_{st} = \{(\dot{m}_1 h_1)_A + (\dot{m}_1 h_1)_B\}_{st}$$

$$\therefore \{h_2\}_{st} = \frac{4 \times 2538 + 10 \times 3065}{14} = 2914 \text{ kJ/kg}$$

Steam A and steam B are at the same pressure. The pressure of the mixture will also be the same. At 500 kPa pressure and with a specific enthalpy of 2914 kJ/kg, the mixture is superheated. Interpolating from the superheat tables, the temperature of the mixture $\{T_2\}_{st}$ is

$$\{T_2\}_{st} = 227.1 \,°C$$

The pressure of the mixture must be the same as the pressures of the two fluids initially, otherwise a pressure discontinuity will occur in the mixing chamber.

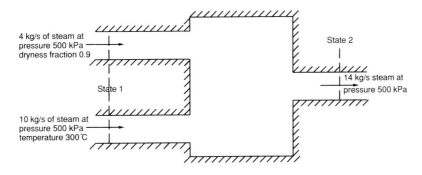

Fig. 8.27 Example 8.12—a mixing process in which two streams of steam are mixed.

8.9 THE SFEE APPLIED TO PIPE/DUCTWORK

This is a very important area and is examined in detail in Chapters 9, 10 and 11.

8.10 SOME COMMON THERMAL SYSTEMS

This book is restricted to a consideration of thermal systems which contain only the components studied above, together with piston cylinder mechanisms, and which would all appear in the energy triangle of Figure 1.3. Fortunately, this covers most of the existing practical systems in use today. For example, a steam turbine plant, a refrigeration plant, a heat pump and a closed cycle gas turbine plant were described in Chapter 4 and can be seen to be composed of the components discussed in this chapter. An open cycle gas turbine plant is like a closed cycle with no cooler. A central heating system consists of a boiler, pipework or ductwork, and heat exchangers, called radiators if the fluid is water. An air-conditioning system consists of a fan, ductwork, heaters and refrigerators, called coolers, chillers or dehumidifiers. The significance and importance of the SFEE is that it can be applied to all these devices and systems. It is only the piston cylinder mechanism that cannot be analysed from a consideration of the SFEE.

EXAMPLE 8.13

Steam enters a horizontal pipe at a pressure of 1 MPa. It passes under steady flow conditions along the pipe at a constant specific volume and constant velocity, undergoing a pressure drop of 100 kPa, and ending in a dry saturated condition. It is throttled through a valve to a pressure of 700 kPa. Thereafter, it is expanded in a rotary turbine of efficiency 85% to a pressure of 100 kPa (Figure 8.28).

Determine, per kilogram of steam, the heat transferred from the steam in passing along the pipe, the change in specific entropy in passing through the valve and the work transfer during the expansion in the turbine.

If the rotary turbine was replaced by a piston cylinder mechanism working between the same pressure limits, what would be the difference in the work transfer? Assume that the fluid expansion in the cylinder is a polytropic non-flow process with an index of expansion of 1.3.

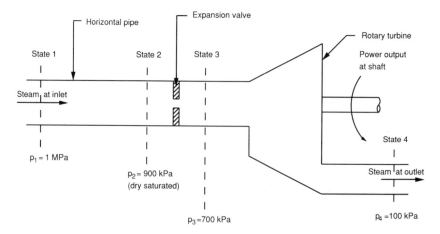

Fig. 8.28 *Example 8.13—steam flows along a pipe, through an expansion valve and into a rotary turbine.*

Solution

With reference to Figure 8.27, for the steam in the pipe, at exit of the pipe where the pressure is 900 kPa, state 2, the steam is saturated. From the saturated water and steam tables, the pipe exit specific enthalpy, specific volume and specific entropy, $\{h_2\}_{st}$, $\{v_2\}_{st}$ and $\{s_2\}_{st}$ respectively, are

$$\{h_2\}_{st} = h_g = 2774 \text{ kJ/kg}$$

$$\{v_2\}_{st} = v_g = 0.2149 \text{ m}^3/\text{kg}$$

$$\{s_2\}_{st} = s_g = 6.623 \text{ kJ/kg K}$$

It is given that the specific volume of the steam in the pipe is constant. Therefore, the specific volume of the steam at the entry to the pipe at state 1, $\{v_1\}_{st}$, is

$$\{v_1\}_{st} = \{v_2\}_{st} = 0.2149 \text{ m}^3/\text{kg}$$

At entry to the pipe where the steam pressure is 1 MPa and specific volume 0.2149 m³/kg, the steam is superheated. Interpolating from the superheat tables gives the specific enthalpy of the steam at inlet to the pipe $\{h_1\}_{st}$ as

$$\{h_1\}_{st} = 2867 \text{ kJ/kg}$$

Considering the SFEE, as in Equation (8.7), applied to the steam flow in the pipe:

$$\dot{W}_{12} = 0 \text{ (because there is no pump or turbine in the pipe)}$$

$$\{C_1\}_{st} = \{C_2\}_{st} \text{ (because the velocity of steam in the pipe is constant)}$$

$$z_1 = z_2 \text{ (because the pipe is horizontal)}$$

The SFEE for the pipe becomes:

$$\dot{Q}_{12} = \{\dot{m}(h_2 - h_1)\}_{st}$$

Therefore, the heat transfer from the pipe per kilogram of steam q_{12} is

$$q_{12} = \frac{\dot{Q}_{12}}{\{\dot{m}\}_{st}} = \{h_2 - h_1\}_{st} = 2774 - 2867 = -93 \text{ kJ/kg}$$

The negative sign implies that it is a heat transfer out of the pipe.

For the valve, the steam inlet conditions of specific enthalpy and specific entropy are the steam exit conditions from the pipe. The SFEE for an expansion valve when the steam goes from state 2 to state 3 (Equation (8.19)) is

$$0 = \{h_3 - h_2\}_{st} \quad \therefore \quad \{h_3\}_{st} = \{h_2\}_{st} = 2774 \text{ kJ/kg}$$

At exit from the valve where the pressure is 700 kPa and specific enthalpy 2774 kJ/kg, the steam is superheated. By interpolation from the superheat tables, the specific entropy of the steam $\{s_3\}_{st}$ is

$$\{s_3\}_{st} = 6.731 \text{ kJ/kg K}$$

The change of specific entropy of the steam across the valve $\{\Delta s_{23}\}_{st}$ is

$$\{\Delta s_{23}\}_{st} = \{s_3 - s_2\}_{st} = 6.731 - 6.623 = 0.108 \text{ kJ/kg K}$$

The entropy of the steam increases, as expected, it being a highly irreversible process. For the turbine, the inlet specific enthalpy and specific entropy of the steam are the valve outlet specific enthalpy and specific entropy at

224 STEADY FLOW PROCESSES

state 3. Assuming, in the first instance, that the expansion in the turbine from state 3 to state 4 is isentropic, the turbine outlet specific entropy $\{s_4\}_{st}$ is

$$\{s_4\}_{st} = \{s_3\}_{st} = 6.731 \text{ kJ/kg K}$$

At outlet from the turbine where the pressure is 100 kPa and specific entropy 6.731 kJ/kg K, the steam is a mixture. From the saturated water and steam tables, the isentropic value of the dryness fraction at the turbine outlet $(X_4)_{isen}$ is

$$\{s_4\}_{st} = (X_4)_{isen} s_g + (1 - (X_4)_{isen}) s_f$$
$$\therefore \quad 6.731 = (X_4)_{isen} \times 7.359 + (1 - (X_4)_{isen}) \times 1.303$$
$$\therefore \quad (X_4)_{isen} = 0.9$$

The isentropic value of the steam specific enthalpy at outlet from the turbine $\{(h_4)_{isen}\}_{st}$, therefore, is

$$\{(h_4)_{isen}\}_{st} = (X_4)_{isen} h_g + (1 - (X_4)_{isen}) h_f$$
$$\therefore \quad \{(h_4)_{isen}\}_{st} = 0.9 \times 2675 + (1 - 0.9) \times 417 = 2449.2 \text{ kJ/kg}$$

Substituting in the SFEE for a turbine (Equation (8.15)) gives the isentropic turbine specific work transfer per kilogram of steam $(w_{34})_{isen}$ as

$$(w_{34})_{isen} = \frac{(\dot{W}_{34})_{isen}}{\{\dot{m}\}_{st}} = \{(h_4)_{isen} - h_3\}_{st}$$
$$\therefore \quad (w_{34})_{isen} = 2449.2 - 2774 = -324.8 \text{ kJ/kg}$$

For a turbine efficiency of 85%, Equation (8.16) gives the actual work transfer per kilogram of steam $(w_{34})_{act}$ as

$$\eta_T = \frac{(\dot{W}_{34})_{act}}{(\dot{W}_{34})_{isen}} = \frac{(w_{34})_{act}}{(w_{34})_{isen}}$$
$$\therefore \quad (w_{34})_{act} = 0.85 \times (-324.8) = -276.1 \text{ kJ/kg}$$

The negative sign shows that it is a work transfer out of the system, as expected for a turbine.

If the rotary turbine is replaced by a piston cylinder mechanism, the steam will undergo a non-flow polytropic process. At entry to the piston cylinder mechanism, the steam is superheated at a pressure $\{p_3\}_{st}$ of 700 kPa and specific enthalpy $\{h_3\}_{st}$ of 2774 kJ/kg. From the superheat tables, the specific volume of the steam $\{v_3\}_{st}$ is

$$\{v_3\}_{st} = 0.2761 \text{ m}^3/\text{kg}$$

Assuming that the steam goes from state 3 to state 5 in following the polytropic relationship:

$$\{p_3 v_3^{1.3}\}_{st} = \{p_5 v_5^{1.3}\}_{st}$$

Hence, the specific volume of the steam at outlet from the piston cylinder mechanism $\{v_5\}_{st}$ is

$$\{v_5\}_{st} = \frac{7^{1/1.3} \times 0.2761}{1^{1/1.3}} = 1.234 \text{ m}^3/\text{kg}$$

Equation (5.12) gives the work transfer per kilogram of fluid from a non-flow polytropic process w_{35} as

$$w_{35} = \frac{\dot{W}_{35}}{\{\dot{m}\}_{st}} = -\frac{\{p_5 v_5 - p_3 v_3\}_{st}}{1 - n}$$

Hence, the work transfer per kilogram of steam from the piston cylinder mechanism w_{35} is

$$w_{35} = -\frac{10^5 \times 1.234 - 7 \times 10^5 \times 0.2761}{1 - 1.3} = -232.9 \text{ kJ/kg}$$

This answer does not include any efficiency terms. If it is compared with the isentropic work transfer from the rotary turbine, it can be seen that it is about 30% less. In other words, the rotary turbine will produce more work transfer than the piston cylinder mechanism when operating between the same pressure limits. But this depends upon the index of expansion.

EXAMPLE 8.14

Air, assumed to be a perfect gas, flows under steady conditions at the rate of 10 kg/s into a horizontal heat exchanger with a velocity of 10 m/s, a pressure of 190 kPa and a temperature of 400 K. The air is heated at constant pressure to a temperature of 1000 K and leaves the heat exchanger at the same velocity (Figure 8.29). What is the heat input to the heat exchanger? Assume that the specific heat at constant pressure of the air is constant and equal to 1.005 kJ/kg K, and that the ratio of the specific heats is 1.4.

The air enters a nozzle where it is expanded isentropically to a pressure of 150 kPa. What is the air exit velocity from the nozzle and the cross-sectional area of the nozzle at outlet?

If, in fact, the nozzle has an efficiency of 90%, what is the actual air exit velocity and temperature?

Solution

The SFEE for a heat exchanger when the fluid is a perfect gas going from state 1 to state 2, Equation (8.10), gives the heat transfer $\dot{Q}_{12}$ as

$$\dot{Q}_{12} = \{\dot{m}c_p(T_2 - T_1)\}_{\text{air}} = 10 \times 1.005 \times (1000 - 400) = 6030 \text{ kW}$$

Assume that the air goes from state 2 to state 3 in the nozzle. The process is initially isentropic. The isentropic value of the nozzle exit temperature $\{(T_3)_{\text{isen}}\}_{\text{air}}$ can be found from Equation (6.17), as follows:

$$\frac{\{(T_3)_{\text{isen}}\}_{\text{air}}}{\{T_2\}_{\text{air}}} = \frac{\{p_3^{(\gamma-1)/\gamma}\}_{\text{air}}}{\{p_2^{(\gamma-1)/\gamma}\}_{\text{air}}}$$

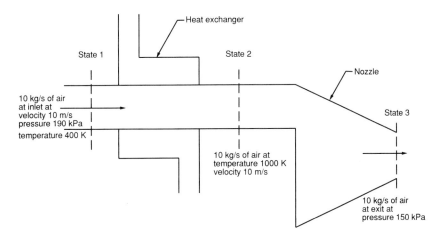

Fig. 8.29 *Example 8.14—air is heated in a heat exchanger before being expanded in a nozzle.*

$$\therefore \quad \{(T_3)_{\text{isen}}\}_{\text{air}} = \frac{1000 \times 1.5^{(1.4-1)/1.4}}{1.9^{(1.4-1)/1.4}} = 934.7 \text{ K}$$

The SFEE for a nozzle when the fluid is a perfect gas (Equation (8.14)) gives the isentropic value of the air velocity at exit from the nozzle $\{(C_3)_{\text{isen}}\}_{\text{air}}$ as

$$0 = \{c_p(T_3 - T_2)_{\text{isen}} + 0.5(C_3^2 - C_2^2)_{\text{isen}}\}_{\text{air}}$$

$$\therefore \quad 0 = 1.005 \times 10^3 \times (934.7 - 1000) + 0.5(\{(C_3)_{\text{isen}}\}_{\text{air}}^2 - 10^2)$$

$$\therefore \quad \{(C_3)_{\text{isen}}\}_{\text{air}} = 362.4 \text{ m/s}$$

The density of the air at exit from the nozzle $\{\rho_3\}_{\text{air}}$ is given by the equation of state (Equation (6.1)) as

$$\{p_3\}_{\text{air}} = \{\rho_3 R T_3\}_{\text{air}}$$

$$\therefore \quad \{\rho_3\}_{\text{air}} = \frac{1.5 \times 10^5}{287 \times 934.7} = 0.56 \text{ kg/m}^3$$

The continuity equation (Equation (8.3)) gives the cross-sectional area of the nozzle at outlet $(A_{xs})_3$ as

$$\{\dot{m}_3\}_{\text{air}} = \{\rho_3 C_3\}_{\text{air}} (A_{xs})_3$$

$$\therefore \quad (A_{xs})_3 = \frac{10}{0.56 \times 362.4} = 0.049 \text{ m}^2$$

Equation (8.12) for the nozzle efficiency allows the actual velocity of the air at exit from the nozzle $\{(C_3)_{\text{act}}\}_{\text{air}}$ to be calculated as follows:

$$\eta_N = \frac{\{(C_3)_{\text{act}}^2\}_{\text{air}}}{\{(C_3)_{\text{isen}}^2\}_{\text{air}}}$$

$$\therefore \quad \{(C_3)_{\text{act}}\}_{\text{air}} = 0.9^{0.5} \times 362.4 = 343.8 \text{ m/s}$$

The SFEE for a nozzle when the fluid is a perfect gas (Equation (8.14)) gives the actual temperature of the air at exit from the nozzle $\{(T_3)_{\text{act}}\}_{\text{air}}$ as

$$0 = \{c_p(T_3 - T_2)_{\text{act}} + 0.5(C_3^2 - C_2^2)_{\text{act}}\}_{\text{air}}$$

$$\therefore \quad 0 = 1.005 \times 10^3 (\{(T_3)_{\text{act}}\}_{\text{air}} - 1000) + 0.5 \times (343.8^2 - 10^2)$$

$$\therefore \quad \{(T_3)_{\text{act}}\}_{\text{air}} = 941.2 \text{ K}$$

The actual temperature of the air at exit from the nozzle is greater than the isentropic, as expected, because the entropy of the air has increased.

8.11 SIMPLIFIED FORMS OF THE SFEE

Whenever a fluid flows over a surface, friction effects act to slow it down, just as the friction force opposes the motion of two solid surfaces sliding together. In fluids, a pressure force in the direction of motion is used to overcome the friction force and maintain the flow, in other words the pressure of the fluid decreases in the direction of the flow. The main effect of the friction force opposing the motion is to increase the internal energy of the fluid, its temperature and the amount of heat transfer. If it is assumed that there is no friction, the effects of heat transfer and internal energy in the fluid can be ignored. If the fluid flow is considered independently of any pump that makes the fluid move, the work transfer term is zero. Hence, with no friction:

$$\dot{Q}_{12} = 0 \quad \dot{W}_{12} = 0 \quad u_1 = u_2$$

The SFEE, as in Equation (8.5), becomes

$$p_2 v_2 + 0.5 C_2^2 + g z_2 = p_1 v_1 + 0.5 C_1^2 + g z_1 \quad (8.23)$$

This is known as Bernoulli's Equation, after Bernoulli, and is an ideal form of the SFEE assuming no friction. All real fluids suffer the effects of friction, but it is sometimes possible to obtain reasonable answers to a problem by assuming them to be ideal and using Bernoulli's equation.

In cases where the flow is horizontal, $z_1 = z_2$, and the equation becomes

$$pv + 0.5 C^2 = \text{const}.$$

This is more commonly written:

$$p + 0.5 \rho C^2 = \text{const}.$$

It is the equation which links the static pressure with the dynamic pressure, considered before as Equation (2.3), namely:

Static pressure + dynamic pressure = total pressure

$$p + 0.5 \rho C^2 = p_0$$

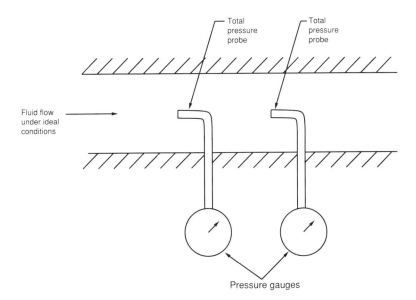

Fig. 8.30 *In an ideal fluid flow, the total pressure of the fluid remains constant.*

228 STEADY FLOW PROCESSES

It can now be seen that, in ideal fluid flow, the total pressure will remain constant from place to place, but in real fluid flow, the total pressure varies from place to place (Figure 8.30).

Note also that, taking a perfect gas for illustration, for the horizontal flow of the gas with no heat and work transfer, Equation (8.7) becomes

$$0 = c_p(T_2 - T_1) + 0.5(C_2^2 - C_1^2)$$

$$\therefore \quad T + \frac{0.5C^2}{c_p} = \text{const.} = T_0 \tag{8.24}$$

where T is the static temperature of the fluid and T_0 is the total or stagnation temperature. As mentioned in Section 2.4, it is the static temperature that really should be substituted in most of the formulae derived, but the total temperature that is measured and actually used. Taking air as an example with $c_p = 1.005$ kJ/kg K, the difference in temperature between the two from Equation (8.24) is only one degree when the velocity is approximately 45 m/s. Given the inaccuracy involved in measuring temperature anyway, this justifies the use of the total temperature in the equations for most applications in thermal systems.

9 Effects of a Fluid in Motion

It is now possible to predict theoretically the work and heat transfer in a non-flow process, a flow process and when the fluid operates in a system completing a cycle made up of both types of process. And this can be done when the fluid is either a perfect gas or water/steam. However, it is important to know what happens to the fluid as it flows from either one component to another, as when it completes a cycle, for example in a steam plant or a refrigerator, or when it flows into or out of a component as in a reciprocating piston cylinder mechanism or a rotary turbine in a hydroelectric scheme. This is because the analysis for the work and heat transfer depends upon the property values of the fluid at entry and exit of a process or component. Generally, the fluid flows in pipe- or ductwork, but before its behaviour in these conduits can be examined, it is useful to highlight a few general characteristics of fluids, and that is attempted in this chapter. Also presented are methods for measuring the rate of flow of the fluid. Chapter 10 deals with the forces that are generated as the fluid is flowing from place to place, and Chapter 11 analyses the consequences in terms of the change in the property values, in particular the decrease in pressure due to friction forces opposing the motion.

In Chapter 2, the property viscosity was introduced as a measure of the resistance to the flow of a fluid. In fact this resistance, which manifests itself as a friction force opposing the motion of the fluid whenever the fluid is flowing past a solid surface, has some fundamental effects upon the behaviour of the fluid.

Consider the situation of a fluid approaching a stationary solid boundary with a uniform velocity, called the Free Stream Velocity C_{FS}. When the fluid flows over the solid surface, the very thin layer of fluid next to the surface adheres to it and has zero velocity. But some distance away from the surface, the fluid is unaffected by the presence of the surface and continues to move at the free stream velocity. In between, the layers of fluid slide past each other at different velocities. A region of velocity change or velocity gradient is created in which the velocity of the fluid varies from zero at the surface to C_{FS}, called the Boundary Layer (Figure 9.1). It is in the boundary layer that shear stresses are generated, caused by the differential velocities of each layer of fluid. The resultant friction force acts on the surface. Outside the boundary layer is called the Free Stream and, here, viscosity effects are negligible, and the fluid may be treated as if it is ideal.

9.1 FLUID FLOW IN A PIPE OR DUCT

A boundary layer forms inside a pipe or duct whenever a fluid is flowing through it. Initially it is very thin, but it increases in thickness with distance downstream. Eventually the boundary layer from the sides of the pipe or duct will meet at the centre so that the

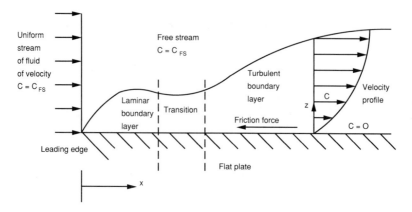

Fig. 9.1 *A boundary layer, in which the fluid velocity varies from zero at the surface to the free stream velocity C_{FS}, forms whenever a fluid flows over a solid surface.*

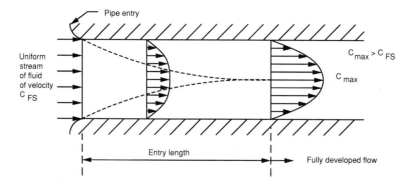

Fig. 9.2 *The boundary layer in a pipe, which begins at the entrance to the pipe next to the pipe wall, eventually fills the whole pipe and the flow becomes fully developed.*

whole of the flow is filled with the boundary layer. This is referred to as Fully Developed Flow in a pipe or duct, and the front piece of the pipe where the boundary layers have not met as the Entry Length, as shown in Figure 9.2.

The velocity profile of the fluid, initially uniform and equal to C_{FS} at entry to the pipe, adopts a parabolic shape in the fully developed flow region. Clearly, as the fluid layer next to the surface of the pipe is stationary and layers of fluid in the vicinity are retarded, the maximum velocity of fluid in the centre of the pipe C_{max} must be greater than C_{FS} if a steady mass flow rate of fluid in the pipe is to be maintained.

In fully developed incompressible pipe flow with a constant mass flow rate, the continuity equation (Equation (8.3)) applied between two points in the pipe, representing two fluid states, is

$$\dot{m}_1 = \dot{m}_2 = \rho_1 (A_{xs})_1 C_1 = \rho_2 (A_{xs})_2 C_2$$

But for the incompressible flow of a fluid in a straight pipe:

$$(A_{xs})_1 = (A_{xs})_2 \quad \text{and} \quad \rho_1 = \rho_2 \quad \therefore \quad C_1 = C_2$$

FLUID FLOW IN A PIPE OR DUCT

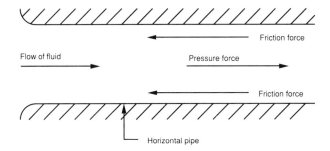

Fig. 9.3 *For the steady incompressible flow of a fluid through a straight pipe, constant velocity and mass flow rate of fluid is maintained when the friction force opposing the motion is balanced by a pressure drop in the fluid.*

The friction force is trying to retard the flow. In order to maintain a constant flow rate and constant velocity, there must be an equal and opposite force to the friction force in the direction of motion of the fluid. This must be a pressure force, in other words there must be a decrease in the pressure of the fluid as it flows through the pipe which exactly counteracts the friction force, as shown in Figure 9.3.

Inside the pipe but near the entry, where the boundary layer is thin, the velocity gradient at the surface is far greater than the velocity gradient downstream where fully developed flow is established. The friction force in the vicinity of the surface must also be greater than that downstream and, therefore, the pressure drop of the fluid is greater over the entry length than over the same length of fully developed flow.

When the Reynolds number of the fluid, defined by Equation (2.15), is less than 2000 and the flow is laminar, the entry length EL in a pipe of diameter D, has been shown by experiment to be given by

$$EL = 0.065 Re D \qquad (9.1)$$

When the Reynolds number of the fluid is greater than 3500 and the flow is turbulent, the conditions near the entry are qualitatively similar to those for laminar flow, but fully developed conditions are established quicker, and the entry length is shorter. It is not possible to predict the entry length exactly because any turbulence in the fluid at entry to the pipe, or roughness in the pipe surface, quickens the establishment of the fully developed flow region.

Flow in ducts is similar to flow in pipes though three-dimensional effects in the corners make it slightly more complicated. In general, however, there is much information available about the flow of fluids in pipes and ducts, both analytical and experimental. Of particular relevance are plots of the log of the friction factor f against the log of the Reynolds number Re, which identifies the laminar and turbulent regions of flow. This type of graph will be considered in more detail in Chapter 11, because the effects of friction in pipes and ducts is extremely important.

9.1.1 Equivalent diameter of a duct

Most liquid flows in thermal systems are in pipes of circular cross-section. Sometimes gas flows, as in air-conditioning systems, are in rectangular or square ducts. Fortunately experiment shows that the relationships developed for circular cross-sections can be applicable

232 EFFECTS OF A FLUID IN MOTION

to cross-sections other than circular if the 'equivalent diameter' is used. The equivalent diameter of a duct with a cross-section other than circular is defined as 'the cross sectional area of the duct multiplied by a factor of 4 and divided by the duct perimeter'.

The equivalent diameter may not be used for duct cross-sections which are highly distorted or irregular, and it is only applicable to turbulent flows. Many fluid flows in a thermal system are turbulent, as discussed in Example 2.8, and if they take place in a duct other than circular, that duct is often square or mildly rectangular, so the concept of the equivalent diameter is most suitable.

9.2 FLUID FLOW OVER A FLAT PLATE

When a uniform stream of fluid of velocity C_{FS} flows over a flat plate, the boundary layer begins at the leading edge of the plate and increases in thickness with distance down the plate, as shown in Figure 9.4. The boundary layer thickness δ occurs when the velocity of the fluid C is 99% of C_{FS}.

According to Equation (2.16), the Reynolds number of the flow is dependent upon the characteristic dimension, which is the length of surface that the fluid maintains contact with. Initially, therefore, the Reynolds number of the flow in the boundary layer at the beginning of the plate, based upon the free stream velocity C_{FS}, will be very small and the boundary layer will be laminar. However, the greater the distance travelled over the plate, the more the Reynolds number will increase. Eventually a Reynolds number will be achieved, the critical Reynolds number, at which the flow will change in a transition region to turbulent conditions in accordance with Section 2.5.5. In practice, this is accompanied by an increase in the boundary layer thickness because there is more mixing in turbulent flow. However, even in the turbulent part of the boundary layer, the random movements of fluid must die out near the surface because the velocity there tends to zero and the resulting velocity gradient and, therefore, the friction force, is so large. Hence, an even thinner Laminar or viscous Sub-layer is formed, in which the friction forces prevent turbulence from occurring.

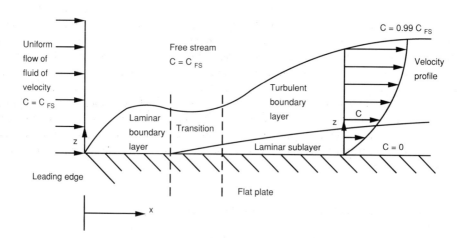

Fig. 9.4 *The boundary layer on a flat plate is initially laminar but changes through a transition region to turbulent, depending upon the Reynolds number of the flow.*

It is possible for the boundary layer to be turbulent almost from the beginning of the plate if the leading edge is sharp, if the plate is rough or if the fluid contains considerable turbulence initially.

Only the region of the boundary layer, by definition, is affected by the presence of the plate surface. The uniform stream of fluid approaching the plate with velocity C_{FS} carries on flowing past the plate in the region outside the boundary layer oblivious of the presence of the plate. It is only in the boundary layer that interest in the type of flow, whether it be laminar or turbulent, is important, because it is only in the boundary layer that the shear stresses are formed and the viscous forces which act on the plate surface. This leads to the analysis of fluid flow past a surface being divided into two regions, the boundary layer where viscosity is important and friction forces are developed, and the free stream where viscosity is not important and there are no friction forces, and where the fluid can be treated as being ideal.

In many applications, such as with aircraft or motor cars, the boundary layer is very thin, of the order of a few millimetres. But this only increases its importance because, in a very short distance away from a surface, the velocity can change considerably from zero at the surface to the free stream value, leading to large friction forces which have to be accommodated in the design.

The analysis of boundary layer flow past a surface is very complicated and relies heavily upon experimental data to support the theoretical equations. Laminar flow is easier because it is possible to use Equation (2.13), but this is not applicable to turbulent flow. In this case, recourse is made to empirical equations based upon Equation (2.19).

A typical equation which describes the thickness of a laminar boundary layer δ at any distance x down the plate is

$$\delta = \frac{5.48x}{Re^{0.5}} \quad (9.2)$$

where the Reynolds number is based upon the distance x. Similarly, an equation for the thickness of a turbulent boundary layer δ at any distance x down the plate is

$$\delta = \frac{0.37x}{Re^{0.2}} \quad (9.3)$$

with the Reynolds number similarly defined. At large Reynolds numbers, the distance down the plate is much greater than the boundary layer thickness and Figure 9.4 is shown greatly exaggerated on the z scale.

In a turbulent flow, there is considerable mixing of the fluid particles, resulting in a more uniform velocity profile, although the velocity must still decrease to zero at the surface. But in laminar flow, only the molecules move from layer to layer, and the velocity profile changes gradually. Typical velocity profiles for a laminar and turbulent boundary layer on a flat plate are shown in Figure 9.5.

A common equation used to describe a laminar velocity profile on a flat plate is

$$\frac{C}{C_{FS}} = \frac{2z}{\delta} - \frac{z^2}{\delta^2} \quad (9.4)$$

and a common equation used to describe a turbulent velocity profile on a flat plate is

$$\frac{C}{C_{FS}} = \frac{z^{1/7}}{\delta^{1/7}} \quad (9.5)$$

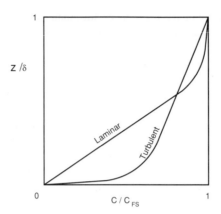

Fig. 9.5 *Typical boundary layer laminar and turbulent velocity profiles for the flow of a fluid over a flat plate.*

As the surface shear stress τ_s (i.e. the value of the shear stress τ at the surface where $z = 0$) is proportional to dC/dz, values of τ_s are greater for turbulent boundary layers than for laminar boundary layers because the velocity gradient is so much steeper near the surface.

To obtain the Friction Force F_F on the plate, as the velocity at any height in the boundary layer varies with distance down the plate, for a plate of length L and width b this must be written:

$$F_F = b \int_0^L \tau_s dx \tag{9.6}$$

In laminar flow, the substitution for τ_s can be made from Equation (2.13), but in turbulent flow it is necessary to use the definition of friction factor f in Equation (2.19), and to know the empirical relationship between the friction factor and the Reynolds number, as mentioned in Section 2.5.6. A typical equation for a flat plate is

$$f = \frac{0.058}{Re^{0.2}} \tag{9.7}$$

EXAMPLE 9.1

An electric plate heater is immersed in an oil tank and is used to heat oil before it enters a particular process. In a first analysis, the heater may be considered to be a flat plate 2 m long by 1 m wide. The oil, of density 800 kg/m³ and dynamic viscosity 0.9 kg/m s, flows parallel to the plate heater, on both sides, with a velocity of 5 m/s, as shown in Figure 9.6. Assuming that the boundary layer of the oil, which forms as the oil passes over the plate, is completely laminar, determine the friction force on the heater.

Solution

For a free stream velocity of oil $\{C_{FS}\}_{oil}$ of 5 m/s, the Reynolds number of the oil at the end of the plate Re of length L is given by Equation (2.16) as

$$Re = \frac{\{\rho C_{FS}\}_{oil} L}{\{\mu\}_{oil}} = \frac{800 \times 5 \times 2}{0.9} = 8888.9$$

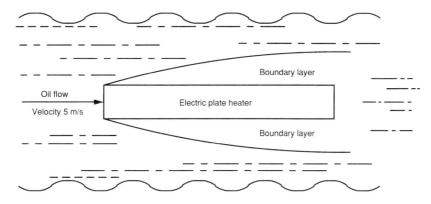

Fig. 9.6 *Example 9.1— as the oil flows past the heater plate, a boundary layer forms on both sides of the plate producing a friction force on the plate.*

As the Reynolds number is less than 10^5 at the end of the plate, this confirms that the fluid boundary layer will be laminar for its entire length as it passes over the heater. The velocity profile of the oil in the boundary layer may be assumed to be given by Equation (9.4) as

$$\frac{\{C\}_{oil}}{\{C_{FS}\}_{oil}} = \frac{2z}{\delta} - \frac{z^2}{\delta^2}$$

$$\therefore \quad \frac{\{dC\}_{oil}}{dz} = \frac{\{C_{FS}\}_{oil} \times 2}{\delta} - \frac{\{C_{FS}\}_{oil} \times 2z}{\delta^2}$$

At the surface, $z = 0$. Hence:

$$\frac{\{dC\}_{oil}}{dz} = \frac{\{C_{FS}\}_{oil} \times 2}{\delta} \quad \text{(at the surface)}$$

From Equation (2.17), the shear stress at the surface produced by the oil τ_s is given by

$$\tau_s = \frac{\{\mu \, dC\}_{oil}}{dz} = \frac{\{\mu C_{FS}\}_{oil} \times 2}{\delta}$$

The friction force on the plate caused by the oil F_F is given by Equation (9.6) as

$$F_F = b \int_0^L \tau_s \, dx = b \int_0^L \frac{\{\mu C_{FS}\}_{oil} \times 2 \, dx}{\delta}$$

For a laminar boundary layer, at a distance x down the plate, the boundary layer thickness of the oil δ is given by Equation (9.2) as

$$\delta = \frac{5.48x}{Re^{0.5}} = \frac{5.48x\{\mu\}_{oil}^{0.5}}{(\{\rho C_{FS}\}_{oil} x)^{0.5}}$$

Substituting in the equation for the friction force gives

$$F_F = b \int_0^L \frac{0.365\{\mu^{0.5} C_{FS}^{1.5} \rho^{0.5}\}_{oil} \, dx}{x^{0.5}}$$

$$\therefore \quad F_F = b \times 0.73 \{\mu^{0.5} C_{FS}^{1.5} \rho^{0.5}\}_{oil} L^{0.5}$$

236   EFFECTS OF A FLUID IN MOTION

$$\therefore \quad F_F = 1 \times 0.73 \times 0.9^{0.5} \times 5^{1.5} \times 800^{0.5} \times 2^{0.5} = 309.7 \text{ N}$$

As the oil flows past the plate on both sides, the total friction force on the heater plate due to the oil F'_F is

$$F'_F = 2F_F = 2 \times 309.7 = 619.4 \text{ N}$$

A restraining force in the opposite direction to the flow of the oil must be provided to hold the plate in place and to counteract the friction force. This is in addition to any support required to hold the plate up against the effects of gravity.

EXAMPLE 9.2

A train may be simply considered as being 100 m long, 3 m wide and 3.75 m high, as shown in Figure 9.7. As a first approximation, the friction force on the sides and top surface of the train may be taken as that acting on one side of a flat plate 10.5 m wide and 100 m long. If the train is moving through air of density 1.2 kg/m³ and dynamic viscosity 1.8×10^{-5} kg/m s with a velocity of 50 m/s, determine the power required to overcome the friction force opposing the motion of the train.

If the train is electric driven, with electricity costing $0.12/kWh, what is the cost incurred in overcoming the friction force on a 2-hour train journey?

Solution

In this example, the air is stationary and the train is moving. But equal and opposite forces would arise if the train was stationary and the air was moving. Consider this to be the case.

If the boundary layer that formed on the surface of the train due to air moving with a free stream velocity $\{C_{FS}\}_{air}$ became turbulent at a Reynolds number Re of 2×10^6 as given in Section 2.5.5, this would occur at a distance L' from the front of the train, and given by Equation (2.16) as

$$Re = \frac{\{\rho C_{FS}\}_{air} L_1}{\{\mu\}_{air}}$$

$$\therefore \quad L' = \frac{2 \times 10^6 \times 1.8 \times 10^{-5}}{1.2 \times 50} = 0.6 \text{ m}$$

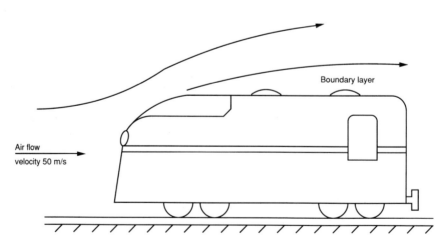

Fig. 9.7 Example 9.2—when a train moves through air, a boundary layer forms at the front of the train which results in a friction force opposing the motion of the train.

As the train is 100 m long, it is a reasonable assumption to consider the boundary layer to be turbulent right from the front of the train. The friction force F_F due to the air is given by Equation (9.6) as

$$F_F = b \int_0^L \tau_s \, dx$$

For a turbulent boundary layer, the shear stress at the surface τ_s is given by Equation (2.19) as

$$\tau_s = f \times 0.5\{\rho C_{FS}^2\}_{air}$$

And, again for turbulent flow, the relationship between the friction factor and Reynolds number may be assumed as that given in Equation (9.7), namely:

$$f = \frac{0.058}{Re^{0.2}} = \frac{0.058\{\mu^{0.2}\}_{air}}{(\{\rho C_{FS}\}_{air} x)^{0.2}}$$

Substituting for τ_s and f in the equation for the friction force F_F gives

$$F_F = b \int_0^L \frac{0.029\{\rho^{0.8} C_{FS}^{1.8} \mu^{0.2}\}_{air} \, dx}{x^{0.2}}$$

$$\therefore \quad F_F = b \times 0.036\{\rho^{0.8} C_{FS}^{1.8} \mu^{0.2}\}_{air} L^{0.8}$$

$$\therefore \quad F_F = 10.5 \times 0.036 \times 1.2^{0.8} \times 50^{1.8} \times (1.8 \times 10^{-5})^{0.2} \times 100^{0.8} = 2238.9 \text{ N}$$

The power required to overcome the friction force $\dot{W}$ is

$$\dot{W} = F_F\{C_{FS}\}_{air} = 2238.9 \times 50 = 112 \text{ kW}$$

For a 2-hour journey, the cost = $112 \times 2 \times 0.12 = $26.9.

Here $\dot{W}$ is only the power needed to overcome the friction forces that arise as the train passes through the stationary air. In addition, power is required to overcome friction between the wheels of the train and the track, and to accelerate the train from a stationary position to full speed.

EXAMPLE 9.3

If the flow of air of density 1.2 kg/m³ and dynamic viscosity 1.8×10^{-5} kg/m s over an aerodynamically shaped motor car, as in Figure 9.8, can be considered to be similar to that over one side of a flat plate of width 4 m (to include the sides of the car) and length 6 m, what is the maximum speed that the car can travel through the air for the boundary layer that forms on it to be laminar throughout? It is known that the boundary layer will remain laminar provided the Reynolds number of the flow is less than 1.5×10^6. What will be the thickness of the boundary layer at the rear of the car under these conditions?

When the car is travelling at 70 m/s, what is now the thickness of the boundary layer at the rear of the car?

Solution

As with Example 9.2, consider the air to be moving with a free stream velocity $\{C_{FS}\}_{air}$ over the stationary car.

The Reynolds number Re for the flow of air of velocity $\{C_{FS}\}_{air}$ over a plate of length L is defined by Equation (2.16) as

$$Re = \frac{\{\rho C_{FS}\}_{air} L}{\{\mu\}_{air}}$$

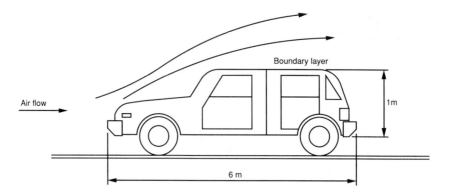

Fig. 9.8 *Example 9.3—when a motor car moves through air, a boundary layer forms at the front of the car which results in a friction force opposing the motion of the car.*

For a maximum Reynolds number of 1.5×10^6 in order to maintain a laminar boundary layer, the air would have to move at a free stream velocity $\{C_{FS}\}_{air}$ of

$$\{C_{FS}\}_{air} = \frac{Re\{\mu\}_{air}}{\{\rho\}_{air}L} = \frac{1.5 \times 10^6 \times 1.8 \times 10^{-5}}{1.2 \times 6} = 3.75 \text{ m/s}$$

Hence, the car must travel at a velocity of 3.75 m/s or less for the boundary layer to be laminar throughout!

The boundary layer thickness of the air δ at the rear of the car for laminar flow is given by Equation (9.2) as

$$\delta = \frac{5.48L}{Re^{0.5}} = \frac{5.48 \times 6}{(1.5 \times 10^6)^{0.5}} = 26.8 \text{ mm}$$

If the car is actually travelling at a velocity of 70 m/s, the initial laminar boundary layer will be of length L', given by Equation (2.16) as

$$Re = \frac{\{\rho C_{FS}\}_{air}L'}{\{\mu\}_{air}}$$

$$\therefore L' = \frac{Re\{\mu\}_{air}}{\{\rho C_{FS}\}_{air}} = \frac{1.5 \times 10^6 \times 1.8 \times 10^{-5}}{1.2 \times 70} = 0.32 \text{ m}$$

With the car being 6 m long, it is reasonable to assume that the boundary layer is turbulent from the front of the car. The Reynolds number of the air at the rear of the car is given by Equation (2.16) as

$$Re = \frac{\{\rho C_{FS}\}_{air}L}{\{\mu\}_{air}} = \frac{1.2 \times 70 \times 6}{1.8 \times 10^{-5}} = 28 \times 10^6$$

Therefore, the boundary layer thickness δ of the air at the rear of the car for turbulent flow is given by Equation (9.3) as

$$\delta = \frac{0.37L}{Re^{0.2}} \quad \therefore \quad \delta = \frac{0.37 \times 6}{(28 \times 10^6)^{0.2}} = 71.9 \text{ mm}$$

It can be seen that the boundary layer is not very thick, but it is within this region that the velocity changes from 0 m/s at the surface of the car to the 70 m/s of the free stream, the change in velocity giving rise to the friction forces which hinder the car's progress.

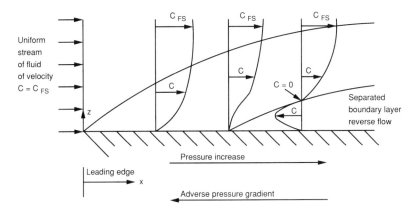

Fig. 9.9 *The boundary layer on a flat plate will separate from the surface of the plate in the presence of an adverse pressure gradient and the flow will be in the opposite direction.*

9.3 SEPARATION OF THE BOUNDARY LAYER ON A FLAT PLATE

Any boundary layer is considerably influenced by the presence of pressure gradients set up in the fluid. Separation of the boundary layer from the surface can occur when there is an Adverse Pressure Gradient, in other words, when there is a pressure force opposing the motion. Typically this occurs when the fluid is flowing in a diffuser or on a curved surface. Consider the following boundary layer on a flat plate in the presence of an adverse pressure gradient, as shown in Figure 9.9.

The fluid in the boundary layer close to the wall has only a small inertia as it is slowed down by the friction forces opposing its motion. If, in addition, there is a pressure gradient acting in the opposite direction to the flow, it is soon brought to rest. This causes the fluid close to the wall but further downstream actually to flow in the reverse direction. The boundary layer is said to have separated from the surface.

9.4 DRAG ON A BLUFF BODY

9.4.1 Ideal fluid theory

An ideal fluid has no viscosity, no boundary layers and no friction forces trying to slow it down. Consider the flow of an ideal fluid over an infinitely long circular cylinder, a bluff body, as shown in Figure 9.10.

The flow will obey the simplified form of Bernoulli's equation relating the total pressure p_0 and the static pressure p of the fluid, as in Equation (2.3), namely:

$$p + 0.5\rho C^2 = p_0 = \text{const}.$$

At positions W and Y, there are Stagnation Points where the fluid is brought to rest, and the pressure there is the total pressure p_0. Over the cylinder, the flow speeds up to reach maximum velocity. As the total pressure remains constant, the fluid static pressure must

240 EFFECTS OF A FLUID IN MOTION

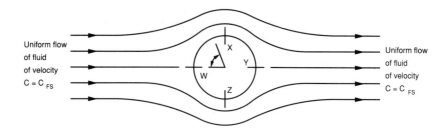

Fig. 9.10 The ideal flow of a fluid over an infinitely long cylinder obeys Bernoulli's equation, giving rise to a symmetrical flow pattern for the streamlines.

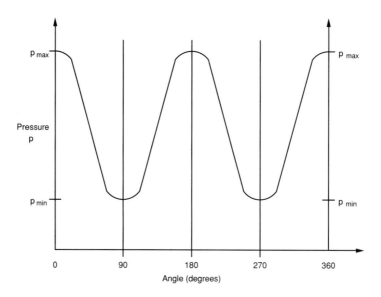

Fig. 9.11 The pressure distribution for the ideal flow of a fluid over an infinitely long cylinder is symmetrical, giving rise to no net forces on the cylinder.

decrease and reach a minimum at positions X and Z. The pressure distribution around the cylinder is symmetrical, as shown in Figure 9.11, and no resultant forces are predicted.

9.4.2 Real fluid

The flow over the front of the cylinder will be similar to that for an ideal fluid with a stagnation point at position W, and the fluid accelerating with static pressure decreasing to points X and Z. However, an important difference is that, because a real fluid has viscosity, a boundary layer forms from the front stagnation point. When the fluid is required to decelerate between points X and Y, and Z and Y (a region of increasing pressure), it has less inertia than it would have had (if there was no boundary layer present) with which to force its way forward against the adverse pressure gradient. The friction forces on the cylinder surface are also opposing the flow, and the net effect is that the fluid separates from the surface of the cylinder and forms a low-pressure wake which fills with eddies, as shown in Figure 9.12.

DRAG ON A BLUFF BODY   241

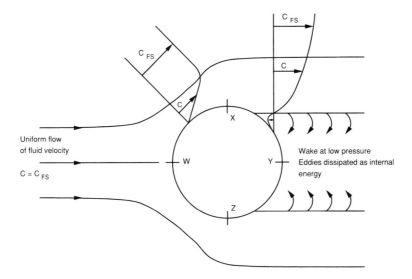

Fig. 9.12 *In the flow of a real fluid over an infinitely long cylinder, the boundary layer separates from the cylinder surface, giving rise to a low pressure wake downstream of the cylinder.*

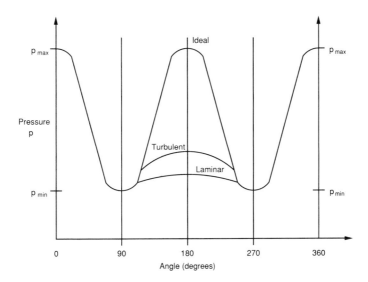

Fig. 9.13 *The pressure distribution of a real fluid over an infinitely long circular cylinder is no longer symmetrical, giving rise to a drag force on the cylinder in the direction of motion of the fluid.*

The streamlines are no longer symmetrical about the z-axis and this gives rise to a force in the direction of motion, known as a drag force and called Form Drag, due to the difference in pressure between the front and back of the cylinder. The pressure distribution is as shown in Figure 9.13.

Due to the change of flow pattern resulting from the formation of the wake, the position of minimum pressure may be altered, for example from 90° to 80°.

In addition to the form drag, there is also the friction force and hence a Friction Drag acting wherever the fluid is in contact with the cylinder surface (before the separation point). The Total Drag F_D is the sum of the form and friction drags. When the cylinder is moving through the fluid, both drags oppose the motion of the cylinder.

A turbulent boundary layer has faster-moving fluid near the cylinder surface and greater inertia with which to resist the adverse pressure gradient and the friction forces. The point of separation of the fluid from the cylinder surface is delayed, to angles typically approaching 130°. The pressure of the fluid is able to rise somewhat before separation and formation of the wake, and the difference in pressure between the front and back of the cylinder is not so great. Form drag is reduced (the size of the wake is correspondingly decreased and this may be taken as an indication of the form drag). The friction drag is increased because the fluid is in contact with the cylinder surface longer but the total drag is less.

When the cylinder is not infinitely long, there are some additional effects at the ends where the fluid pressure on the top tries to equalise itself with the fluid pressure on the bottom, and the total drag becomes more difficult to predict.

9.5 STREAMLINE BODIES

Consider the example of an aerofoil, as shown in Figure 9.14.

The tapered shape of the aerofoil past the position of minimum pressure reduces the rate of pressure rise, and boundary layer separation is delayed until almost the tail of the aerofoil. The pressure nearly increases to the stagnation pressure before separation, the wake is very small and form drag is negligible. Friction drag is increased because the fluid is in contact with greater aerofoil surface area, but the total drag is considerably reduced.

9.6 LIFT AND DRAG COEFFICIENTS

9.6.1 Bluff bodies

On a bluff body such as the circular cylinder, the total drag force F_D, which is in the direction of motion, can be expressed as a non-dimensional number called the Drag Coefficient C_D,

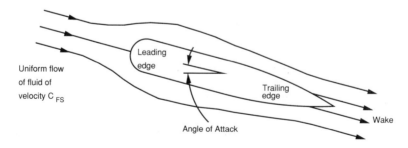

Fig. 9.14 *The lower rate of fluid pressure rise towards the trailing edge of an aerofoil delays separation of the boundary layer, leading to a small wake and minimal drag forces.*

as follows:

$$C_D = \frac{\text{Drag force per unit area}}{\text{Dynamic pressure}}$$

$$\therefore \quad C_D = \frac{F_D}{A \times 0.5\rho C^2} \tag{9.8}$$

where F_D is the total drag force on bluff body = form drag + friction drag force, A is the characteristic area, usually the cross-sectional area of the bluff body, ρ is the fluid density and C is the velocity of the bluff body, equivalent to the free stream velocity of the fluid.

Graphs of drag coefficient against Reynolds number can be plotted for various bluff body shapes, such as a circular cylinder or sphere. The characteristic dimension in the Reynolds number depends upon the bluff body shape. For a circular cylinder or sphere, it is the diameter.

EXAMPLE 9.4

During a famine relief operation, it becomes necessary to drop parcels of food and drugs from an aeroplane into an inaccessible area. Each parcel weighs 300 kg and the parachute which is to be used has a diameter of 8 m, as shown in Figure 9.15. A model test in a wind tunnel on half a hollow sphere, placed such as to simulate the airflow past a parachute, reveals that it has an approximately constant drag coefficient of 1.3 when the Reynolds number of the airflow is greater than 10^3. If the air density and air viscosity can be assumed constant at 1.2 kg/m^3 and 1.8×10^{-5} kg/m s respectively, at what velocity will each parcel hit the ground?

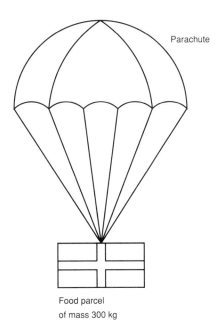

Fig. 9.15 *Example 9.4—a food parcel weighing 300 kg, supported by a parachute, descending in air after being dropped from an aeroplane.*

244 EFFECTS OF A FLUID IN MOTION

Solution

The cross-sectional area A_{xs} of the parachute of diameter D facing the flow is

$$A_{xs} = 0.25\pi D^2 = 0.25\pi \times 8^2 = 50.3 \text{ m}^2$$

The drag force F_D acting vertically downwards due to the weight of the parcel is

$$F_D = 300 \times 9.81 = 2943 \text{ N}$$

The drag coefficient C_D is given by Equation (9.8) as

$$C_D = \frac{F_D}{A \times 0.5\rho C^2}$$

Taking the characteristic area as being the cross-sectional area of the parachute, gives the parcel velocity C as

$$1.3 = \frac{2943}{0.5 \times 1.2 C^2 \times 50.3} \qquad \therefore \quad C = 8.7 \text{ m/s}$$

If the circumstances are reversed such that the air is moving past the stationary parcel, the Reynolds number of the air at a free stream velocity $\{C_{FS}\}_{air}$ of 8.7 m/s is given by Equation (2.15) as

$$Re = \frac{\{\rho C_{FS}\}_{air} D}{\{\mu\}_{air}} = \frac{1.2 \times 8.7 \times 8}{1.8 \times 10^{-5}} = 46.4 \times 10^5$$

This is well above the lower limit at which the drag coefficient becomes constant.

9.6.2 Streamline bodies

On a streamline body such as an aircraft wing, the design is such that the air has to travel over a greater distance on the top of the wing compared to the bottom, between the leading and trailing edges. The velocity of the fluid is greater over the top surface than the bottom, and the pressure on the top wing is less than on the bottom wing. This gives rise to a Lift Force F_L perpendicular to the direction of motion (Figure 9.16).

For an aircraft wing, there is both a drag force (due to the wake formation) and a lift force F_L, and there is both a drag coefficient and a lift coefficient. The drag coefficient is defined as above, and the lift coefficient C_L for a wing velocity C is defined as

$$C_L = \frac{\text{Lift force per unit area}}{\text{Dynamic pressure}}$$

$$\therefore \quad C_L = \frac{F_L}{A \times 0.5\rho C^2} \qquad (9.9)$$

The area A is usually taken to be the plan area of the aerofoil, which is the product of its span (the width) and its chord (the mean straight line distance from the leading to the trailing edge).

A well-designed aerofoil has a high lift and low drag. But as the angle of attack increases, the flow on the upper surface separates not at the trailing edge but more towards the leading edge, resulting in a large wake and big drag force. Eventually the point is reached (called Stalling) when the lift coefficient falls (Figure 9.17).

LIFT AND DRAG COEFFICIENTS

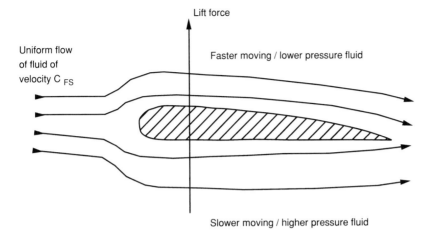

Fig. 9.16 *The fluid streamline pattern above an aerofoil is not the same as the streamline pattern below the aerofoil, giving rise to a differential fluid pressure and a lift force.*

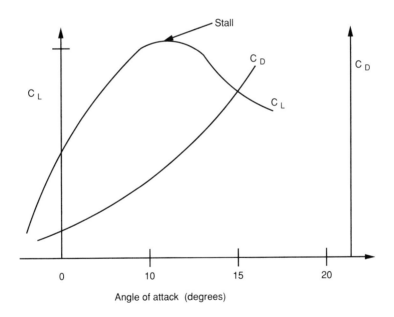

Fig. 9.17 *A plot of the lift coefficient and drag coefficient on an aerofoil versus the angle of attack.*

EXAMPLE 9.5

A small aeroplane has a mass of 10 000 kg and a total wing plan area of 35 m² (including the tail), as shown in Figure 9.18. What is the lift coefficient when the aeroplane is flying at 120 m/s in air of density 1.15 kg/m³?

If the ratio of the lift to drag coefficient is 12 to 1 at this operating condition, what engine power is required to provide the forward thrust?

246 EFFECTS OF A FLUID IN MOTION

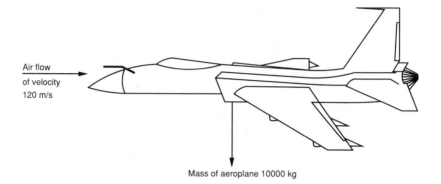

Fig. 9.18 *Example 9.5—determination of the lift coefficient and forward thrust of an aeroplane.*

Solution

The lift coefficient C_L is given by Equation (9.9) as

$$C_L = \frac{F_L}{A \times 0.5\rho C^2}$$

$$\therefore \quad C_L = \frac{10\,000 \times 9.81}{0.5 \times 1.15 \times 120^2 \times 35} = 0.34$$

The ratio of the lift coefficient C_L to the drag coefficient C_D is given by the ratio of Equation (9.9) to Equation (9.8) as

$$\frac{C_L}{C_D} = \frac{F_L}{F_D} \quad \therefore \quad F_D = \frac{10\,000 \times 9.81}{12} = 8175 \text{ N}$$

The power required to overcome this drag force $\dot{W}$ is

$$\dot{W} = 8175 \times 120 = 981 \text{ kW}$$

Once the aeroplane has been accelerated to a velocity of 120 m/s, if there were no friction forces opposing its motion because the air had no viscosity, no forward thrust would be necessary, provided it is maintained in flight. In practice, the more aerodynamic the shape of the aeroplane, the lower the drag force, resulting in a reduced power output being required from the engines.

9.7 FLOW MEASUREMENT

The mass flow rate of fluids $\dot{m}$ in kilograms per second can be measured exactly by two methods. For liquids, collect the liquid in a weighing tank and determine the mass of liquid accumulated in a given time (Figure 9.19).

The same method can be used for gases, but it takes some time to get a sensible reading. An alternative for gases is to collect the gas in a container of known volume, measure the pressure and temperature at a given time, measure the pressure and temperature again after a certain interval of time has elapsed, and calculate the mass flow rate from the equation of state (Figure 9.20).

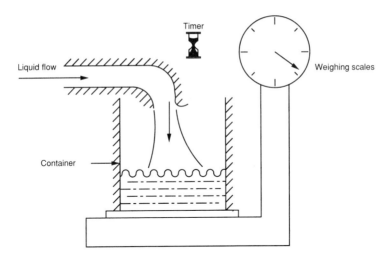

Fig. 9.19 *The measurement of the mass flow rate of liquids can be achieved by collecting the liquid in a weighing tank and determining the mass accumulated in a given time.*

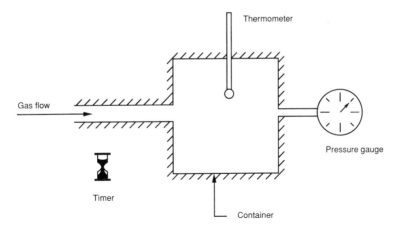

Fig. 9.20 *The measurement of the mass flow rate of gases can be achieved by collecting the gas in a container, measuring the change in the pressure and temperature in a given time, and using the equation of state.*

However, both these methods are awkward and imply that the flow of the fluid must be interrupted, which may not be convenient. Other methods of flow measurement, which can be calibrated against the absolute methods described above, are required.

9.8 FLOW-MEASURING DEVICES BASED UPON THE CONTINUITY AND BERNOULLI EQUATIONS

The continuity equation, described in Chapter 8, relates the mass flow rate of fluid $\dot{m}$ with its density ρ, velocity C and the cross-sectional area of the flow A_{xs}. For a steady mass flow rate of fluid between the inlet (state 1) and outlet (state 2) of a thermal system, Equation (8.3) can be written:

$$\dot{m}_1 = \rho_1(A_{xs})_1 C_1 = \dot{m}_2 = \rho_2(A_{xs})_2 C_2 = \dot{m} \tag{8.3}$$

The Bernoulli equation was also derived in Chapter 8. It is an ideal form of the SFEE assuming no losses due to friction and turbulence. For a steady mass flow rate of fluid between the inlet and outlet of a thermal system, Equation (8.20) can be written:

$$p_2 + 0.5\rho_1 C_1^2 + \rho_1 g z_1 = p_2 + 0.5\rho_2 C_2^2 + \rho_2 g z_2 \tag{8.20}$$

From Equation (8.3), the outlet velocity C_2 is

$$C_2 = \frac{\rho_1(A_{xs})_1 C_1}{\rho_2(A_{xs})_2}$$

Substituting for C_2 in Equation (8.20) gives

$$C_1 = \frac{(2g((p_1/\rho_1 g) - (p_2/\rho_2 g) + (z_1 - z_2)))^{0.5}}{(((\rho_1(A_{xs})_1)/(\rho_2(A_{xs})_2))^2 - 1)^{0.5}} \tag{9.10}$$

Substituting for C_1 from Equation (9.10) in Equation (8.3) gives

$$\dot{m} = \frac{\rho_1(A_{xs})_1 (2g((p_1/\rho_1 g) - (p_2/\rho_2 g) + (z_1 - z_2)))^{0.5}}{(((\rho_1(A_{xs})_1)/(\rho_2(A_{xs})_2))^2 - 1)^{0.5}} \tag{9.11}$$

Equation (9.11) links the mass flow rate of the fluid with:

1. Fluid densities which can be determined easily;
2. Cross-sectional areas which can be measured easily;
3. Vertical heights which can be measured easily;
4. A fluid pressure difference $(p_1 - p_2)$ which is caused by an area change and which again can be measured easily by a manometer or pressure gauges.

Equation (9.11) can be the basis of a measuring instrument to determine the mass flow rate of a fluid. What is required is a device in which the cross-sectional area of flow changes. This will result in a pressure change which can be measured. A pipe in which the diameter decreases would be a suitable device.

Equation (9.11) can be made more simple if the measuring device is horizontal, in which case $z_1 = z_2$. Also, if the fluid were incompressible, $\rho_1 = \rho_2 = \rho$. Even if the fluid were compressible, if the changes in pressure were small, the changes in density would also be small and it could be assumed that $\rho_1 = \rho_2$. Under these circumstances, Equation (9.11) reduces to

$$\dot{m} = \frac{\rho(A_{xs})_1 (2((p_1/\rho) - (p_2/\rho)))^{0.5}}{(((A_{xs})_1/(A_{xs})_2)^2 - 1)^{0.5}} \tag{9.12}$$

Because Equation (9.11) was derived using Bernoulli's equation which is ideal, the mass flow rate so determined will be the theoretical value $\dot{m}_{theor}$. The actual flow rate will be

FLOW-MEASURING DEVICES BASED UPON BERNOULLI EQUATIONS

different to the theoretical flow rate because of the effects of friction in a real fluid. The actual mass flow rate of the fluid $\dot{m}_{act}$ is

$$\dot{m}_{act} = C_d \dot{m}_{theor} \quad (9.13)$$

where $\dot{m}_{theor}$ is the value given by Equation (9.11).

Here C_d is called the Coefficient of Discharge and is a factor representing the losses incurred in a flow-measuring device based upon Equation (9.11). In other words, any meter designed using Equation (9.11) must be calibrated against an absolute measurement of the flow rate in order to determine the coefficient of discharge. Three such meters are commonly used and, because they are considered standard designs, it is agreed that they be made according to a National Standard. If the accepted rules are followed, the value of C_d can be obtained from standard graphs and tables. There is no need to calibrate each meter in order to find out its coefficient of discharge. The three common meters in use are the Venturi meter, the Nozzle meter and the Orifice Plate meter.

One disadvantage of Equation (9.11) is that the mass flow rate of the fluid $\dot{m}$ is proportional to the square root of the pressure difference. If a calibration graph of $\dot{m}_{act}$ against $(p_1 - p_2)$ is drawn, it is a curve. Errors can occur at either extreme of the curve where small changes in one value can produce large changes in the other value. A graph of $\dot{m}_{act}$ against $(p_1 - p_2)^{0.5}$ should be a straight line, which is better for calibration purposes except that it requires a calculation to be carried out, rather than a direct reading to be made.

EXAMPLE 9.6

Water of density 1000 kg/m³ is supplied to a natural gas-fired water heater through a 50 mm diameter horizontal pipe. Situated in the pipe is a nozzle section of outlet diameter 20 mm followed by a diffusing section. The objective of the nozzle is to create a pressure difference in the water between the nozzle inlet and outlet and to utilise this pressure difference to move a 25 mm diameter piston which is connected to a valve in the natural gas supply line to the water heater, as shown in Figure 9.21. The piston operates against a spring force and will only move when there is sufficient water flow rate through the pipe to the heater. In other words, the valve in the natural gas supply line will only open when the force on the piston due to the pressure difference across the nozzle is great enough to overcome the opposing spring force and move the piston. The pressure difference increases with water flow rate and the device acts as a safety mechanism by not allowing the natural gas to be supplied to the heater until there is a suitable flow of water. If the minimum flow rate of water specified by the natural gas heater manufacturer is 2.4 kg/s, what is the force on the piston assuming frictionless conditions?

Solution

Let the entry to the nozzle be state 1 and the exit from the nozzle state 2.

The cross-sectional area at entry to the nozzle $(A_{xs})_1$, where the diameter is D_1, is

$$(A_{xs})_1 = 0.25\pi D_1^2 = 0.25\pi \times 0.05^2 = 1.96 \times 10^{-3} \text{ m}^2$$

The cross-sectional area at exit of the nozzle $(A_{xs})_2$, where the diameter is D_2, is

$$(A_{xs})_2 = 0.25\pi D_2^2 = 0.25\pi \times 0.02^2 = 3.14 \times 10^{-4} \text{ m}^2$$

If the density of the incompressible water is $\{\rho\}_w$, the continuity equation (Equation (8.3)) gives the velocity of the water at entry to the nozzle $\{C_1\}_w$ as

$$\{C_1\}_w = \frac{\{\dot{m}\}_w}{\{\rho\}_w (A_{xs})_1} = \frac{2.4}{1000 \times 1.96 \times 10^{-3}} = 1.22 \text{ m/s}$$

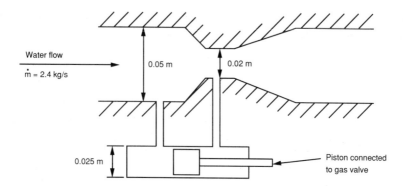

Fig. 9.21 *Example 9.6—a natural gas water heater control valve which only opens when there is sufficient water flow in the pipe.*

and the velocity of the water at exit from the nozzle $\{C_2\}_w$ as

$$\{C_2\}_w = \frac{\{\dot{m}\}_w}{\{\rho\}_w (A_{xs})_2} = \frac{2.4}{1000 \times 3.14 \times 10^{-4}} = 7.64 \text{ m/s}$$

Assuming frictionless flow of water through the horizontal nozzle, Bernoulli's equation (Equation (8.20)) gives the pressure difference of the water across the nozzle $\{p_1 - p_2\}_w$ as

$$\{p_1\}_w + 0.5\{\rho C_1^2\}_w = \{p_2\}_w + 0.5\{\rho C_2^2\}_w$$

$$\therefore \quad \{p_1 - p_2\}_w = 0.5 \times 1000 \times (7.64^2 - 1.22^2) = 28\,440.6 \text{ Pa}$$

The cross-sectional area of the piston A'_{xs} of diameter D' is

$$A'_{xs} = 0.25\pi D'^2 = 0.25\pi \times 0.025^2 = 4.91 \times 10^{-4} \text{ m}^2$$

Assuming the movement of the piston to be frictionless, the force F' on the piston is

$$F' = \{p_1 - p_2\}_w A'_{xs} = 28\,440.6 \times 4.91 \times 10^{-4} = 14 \text{ N}$$

In practice, the spring force is slightly different to this because the flow of the water through the restriction has been assumed to be frictionless, and the ideal Bernoulli's equation has been used to determine the pressure differential.

9.8.1 Venturi meter

A venturi meter is shown in Figure 9.22. It consists of a nozzle section in which the cross-sectional area of flow decreases, followed by a throat section, and then a diffuser section in which the cross-sectional area increases. If the flow rate of the fluid is steady, when the cross-sectional area decreases in the nozzle, the velocity of the fluid increases, as predicted by the continuity equation. Therefore, the pressure of the fluid decreases, according to the Bernoulli equation. A manometer, or pressure gauges, can be used to measure the pressure drop between the entrance to the nozzle section and the throat of the venturi, and substitution in Equation (9.11) will give the theoretical mass flow rate. Knowing the coefficient of discharge of the meter will enable the real mass flow rate to be determined.

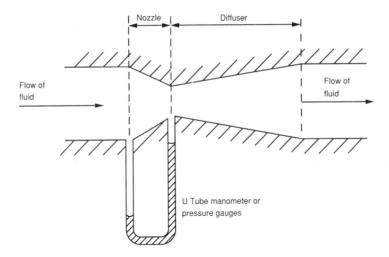

Fig. 9.22 *A venturi meter.*

The nozzle section and throat are followed by a diffusing section in which the cross-sectional area of flow increases, the velocity of the fluid decreases, and, therefore, the pressure of the fluid rises. Hence, the pressure drop in the nozzle is recovered in the diffuser. If there were no losses due to friction and turbulence in the venturi, the pressure of the fluid at exit of the diffuser would be the same as at inlet to the nozzle. However, due to losses, the pressure at exit is always less than at inlet, and some of the kinetic energy of the fluid is converted to internal energy in overcoming friction and turbulence. In order to reduce the losses to a minimum, the diffuser has a maximum included angle of 8 degrees, which is the angle the fluid would come out at if there were no physical boundary present. If the included angle of the physical boundary were bigger, the fluid would not follow its shape but, instead, would form vortices or eddies in the dead space, leading to an even greater conversion of kinetic energy to internal energy, resulting in minimal fluid pressure recovery at exit from the diffuser. This means that the diffuser section of a venturi is quite long which increases its expense. Also, the longer the diffuser, the greater are the surface friction forces but overall the venturi is a very efficient device.

Values of C_d of about 0.96 are typical. This implies minimal use of available energy to overcome friction between the inlet and the throat. In addition, most of the fluid pressure is recovered in the diffuser and the running costs of a venturi meter are small. Note that, in the nozzle, the fluid has no option but to follow the physical boundary because the shape is of decreasing area.

The length of the diffuser can be made shorter by having a smaller area ratio in the nozzle, although this may affect the accuracy of the pressure measurement. National Standards provide all the necessary instructions to design a venturi meter with a C_d of 0.96 for given flow rates and densities of fluid. The coefficient of discharge is not necessarily constant, but it does tend to have a fixed value as the Reynolds number of the flow increases.

Other advantages of the venturi meter are that it has no moving parts, it does not interrupt the flow, it can be made of a range of materials depending upon the fluid, it is easy to install in a pipeline and the pressure measurement is cheap. Once installed, a calibration chart of flow rate against pressure drop can be used. This is a curve because the flow rate

252 EFFECTS OF A FLUID IN MOTION

is proportional to the square root of the pressure difference between the entry to the nozzle and the throat, and this can lead to errors in measurement at both low and high fluid flow rates. In terms of disadvantages, as well as the length of the diffuser which can make the venturi slightly expensive to make, in order to obtain a C_d of 0.96, it must be installed in a pipeline with a given length of straight pipe both upstream and downstream of the meter. The actual length required is specified in the National Standards.

EXAMPLE 9.7

A venturi meter, placed horizontally in a pipeline, is being tested with water of density 1000 kg/m³ in the laboratory, as shown in Figure 9.23. The inlet diameter and throat diameter of the venturi are 0.026 and 0.016 m respectively. Six different water flow rates are examined and, at each flow, the pressure difference of the water between the inlet to the venturi and the throat is measured on a manometer, along with the time taken for the water to fill a container of volume 0.0068 m³. The results are shown in Table 9.1.

Determine the coefficient of discharge of the venturi meter for each water flow rate.

Solution

Let the inlet to the venturi meter be state 1 and the throat of the venturi meter be state 2.

The cross-sectional area at entry to the venturi $(A_{xs})_1$, where the diameter is D_1, is

$$(A_{xs})_1 = 0.25\pi D_1^2 = 0.25\pi \times 0.026^2 = 5.31 \times 10^{-4} \text{ m}^2$$

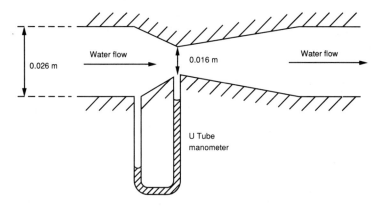

Fig. 9.23 Example 9.7—determining the coefficient of discharge of a Venturi meter.

Table 9.1 Example 9.7.

Test number	Pressure difference of water between entry and throat (mm water)	Time for water to fill 0.0068 m³ container (s)
1	236	15.1
2	228	15.4
3	207	16.1
4	184	17.1
5	110	22.1
6	67	28.3

Table 9.2 Example 9.7 — values of $\{\dot{m}_{theor}\}_w$, $\{\dot{m}_{act}\}_w$ and C_d.

Test number	$\{\dot{m}_{theor}\}_w$ (kg/s)	$\{\dot{m}_{act}\}_w$ (kg/s)	C_d
1	0.468	0.45	0.962
2	0.46	0.442	0.961
3	0.438	0.422	0.963
4	0.413	0.4	0.969
5	0.319	0.308	0.966
6	0.249	0.24	0.964

The cross-sectional area of the throat of the venturi $(A_{xs})_2$, where the diameter is D_2, is

$$(A_{xs})_2 = 0.25\pi D_2^2 = 0.25\pi \times 0.016^2 = 2.01 \times 10^{-4} \text{ m}^2$$

Equation (2.8) converts the pressure difference between the inlet and throat of the venturi from millimetre of water to pascals. As the water is incompressible and the venturi meter horizontal, Equation (9.10) allows the theoretical mass flow rate $\{\dot{m}_{theor}\}_w$ of the water through the venturi to be calculated. The actual mass flow rate $\{\dot{m}_{act}\}_w$ of the water is determined by dividing the volume of water collected in the container by the time taken to fill the container, and multiplying by the density of water. The coefficient of discharge C_d of the venturi is given by Equation (9.13) as the ratio of $\{\dot{m}_{act}\}_w$ to $\{\dot{m}_{theor}\}_w$. The results, therefore, are as shown in Table 9.2. The average value of C_d is 0.964. The result, that C_d is constant over a large flow range of water flow, is typical for most venturi meters.

9.8.2 Nozzle meter

This simply consists of the upstream part of the venturi meter. The coefficient of discharge is similar to a venturi meter but, of course, there is no diffusing section to recover the fluid pressure, so the running costs of the meter are increased. But it is cheaper to make and occupies less space than a venturi meter. The point of minimum pressure is a little downstream of the nozzle meter exit, and is known as the vena contracta. The positions of the static pressure tappings for pressure measurement of the fluid, at entry to the nozzle meter and at the vena contracta, are specified in the National Standards. A nozzle meter is shown in Figure 9.24.

EXAMPLE 9.8

A technician is asked to install a nozzle meter of throat diameter 50 mm in a horizontal length of pipeline of diameter 100 mm, as shown in Figure 9.25. The pipeline is used to transport water of density 1000 kg/m³. The technician is instructed to adjust the actual volume flow rate of water in the pipeline to be 0.0067 m³/s. There is a calibration chart available for the nozzle meter which shows that, at this flow rate, there should be a 52 mm difference in the height of the mercury contained in the U-tube manometer which is connected between the inlet and throat of the nozzle meter. If the density of the mercury is 13 600 kg/m³, what is the coefficient of discharge of the nozzle meter?

At a later date, the pipeline is rerouted such that the nozzle meter ends up in a vertical length with the water flow going downwards. If the coefficient of discharge remains the same and the vertical distance between the static pressure tapping at the nozzle meter inlet and nozzle meter throat is 0.1 m, what is the flow rate of the water if the difference of the height of the mercury in the U-tube manometer is again 52 mm?

254   EFFECTS OF A FLUID IN MOTION

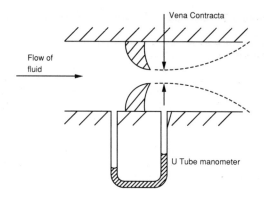

Fig. 9.24 A nozzle meter.

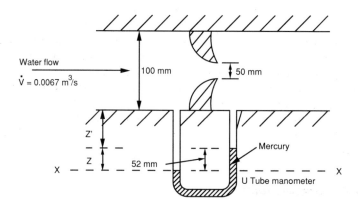

Fig. 9.25 Example 9.8—the nozzle meter in the horizontal length of pipe.

Solution

Let the inlet to the nozzle meter be state 1 and the outlet from the nozzle meter be state 2.
The cross-sectional area at the nozzle meter inlet $(A_{xs})_1$, where the diameter is D_1, is

$$(A_{xs})_1 = 0.25\pi D_1^2 = 0.25\pi \times 0.1^2 = 0.007\,85 \text{ m}^2$$

The cross-sectional area at the nozzle meter outlet $(A_{xs})_2$, where the diameter is D_2, is

$$(A_{xs})_2 = 0.25\pi D_2^2 = 0.25\pi \times 0.05^2 = 0.001\,96 \text{ m}^2$$

Assume that the static pressure of the water is uniform over the cross-section at entry and at the throat of the nozzle meter. Let the density of the incompressible water be $\{\rho\}_w$ and the density of the incompressible mercury be $\{\rho\}_m$. If the pressure of the mercury either side of the U-tube manometer is the same at level XX in Figure 9.25, equating the pressures gives

$$\{p_1\}_w + \{\rho\}_w g(Z' + Z) = \{p_2\}_w + \{\rho\}_w g Z' + \{\rho\}_m g Z$$

$$\therefore \{p_1 - p_2\}_w = Zg(\{\rho\}_m - \{\rho\}_w)$$

$$\therefore \{p_1 - p_2\}_w = 0.052g(13\,600 - 1000) = 6427.5 \text{ Pa}$$

Substituting in Equation (9.12) gives the theoretical mass flow rate of the water $\{\dot{m}_{theor}\}_w$ as

$$\{\dot{m}_{theor}\}_w = \frac{\{\rho\}_w (A_{xs})_1 (2(\{p_1/\rho\}_w - \{p_2/\rho\}_w))^{0.5}}{(((A_{xs})_1/(A_{xs})_2)^2 - 1)^{0.5}}$$

$$\therefore \{\dot{m}_{theor}\}_w = \frac{1000 \times 0.00785 \times (2 \times 10^{-3} \times 6427.5)^{0.5}}{((7.85/1.96)^2 - 1)^{0.5}}$$

$$\therefore \{\dot{m}_{theor}\}_w = 7.26 \text{ kg/s}$$

If the volume flow rate of the water of density 1000 kg/m³ is 0.0067 m³/s, the actual mass flow rate $\{\dot{m}_{act}\}_w$ is

$$\{\dot{m}_{act}\}_w = 1000 \times 0.0067 = 6.7 \text{ kg/s}$$

Hence, from Equation (9.13), the coefficient of discharge C_d is

$$C_d = \frac{\{\dot{m}_{act}\}_w}{\{\dot{m}_{theor}\}_w} = \frac{6.7}{7.257} = 0.92$$

When the nozzle meter is installed in a vertical length of pipe, as in Figure 9.26, equating the pressures in the mercury manometer at level XX gives

$$\{p'_1\}_w + \{\rho\}_w g (0.1 + Z'' + Z) = \{p'_2\}_w + \{\rho\}_w g Z'' + \{\rho\}_m g Z$$

$$\therefore \{p'_1 - p'_2\}_w = Zg(\{\rho\}_m - \{\rho\}_w) - \{\rho\}_w g \times 0.1$$

$$\therefore \{p'_1 - p'_2\}_w = 0.052g(13\,600 - 1000) - 1000g \times 0.1 = 5446.5 \text{ Pa}$$

Substituting in Equation (9.11) gives the theoretical mass flow rate of the water $\{\dot{m}'_{theor}\}_w$ as

$$\{\dot{m}'_{theor}\}_w = \frac{\{\rho\}_w (A_{xs})_1 (2g(\{p_1/\rho g\}_w - \{p_2/\rho g\}_w + (z_1 - z_2)))^{0.5}}{(((A_{xs})_1/(A_{xs})_2)^2 - 1)^{0.5}}$$

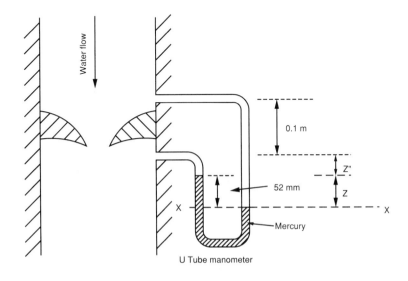

Fig. 9.26 Example 9.8—the nozzle meter in the vertical length of pipe.

$$\therefore \quad \{\dot{m}'_{\text{theor}}\}_w = \frac{1000 \times 0.00785 \times (2 \times 10^{-3} \times 5446.5 + 2g \times 0.1)^{0.5}}{((7.85/1.96)^2 - 1)^{0.5}}$$

$$\therefore \quad \{\dot{m}'_{\text{theor}}\}_w = 7.26 \text{ kg/s}$$

If the coefficient of discharge remains the same, Equation (9.13) gives the actual mass flow rate $\{\dot{m}'_{\text{act}}\}_w$ as

$$\{\dot{m}'_{\text{act}}\}_w = C_d \{\dot{m}'_{\text{theor}}\}_w = 0.92 \times 7.26 = 6.7 \text{ kg/s}$$

The answer is the same because the pressure term due to the vertical distance between the static pressure tappings is accounted for in the manometer.

9.8.3 Orifice plate meter

This is even cheaper than a nozzle meter to make, but the coefficient of discharge is around 0.6 and, with no diffusing section to recover the fluid pressure, the running costs are appreciably higher. An orifice plate meter is shown in Figure 9.27.

EXAMPLE 9.9

An orifice plate meter of throat diameter 50 mm is installed in a length of pipe of diameter 100 mm, which is at an angle of 45 degrees, as shown in Figure 9.28. The pipe is carrying oil of density 800 kg/m³ upwards in the pipe at the rate of 3 kg/s. The coefficient of discharge of the orifice plate meter is 0.6. Static pressure tappings, 0.075 m apart, are situated upstream and downstream of the orifice plate meter and are connected to a U-tube manometer which records the pressure difference. What is the difference in height of the mercury of density 13 600 kg/m³ in the manometer?

Solution
Let the entry to the orifice plate meter be state 1 and the exit from the orifice plate meter be state 2.
The cross-sectional area of the pipe at the orifice plate meter inlet $(A_{xs})_1$, where the diameter is D_1, is

$$(A_{xs})_1 = 0.25\pi D_1^2 = 0.25\pi \times 0.1^2 = 0.00785 \text{ m}^2$$

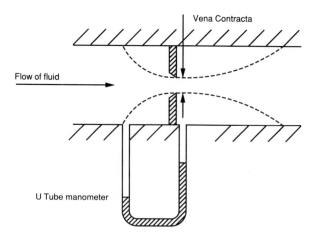

Fig. 9.27 *An orifice plate meter.*

FLOW-MEASURING DEVICES BASED UPON BERNOULLI EQUATIONS

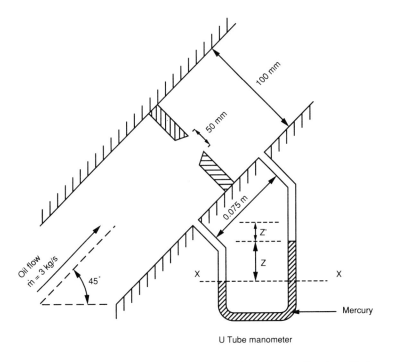

Fig. 9.28 *Example 9.9—the orifice plate meter in a pipe inclined at 45 degrees.*

The cross-sectional area at the orifice plate meter outlet $(A_{xs})_2$, where the diameter is D_2, is

$$(A_{xs})_2 = 0.25\pi D_2^2 = 0.25\pi \times 0.05^2 = 0.00196 \text{ m}^2$$

Substituting in Equation (9.13) gives the theoretical mass flow rate of the oil $\{\dot{m}_{\text{theor}}\}_{\text{oil}}$ as

$$\{\dot{m}_{\text{theor}}\}_{\text{oil}} = \frac{\{\dot{m}_{\text{act}}\}_{\text{oil}}}{C_d} = \frac{3}{0.6} = 5 \text{ kg/s}$$

If the density of the incompressible oil is $\{\rho\}_{\text{oil}}$, substituting in Equation (9.11) gives the pressure drop across the orifice plate $\{p_1 - p_2\}_{\text{oil}}$ as

$$\{\dot{m}_{\text{theor}}\}_{\text{oil}} = \frac{\{\rho\}_{\text{oil}}(A_{xs})_1(2g(\{p_1/\rho g\}_{\text{oil}} - \{p_2/\rho g\}_{\text{oil}} + (z_1 - z_2)))^{0.5}}{(((A_{xs})_1/(A_{xs})_2)^2 - 1)^{0.5}}$$

$$\therefore 5 = \frac{800 \times 0.00785 \times (2\{p_1 - p_2\}_{\text{oil}}/800 - 2g \times 0.075 \sin 45)^{0.5}}{((7.85/1.96)^2 - 1)^{0.5}}$$

$$\therefore \{p_1 - p_2\}_{\text{oil}} = 4230 \text{ Pa}$$

If the density of mercury is $\{\rho\}_m$, and assuming uniform pressures across the cross-section of the pipe at entry and exit of the orifice plate meter, equating the pressures either side of the U-tube manometer on level XX gives

$$\{p_1\}_{\text{oil}} + \{\rho\}_{\text{oil}}g(Z' + Z) = \{p_2\}_{\text{oil}} + \{\rho\}_{\text{oil}}g(Z' + 0.075 \sin 45) + \{\rho\}_m g Z$$

$$\therefore Zg(\{\rho\}_m - \{\rho\}_{\text{oil}}) = \{p_1 - p_2\}_{\text{oil}} - \{\rho\}_{\text{oil}}g \times 0.075 \sin 45$$

$$\therefore \quad Zg(13600 - 800) = 4230 - 800g \times 0.075 \sin 45$$

$$\therefore \quad Z = 30.4 \text{ mm}$$

It would be unfortunate if the orifice plate meter were situated in a pipe at an angle of 45° and it is not recommended practice. However, it may sometimes happen, and the solution shows how to cope with this situation when it does.

9.9 OTHER METHODS OF FLOW MEASUREMENT BASED UPON THE BERNOULLI EQUATION

9.9.1 Pitot static tube

This device consists of a tube which faces into the flow, thereby measuring the total pressure of the fluid p_0, and a tube at right angles to the flow which measures the static pressure p. The two tubes can be combined into one instrument, as shown in Figure 9.29, or the static pressure measurement can be made from a simple static pressure tapping in the side of the duct if the fluid is so enclosed.

The Pitot Static Tube only measures the velocity of the fluid C at a point, as follows:

A simplified form of Bernoulli's equation for a fluid of density ρ is Equation (2.3), namely:

$$p + 0.5\rho C^2 = p_0$$

$$\therefore \quad C = \frac{(2(p_0 - p))^{0.5}}{\rho^{0.5}} \quad (9.14)$$

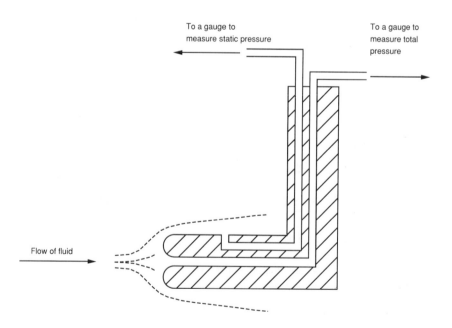

Fig. 9.29 *A pitot static tube.*

OTHER METHODS OF FLOW MEASUREMENT—THE BERNOULLI EQUATION

To find the mass flow rate, the fluid must be flowing in a duct of known diameter and a velocity traverse of the duct carried out. The cross-sectional area of the duct must be divided into a number of annuli of equal area, and the velocity of the fluid measured at the centre of each of the annuli. National Standards specify exactly how the traverse is to be performed. The total mass flow rate of fluid $\dot{m}$ is given by the continuity equation as the sum of each point velocity C multiplied by the cross-sectional area of the annulus in which it is measured A_{xs}, multiplied by the density of the fluid ρ in that annulus. Hence, Equation (8.3) gives

$$\dot{m} = \sum (\rho A_{xs} C) \quad \text{in each annulus} \tag{9.15}$$

The pitot static tube is a simple and cheap method of determining the mass flow rate of a fluid, if a bit laborious. It is also a true method of measurement, provided a few rules are followed such as to point the tube only in the direction of motion of the fluid. If the flow is highly swirling, a pitot static tube cannot be used. It can be utilised for calibrating other devices and is particularly beneficial in gas flows. The assumptions inherent in the use of the simplified Bernoulli equation are justified. However, some sort of velocity traverse is essential unless the fluid velocity is uniform, otherwise a false reading of the mass flow rate will be obtained. For example, the velocity profile of a fluid in a pipe in fully developed flow is parabolic and the centre-line velocity is greater than the average velocity, as explained in Section 9.1.

EXAMPLE 9.10

A horizontal pipe carrying water of density 1000 kg/m^3 tapers from a diameter of 150 mm to a diameter of 100 mm. A pitot static tube, connected to a mercury manometer, is used to measure the velocity of the water after the pipe contraction, as shown in Figure 9.30. There is a difference in the height of the mercury on the manometer of 650 mm. The mercury has a density of 13 600 kg/m^3. If the flow can be assumed frictionless through the contraction and the velocity profile over any cross-section of the duct uniform, what is the mass flow rate of the water?

If the flow rate were maintained constant but the positions of the total pressure and static pressure tubes were reversed, what would now be the reading on the mercury manometer?

Solution

Let the water conditions upstream of the contraction be state 1 and the water conditions downstream of the contraction be state 2.

Let the total pressure of the water, as measured by the total pressure probe, be $\{p_0\}_w$ and the static pressure of the water, as measured by the static pressure tapping, be $\{p\}_w$. If the density of the water is $\{\rho\}_w$ and the density of the mercury $\{\rho\}_m$, equating the pressures of the mercury either side of the U-tube manometer on level XX gives

$$\{p_0\}_w + \{\rho\}_w g(0.075 + Z' + Z) = \{p\}_w + \{\rho\}_w g(0.025 + Z') + \{\rho\}_m g Z$$

$$\therefore \quad \{p_0 - p\}_w = gZ(\{\rho\}_m - \{\rho\}_w) - \{\rho\}_w g \times 0.05$$

$$\therefore \quad \{p_0 - p\}_w = g \times 0.65 \times (13\,600 - 1000) - 1000g \times 0.05 = 79\,853.4 \text{ Pa}$$

If the flow through the contraction is frictionless, Equation (2.3) predicts that the total pressure of the water is constant. Therefore, it does not matter where the total pressure is measured and what the position of the total pressure probe is. The static pressure tapping is situated in the smaller diameter pipe after the contraction, so the pitot static tube, as installed in Figure 9.30, is recording the velocity of the water flow in the smaller diameter pipe.

260 EFFECTS OF A FLUID IN MOTION

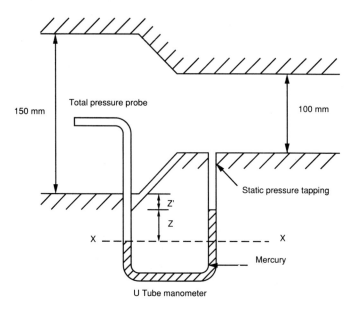

Fig. 9.30 *Example 9.10—a pitot static tube being used to determine the mass flow rate of water in a pipe after a contraction, assuming a uniform velocity profile.*

Substituting in Equation (9.14) gives the centre-line velocity of the water in the smaller diameter pipe $\{C_2\}_w$ as

$$\{C_2\}_w = \frac{(2\{p_0 - p\}_w)^{0.5}}{\{\rho\}_w^{0.5}} = \frac{(2 \times 79\,853.4)^{0.5}}{(1000)^{0.5}} = 12.6 \text{ m/s}$$

The cross-sectional area of the pipe at exit of the contraction $(A_{xs})_2$, where the diameter is D_2, is

$$(A_{xs})_2 = 0.25\pi D_2^2 = 0.25\pi \times 0.1^2 = 7.85 \times 10^{-3} \text{ m}^2$$

The continuity equation (Equation (8.3)) gives the mass flow rate of the water $\{\dot{m}\}_w$ as

$$\{\dot{m}\}_w = \{\rho C_2\}_w (A_{xs})_2 = 1000 \times 7.85 \times 10^{-3} \times 12.6 = 99 \text{ kg/s}$$

If the positions of the total pressure probe and the static pressure tapping are reversed, as in Figure 9.31, the pitot static tube will now be recording the velocity of the water in the larger diameter pipe.

The cross-sectional area of the pipe at inlet to the contraction $(A_{xs})_1$, where the diameter is D_1, is

$$(A_{xs})_1 = 0.25\pi D_1^2 = 0.25\pi \times 0.15^2 = 0.0177 \text{ m}^2$$

As the mass flow rate remains the same, the continuity equation (Equation (8.3)) gives the velocity of the water in the larger diameter pipe $\{C_1\}_w$ as

$$\{C_1\}_w = \frac{\{\dot{m}\}_w}{\{\rho\}_w (A_{xs})_1} = \frac{99}{1000 \times 0.0177} = 5.59 \text{ m/s}$$

Equation (9.14) allows the total pressure and static pressure difference of the water $\{p_0 - p\}_w$ to be determined as

$$\{p_0 - p\}_w = 0.5\{\rho C_1^2\}_w = 0.5 \times 1000 \times 5.59^2 = 15\,624 \text{ Pa}$$

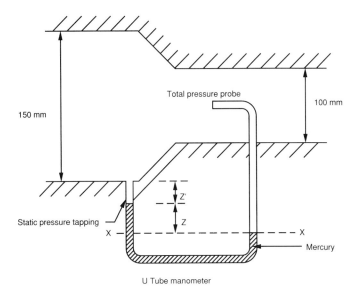

Fig. 9.31 *Example 9.10—the pitot static tube in the pipe before the contraction.*

Referring to Figure 9.31, equating the pressures of the mercury either side of the U-tube manometer on level XX gives

$$\{p\}_w + \{\rho\}_w g Z' + \{\rho\}_m g Z = \{p_0\}_w + \{\rho\}_w g(0.075 + Z' + Z)$$

$$\therefore \quad 13\,600 g Z = 15\,624 + 1000 g(0.075 + Z)$$

$$\therefore \quad Z = 0.132 \text{ m}$$

In large-diameter water pipes where a pitot static tube is employed, for example in those leading to some cooling towers, the tube will have to be considerably strengthened to prevent it from bending. This may result in some blockage to the flow which must be accounted for.

9.9.2 Entry nozzle

This device can be positioned at the entrance to a pipe or duct. It simply consists of a smooth nozzle type entry with a static pressure tapping situated at a certain position downstream, which records the static pressure of the fluid there, relative to the atmospheric pressure. It is shown in Figure 9.32.

Equation (9.14) is valid for the Entry nozzle. The total pressure p_0 is the pressure of the fluid in the reservoir before the nozzle, the atmospheric pressure, and the static pressure of the fluid is that measured at the static pressure tapping. The difference in pressure can be obtained by connecting a manometer, which is open to the atmosphere, to the static pressure tapping. Equation (9.14) determines the average velocity of the fluid in the duct C. If the fluid has a density ρ and the cross-sectional area of the duct at the location of the static pressure tapping is A_{xs}, the continuity equation (Equation (8.3)) gives the mass flow rate of fluid $\dot{m}$ as

$$\dot{m} = \rho A_{xs} C$$

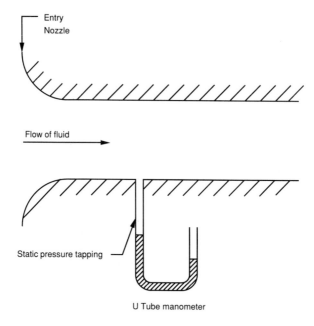

Fig. 9.32 *An entry nozzle.*

In a well-designed entry nozzle, the friction losses at entry are negligible. Otherwise, a C_d of about 0.96 is typical.

EXAMPLE 9.11

A fan draws air, assumed to have a constant density of 1.2 kg/m³, into a 75 mm diameter pipe through an entry nozzle. A manometer records a pressure difference of 55 mm water across the entry nozzle. Some distance downstream of the entry nozzle, where the flow is fully developed, a pitot static tube is used to traverse the airflow across the pipe diameter, as shown in Figure 9.33. The cross-section of the pipe is divided into five annuli of equal area of 8.835×10^{-4} m² and the difference between the total and static pressure of the air $\{p_0 - p\}_{air}$ at the centre of each annulus is recorded from a manometer. The readings are as shown in Table 9.3.

Determine the mass flow rate of air as predicted by the entry nozzle assuming frictionless flow, given that the density of water is 1000 kg/m³.
Determine the actual mass flow rate of air from the pitot static tube traverse.
Determine the C_d value for the entry nozzle.

Table 9.3 *Example 9.11.*

Annulus number	1	2	3	4	5
Radius at which reading taken (mm)	0	20.24	26.38	31.29	35.52
Annulus area (m² × 10⁻⁴)	8.835	8.835	8.835	8.835	8.835
Pressure difference $\{p_0 - p\}_{air}$ (Pa)	745.56	597.28	526.78	445.94	321.73

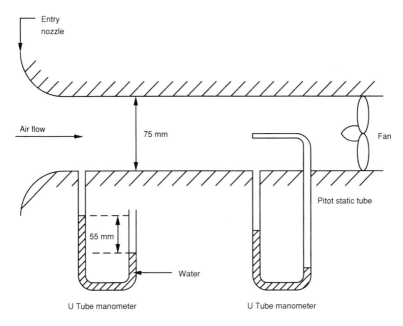

Fig. 9.33 *Example 9.11 — an entry nozzle and a pitot static tube used for determining the mass flow rate of air in a pipe.*

Solution

For the entry nozzle, the theoretical pressure difference $(p_0 - p)_{theor}$ can be converted from millimetres of water into pascals using Equation (2.8). If the density of water is $\{\rho\}_w$:

$$\{(p_0 - p)_{theor}\}_{air} = 55 \times 10^{-3} g \{\rho\}_w = 55 \times 10^{-3} \, g \times 10^3 = 539.55 \text{ Pa}$$

If the density of the air is $\{\rho\}_{air}$, the average theoretical velocity of the airflow predicted by the entry nozzle $\{C_{theor}\}_{air}$ can be determined from Equation (9.14) as follows:

$$\{C_{theor}\}_{air} = \frac{(2\{(p_0 - p)_{theor}\}_{air})^{0.5}}{\{\rho^{0.5}\}_{air}}$$

$$\therefore \{C_{theor}\}_{air} = \frac{(2 \times 539.55)^{0.5}}{1.2^{0.5}} = 29.99 \text{ m/s}$$

The cross-sectional area of the pipe A_{xs}, for a pipe of diameter D, is

$$A_{xs} = 0.25\pi D^2 = 0.25\pi \times 0.075^2 = 4.418 \times 10^{-3} \text{ m}^2$$

The theoretical mass flow rate of the air predicted by the entry nozzle $\{\dot{m}_{theor}\}_{air}$ can be found from the continuity equation (Equation (8.3)) as

$$\{\dot{m}_{theor}\}_{air} = \{\rho C_{theor}\}_{air} A_{xs} = 1.2 \times 29.99 \times 4.418 \times 10^{-3} = 0.159 \text{ kg/s}$$

For the pitot static tube traverse, the air velocity at any point $\{C_{act}\}_{air}$ can be determined from Equation (9.14). Knowing the cross-sectional annulus area A'_{xs} and density of air $\{\rho\}_{air}$, the actual mass flow rate of air in each annulus $\{\dot{m}'_{act}\}_{air}$ can be found from the Continuity equation (Equation (8.3)). The results are as shown in Table 9.4.

Table 9.4 *Example 9.11—results.*

Annulus number	1	2	3	4	5
Velocity of air $\{C_{act}\}_{air}$ (m/s)	35.25	31.55	29.63	27.26	23.16
Annulus area A'_{xs} (m² × 10⁻⁴)	8.835	8.835	8.835	8.835	8.835
Annulus mass flow rate $\{\dot{m}'_{act}\}_{air}$ (kg/s)	0.0374	0.0334	0.0314	0.0289	0.0246

The actual mass flow rate of the air in the pipe $\{\dot{m}_{act}\}_{air}$, as predicted from the pitot static tube traverse, is the sum of the mass flow rates in each annulus. Thus:

$$\{\dot{m}_{act}\}_{air} = \sum \{\dot{m}'_{act}\}_{air} = 0.156 \text{ kg/s}$$

The C_d value for the entry nozzle is given by Equation (9.13) as

$$C_d = \frac{\{\dot{m}_{act}\}_{air}}{\{\dot{m}_{theor}\}_{air}} = \frac{0.156}{0.159} = 0.98$$

C_d values for entry nozzles depend very much upon their shape but they should approach $C_d = 1$ if the profile is smooth and aerodynamic.

9.10 OTHER METHODS OF FLOW MEASUREMENT

9.10.1 Turbine meter

Any physical phenomena which varies in a regular manner with fluid flow rate can be made into a flow meter and correspondingly calibrated. A Turbine Meter relies upon the fact that the turbine will turn faster the greater the mass flow rate. It is shown in Figure 9.34. The number of revolutions of the turbine in a certain time can be calibrated to give a reading of velocity, which is the usual output when the fluid is air. Alternatively, the meter can be calibrated to read the volume of a fluid. To obtain a flow rate, the output change must be recorded over a given interval of time. This is particularly used by water authorities to charge for water over a period of a few months, when the instantaneous flow rate is not required.

9.10.2 Rotameter

This is another simple but accurate device, shown in Figure 9.35. It consists of a tapered tube through which the fluid flows upwards into the increasing area. The float varies its position according to the pressure drop across it, which depends upon the flow rate. By tapering the tube, a linear scale of flow rate can be obtained. The meter must be calibrated against a standard measurement. Each meter is usually only suitable for one fluid under given conditions.

OTHER METHODS OF FLOW MEASUREMENT   265

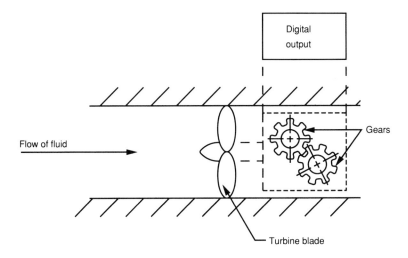

Fig. 9.34 *A turbine meter.*

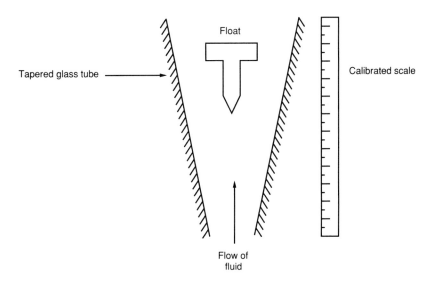

Fig. 9.35 *A rotameter.*

9.10.3 Hot wire anemometer

These are devices which, generally, consist of a piece of wire or metal, heated by an electric current to a certain constant temperature. When placed in a fluid stream, the metal is cooled and the temperature decreases. The amount by which the current has to be increased in order to return the temperature to its original constant value is proportional to the velocity of the fluid stream and can be calibrated accordingly. The velocity can be converted to a mass flow rate through the continuity equation. Such anemometers are most commonly used in gas flows, and can be quite delicate and expensive.

9.11 APPLICATION IN THERMAL SYSTEMS

Apart from the measurement of the mass flow rate of a fluid, which is essential in any thermal system, this chapter has described some of the fundamental characteristics of a fluid which determine its behaviour. The consequences, as far as a thermal system is concerned, rest in the nature of the forces that arise and their effect upon the fluid properties as the fluid flows through the system, either in a component or in the interconnecting pipework. Chapter 10 demonstrates how the magnitude of the forces can be predicted, and Chapter 11 evaluates the changes that occur in the fluid property values as a result of these forces being in action.

10 The Steady Flow Momentum Equation

Whenever a fluid is flowing either within a solid boundary, as in a pipe, or over a solid boundary, as in the external flow over bodies such as cars or aeroplane wings or the blading in fluid machinery, there are a number of forces acting which need to be accounted for. The Steady Flow Momentum Equation (SFME) is an attempt to do just that. Only a very simple approach is adopted, but it is sufficient for a first analysis in many flow situations.

Some forces that arise in fluid flow have already been examined, for example, those due to the properties mass, pressure and viscosity. The SFME considers all the forces that may be exerted by the fluid on a solid object as a consequence of the fluid flowing either within or over the surfaces of the object.

The SFME developed is restricted to steady flow conditions only, as defined in Chapter 8. It is also limited to a one-dimensional approach in that only the average property values such as the velocity and pressure of the fluid at any given state are considered although, of course, the fluid may change direction and so introduce another dimension into the analysis.

10.1 FORCES DUE TO THE FLOW OF A FLUID

The SFME results from a consideration of Newton's laws of motion as applied to the fluid. The laws may be stated as follows:

> A body will remain at rest or in a condition of uniform motion unless acted upon by a force.
>
> The magnitude of the force is proportional to the rate of change of momentum of the body in the direction of the force.
>
> There is an equal and opposite reaction to the force.

In the case of a flowing fluid, it is the force of the fluid upon the solid object which it comes into contact with that is important. Therefore, consider Figure 10.1, in which a fluid passes over a solid object which is blocking its path. The fluid flow is identified by drawing a few streamlines. The fluid will have to undergo a change in direction and velocity in order to get past the solid object. It will experience a change in momentum which will be related to the force acting upon it, caused by the object. The reaction exerted by the fluid on the object will be equal and opposite to the force of the object on the fluid. By examining what happens to the fluid properties, it is possible to determine the size of the force.

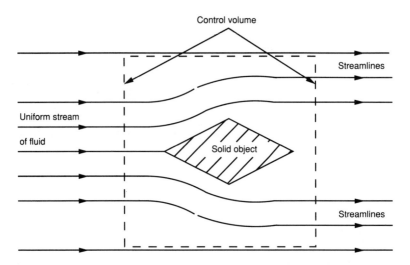

Fig. 10.1 *The control volume for the flow of a fluid past a solid object.*

For the purposes of the analysis, it is necessary to identify a boundary, usually referred to as a Control Volume, associated with the flow. This can be an imaginary, or sometimes real, boundary in which all the relevant forces apply. The beauty of the SFME is that it only considers the forces acting on the boundary, not those apparent during the passage of the fluid either over or through the object. This is because any force within the control volume will be balanced by an equal and opposite reaction. Therefore, the equation is a general one and does not require a knowledge of the exact flow path taken by the fluid. In addition, the fluid may be compressible or incompressible, and friction effects may be included or not, depending upon the application. For example, take the case of a fluid flowing over a curved vane. If the fluid is considered ideal, there are no friction forces exerted upon the fluid as it passes over the surface of the vane, so the outlet velocity of the fluid will be the same as the inlet velocity, albeit in a different direction. If the fluid is not ideal and friction exerts a retarding force, the fluid outlet velocity will be less than the inlet velocity. A knowledge of the velocities at inlet and outlet will account for the effects of the friction force. Again, if the fluid is incompressible, its density will be constant. Otherwise, a knowledge of the density at entry and exit to the control volume will account for forces due to compressibility.

10.1.1 Momentum forces

With reference to Figure 10.2, applying Newton's laws of motion to the control volume at entry and exit of the fluid, identified as state 1 and state 2 respectively, the force exerted by the object is proportional to the rate of change of momentum of the fluid. The rate of change of momentum is the mass of fluid multiplied by its acceleration. However, it is easier to consider the mass flow rate of fluid when the fluid is flowing from one place to another, in which case the rate of change of momentum becomes the mass flow rate multiplied by the change of velocity of the fluid. In other words, in solid body mechanics, Newton's second law of motion is

$$\text{Force} = (\text{mass}) \times (\text{acceleration})$$

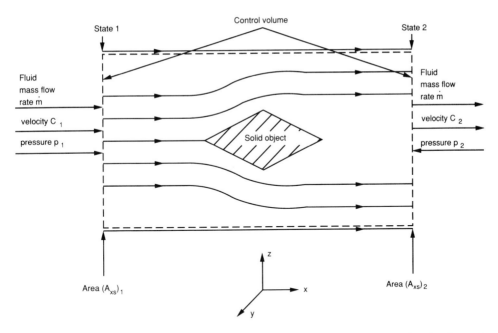

Fig. 10.2 *The fluid properties applicable at the control volume boundary for the flow of a fluid past a solid object.*

In thermofluids, this becomes

$$\text{Force} = (\text{mass flow rate}) \times (\text{change of velocity})$$

For steady flow conditions, the mass flow rate of fluid $\dot{m}$ entering the control volume equals the mass flow rate of fluid leaving the control volume. If the boundaries AB and CD of the control volume are streamlines, there will be no flow across them, so the change in velocity of fluid to be considered is the average velocity of the fluid leaving boundary BC minus the average velocity of the fluid entering boundary DA. However, velocity, like force and momentum, is a vector quantity, and its direction must be taken into account. In Figure 10.2, the velocity at entry C_1 is in the same direction as the velocity at exit C_2. It is possible for a force to be generated even if the inlet and outlet velocities are the same, provided they are in different directions. This is the situation when a fluid passes over a curved vane in rotodynamic machinery.

If the x-, y- and z-directions are identified, the force due to the rate of change of momentum in the x-direction F_{Mx} is given by

$$F_{Mx} = (\dot{m}(C_2 - C_1))_x$$

where the subscript x on the right-hand side of the equation implies that only the components of the fluid velocity at inlet and outlet in the x-direction should be considered. This is the force exerted by the object upon the fluid. From Newton's third law of motion, the force exerted by the fluid on the object is equal and opposite, and given by

$$F_{Mx} = (\dot{m}(C_1 - C_2))_x \qquad (10.1)$$

The momentum force exerted by the fluid on the object in the y- and z-directions, F_{My} and F_{Mz} respectively, is given by

$$F_{My} = (\dot{m}(C_1 - C_2))_y \tag{10.2}$$

$$F_{Mz} = (\dot{m}(C_1 - C_2))_z \tag{10.3}$$

The subscripts y and z on the right-hand side of the equation imply that only those components of the fluid velocity at entry and exit in the y- and z-directions should be considered in Equations (10.2) and (10.3) respectively. In fact, in Figure 10.2, both C_1 and C_2 are drawn in the x-direction only. Therefore, both F_{My} and F_{Mz} will be zero in this case.

10.1.2 Pressure forces

The forces due to the rate of change of momentum are not the only forces acting. If the static pressure of the fluid is not the same at entry to the control volume as at exit, a pressure force will result. Again, this will be directional and it is necessary to consider the x-, y- and z-components. As stated in Chapter 2, pressures are drawn to hold the fluid in place, in the control volume in this case, as shown in Figure 10.2. The force due to pressure is effectively the force exerted on the fluid in the control volume by the fluid outside the control volume. Boundaries AB and CD are streamlines with no flow across them. Any pressure forces acting on that part of the boundary are perpendicular to the flow direction and, when resolved in the direction of flow, become zero. The same will apply to the boundaries of the control volume in the y-direction, not shown in Figure 10.2. But assuming that the average static pressure p_1 on boundary DA of cross-sectional area $(A_{xs})_1$ is not the same as the average static pressure p_2 on boundary BC of cross sectional area $(A_{xs})_2$, the force due to pressure exerted by the fluid on the object in the x-direction F_{Px} is

$$F_{Px} = \sum (p(A_{xs}))_x$$

$$\therefore \quad F_{Px} = (p_1(A_{xs})_1 - p_2(A_{xs})_2)_x \tag{10.4}$$

Here p_2 is negative because its direction is opposite to the x-direction. Again, the subscript x means that only the components of p_1 and p_2 in the x-direction should be used.

In the y- and z-directions, the force exerted by the fluid on the object due to the fluid pressure, F_{Py} and F_{Pz}, respectively is

$$F_{Py} = \sum (p(A_{xs}))_y \tag{10.5}$$

$$F_{Pz} = \sum (p(A_{xs}))_z \tag{10.6}$$

In Figure 10.2, both p_1 and p_2 are entirely in the x-direction, so there will be no force due to pressure in the y- and z-directions, and F_{Py} and F_{Pz} will be zero.

The force due to pressure differences in the fluid needs careful consideration. In some applications, such as when a jet of fluid is striking a plate in the atmosphere, the jet may be considered to be at atmospheric pressure everywhere inside the control volume. As the pressure of the fluid outside the control volume is also atmospheric, the pressure force exerted is zero. In fluid machinery, such as the rotary turbine of Figure 8.20 where the fluid is not exposed to the atmosphere, although the static pressure varies in the direction of motion of the fluid, the force is applied in the rotary direction. At any given radius, the static pressure will be the same and its contribution to the total force exerted on the shaft will be zero. In flow around a bend in a pipe, it is the changes in gauge static pressure that must be considered. This is because the atmosphere is applying a pressure to the boundary, and it is the relative value of the static pressure that gives rise to the force. All these points will be examined further in the examples that follow.

10.1.3 Body forces

In many applications, forces exerted by the fluid on the object due to the weight of the fluid, called Body Forces, are relevant. These are simply given by the mass of fluid in the control volume m, multiplied by the acceleration due to gravity g. Thus, in the x-, y- and z-directions of Figure 10.2, the body forces F_{Bx}, F_{By} and F_{Bz} respectively, are

$$F_{Bx} = (mg)_x \tag{10.7}$$

$$F_{By} = (mg)_y \tag{10.8}$$

$$F_{Bz} = (mg)_z \tag{10.9}$$

Of course, in Figure 10.2, the weight of fluid in the control volume is acting vertically downwards. Therefore, F_{Bx} and F_{By} are zero and F_{Bz} will be negative because its direction is in the negative z-direction. In many applications, such as the flow of a fluid in a horizontal pipe, the body forces only contribute to the resultant force on the pipe in the vertical direction.

The total force exerted by the fluid on the object in each direction, F_x, F_y and F_z respectively, is given by

$$F_x = F_{Mx} + F_{Px} + F_{Bx} \tag{10.10}$$

$$F_y = F_{My} + F_{Py} + F_{By} \tag{10.11}$$

$$F_z = F_{Mz} + F_{Pz} + F_{Bz} \tag{10.12}$$

The resultant force F is

$$F = (F_x^2 + F_y^2 + F_z^2)^{0.5} \tag{10.13}$$

Equation (10.13) is, in effect, the SFME, but it can only be used after Equations (10.1)–(10.12) have been applied to any given flow under consideration.

10.2 THE KINETIC ENERGY AND MOMENTUM CORRECTION FACTOR

In the derivation of the SFME, the average velocity of the fluid at entry and exit to the control volume is used. In practice, the velocity in any duct or pipe will vary from a maximum, usually at the centre, to zero at the surface. Even in the atmosphere, the velocity profile may not be uniform. The kinetic energy and momentum correction factors may be applied to account for this variation in velocity, to kinetic energy and momentum terms

272 THE STEADY FLOW MOMENTUM EQUATION

respectively. The kinetic energy correction factor is usually given the symbol α, and the momentum correction factor β. Values of α and β can be found in standard reference books and, whilst they can be of some consequence, in reasonably streamline flows they are only of the order of a few per cent. They will not be considered further.

10.3 APPLICATIONS OF THE SFME

10.3.1 Jet striking a perpendicular flat plate

Consider initially the simple case of a horizontal jet of incompressible fluid of density ρ striking a stationary circular flat plate held perpendicular to the flow at its centre. The fluid exits from a nozzle of diameter D into the atmosphere with a velocity C and strikes the perpendicular stationary plate held a nominal distance away from it, as shown in Figure 10.3.

When the fluid strikes the plate, it is dispersed outwards in the radial direction. Because the flow over the plate is symmetrical, it is only necessary to look at the flow in two dimensions. If Figure 10.3 is a plan view, consider the x- and y-directions. If the initial mass flow rate of fluid from the nozzle is $\dot{m}$, after striking the plate half the mass flow of fluid will leave in the positive y-direction, and the other half will leave in the negative y-direction. In the first instance, it is a reasonable assumption that the fluid will leave the plate at the same velocity C with which it leaves the nozzle. The control volume is drawn to include all the forces that may be acting.

Consider first the momentum forces. In the x-direction, Equation (10.1) is applicable. The initial velocity of the fluid in the x-direction is C. At exit, all the fluid is travelling in

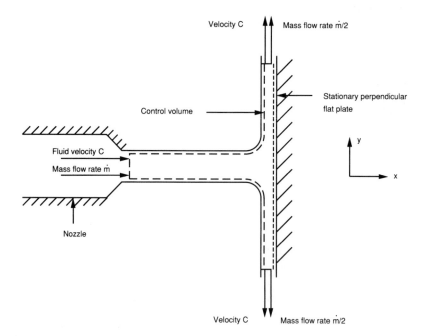

Fig. 10.3 Two-dimensional plan view of the control volume for a horizontal jet of fluid striking a perpendicular stationary plate.

the y-direction, so the exit velocity in the x-direction will be C cos 90, which is zero. In this particular situation, the mass flow rate of fluid entering the control volume is $\dot{m}$, but it is divided in half by the plate and leaves in two places which must be considered separately. Equation (10.1) becomes

$$F_{Mx} = \dot{m}C - (0.5\dot{m}C \text{ cos } 90 + 0.5\dot{m}C \text{ cos } 90) = \dot{m}C$$

In the y-direction, Equation (10.2) is applicable. At entry to the control volume, the fluid velocity is entirely in the x-direction, so its component in the y-direction is C cos 90, zero again. At exit, half the fluid comes out in the positive y-direction, but the other half comes out in the negative y-direction. Equation (10.2) becomes

$$F_{My} = \dot{m}C \text{ cos } 90 - (0.5\dot{m}C + 0.5\dot{m}(-C)) = 0$$

In other words, there is no momentum force in the y-direction because the flow is symmetrical.

Now consider the pressure forces. The fluid leaves the nozzle into the atmosphere. As it is in the form of a compact jet, it is a reasonable assumption that its pressure throughout remains approximately constant and equal to atmospheric. As the atmosphere is at the same pressure, the force due to pressure on the control volume will be zero. Thus:

$$F_{Px} = F_{Py} = 0$$

Finally, consider the body forces. The weight of the fluid in the control volume will be acting vertically downwards in the negative z-direction. But only the x- and y-directions are being considered in Figure 10.3, which is a plan view of the jet. Therefore, in this application, the body forces are zero, and so

$$F_{Bx} = F_{By} = 0$$

Equations (10.10) and (10.11) give the net forces F_x and F_y in the x- and y-directions respectively as

$$F_x = F_{Mx} + F_{Px} + F_{Bx} = \dot{m}C + 0 + 0 = \dot{m}C$$
$$F_y = F_{My} + F_{Py} + F_{By} = 0 + 0 + 0 = 0$$

The resultant force F from Equation (10.13) is

$$F = (F_x^2 + F_y^2)^{0.5} = \dot{m}C \qquad (10.14)$$

and it is in the x-direction. The mass flow rate of the fluid can be determined from the continuity equation applied to the nozzle (Equation (8.3)).

The above analysis is somewhat simplified of course. When the fluid hits the plate, it is unlikely to flow smoothly out in the radial direction. Instead, some fluid will bounce back off the plate into the jet, disturbing the flow pattern. As the fluid flows out over the plate, it will experience a retarding friction force. Therefore, the velocity of the fluid at exit from the control volume will be less than the velocity of the fluid leaving the nozzle. Again, as the fluid passes over the plate, the flow will be three dimensional and body forces may affect the mass flow of fluid leaving the top of the plate as compared to the bottom. Finally, the distance of the plate from the nozzle does affect the force exerted on it. When the fluid emerges from the nozzle into the atmosphere, shear stresses are established between the moving jet and the stationary atmosphere. As the fluid is not in a fixed boundary, it expands in shape, drawing air into it. The velocity of the jet decreases as its cross sectional area increases. The momentum of the fluid including the entrained air may stay approximately constant to begin with, but, eventually, the body forces of the fluid are sufficient to deflect the jet from its horizontal path.

If the plate is held sufficiently close to the jet, but not too close, all the above effects are negligible and the force exerted will be as predicted by Equation (10.14) to a sufficient order of accuracy.

EXAMPLE 10.1

A jet of water of density 1000 kg/m³ issues from a 25 mm diameter nozzle with a velocity of 10 m/s horizontally into the atmosphere. The jet strikes a circular flat plate held perpendicular to the flow at its centre. What is the force on the plate (Figure 10.4)?

If the flat plate is moving with a velocity of 2 m/s in the same direction as the jet, what is now the force on the plate?

The plate is replaced by another which has a 10 mm diameter hole in its centre. What is the force on the plate if it is held stationary in this case (Figure 10.5)?

Solution

The mass flow rate of the water $\{\dot{m}\}_w$ from the nozzle can be determined from the continuity equation. Equation (8.3) gives

$$\{\dot{m}\}_w = A_{xs}\{\rho C\}_w = 0.25\pi D^2\{\rho C\}_w = 0.25\pi \times 0.025^2 \times 1000 \times 10 = 4.91 \text{ kg/s}$$

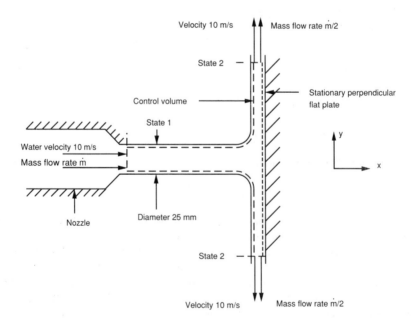

Fig. 10.4 *Example 10.1—two-dimensional plan view of the control volume for a horizontal jet of water striking a perpendicular stationary plate.*

APPLICATIONS OF THE SFME

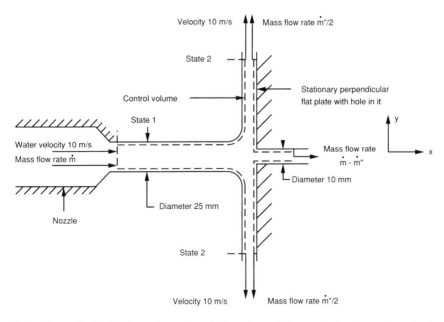

Fig. 10.5 *Example 10.1—two-dimensional plan view of the control volume for a horizontal jet of water striking a perpendicular stationary plate with a hole in it.*

Equation (10.14) gives the resultant force F in the x-direction as

$$F = \{\dot{m}C\}_w = 4.91 \times 10 = 49.1 \text{ N}$$

When the plate is moving away from the jet with a velocity of 2 m/s, it is the relative velocity between the jet and the plate that provides the force. The equations are as above provided the relative velocity of the water $\{C'\}_w$ is used:

$$\{C'\}_w = \text{(velocity of jet)} - \text{(velocity of plate)} = 10 - 2 = 8 \text{ m/s}$$

The relative mass flow rate $\{\dot{m}'\}_w$ from Equation (8.3), the continuity equation, is

$$\{\dot{m}'\}_w = 0.25\pi D^2 \{\rho C'\}_w = 0.25\pi \times 0.025^2 \times 1000 \times 8 = 3.93 \text{ kg/s}$$

The resultant force in the x-direction F' when the plate is moving is

$$F' = \{\dot{m}'C'\}_w = 3.93 \times 8 = 31.4 \text{ N}$$

When the plate is stationary but has a hole in its centre, only the mass flow of fluid which hits the plate will contribute to the force. That which passes through the hole in the plate will exert no force. If the jet is of diameter D and the hole of diameter D'', the mass flow of water which hits the plate $\{\dot{m}''\}_w$ is given by Equation (8.3), the continuity equation, as

$$\{\dot{m}''\}_w = 0.25\pi (D^2 - D''^2)\{\rho C\}_w = 0.25\pi (0.025^2 - 0.01^2) \times 1000 \times 10$$

$$\therefore \quad \{\dot{m}''\}_w = 4.12 \text{ kg/s}$$

The resultant force in the x-direction on the plate F'' is now, from Equation (10.14):

$$F'' = \{\dot{m}''C\}_w = 4.12 \times 10 = 41.2 \text{ N}$$

If the plate with the hole in was moving in the same direction as the jet of water, the relative velocities should be used and the force on the plate would be correspondingly reduced.

10.3.2 Jet striking a plate at an angle

In this case, the plate may be flat or curved as in fluid machinery. Consider the simple case of a jet of fluid of density ρ and velocity C issuing horizontally from a nozzle of diameter D into the atmosphere. The fluid strikes a stationary curved plate tangentially and its direction is turned through an acute angle θ. The fluid leaves the plate with the same velocity C with which it left the nozzle, as shown in Figure 10.6. For simplicity, to ignore body forces, consider the water to be turned through the angle in a horizontal plane.

Figure 10.6 shows the control volume which is drawn to include all the forces and, as with the fluid striking a flat plate perpendicular to the flow, it is a plan view so that the fluid only travels horizontally in the x- and y-directions.

Consider first the momentum forces. The initial fluid velocity C is entirely in the x-direction. The final fluid velocity C can be resolved into both the x- and y-directions. Therefore, if the mass flow rate of the fluid is $\dot{m}$, the momentum force in the x-direction F_{Mx}, from Equation (10.1), is

$$F_{Mx} = \dot{m}(C - C\cos\theta)$$

In the y-direction, the momentum force F_{My} is given by Equation (10.2) as

$$F_{My} = \dot{m}(0 - (-C\sin\theta)) = \dot{m}C\sin\theta$$

Note that the final velocity is negative because the resolved component of C is in the negative y-direction.

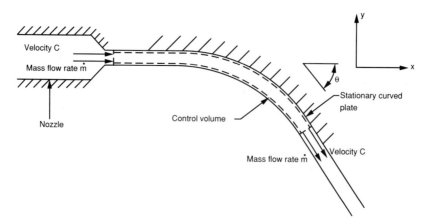

Fig. 10.6 Two-dimensional plan view of the control volume for a horizontal jet of fluid striking a stationary curved plate in the horizontal plane.

Both the pressure and body forces are zero for the same reasons described for a jet striking a stationary flat plate held perpendicular to the flow. Therefore, the total force in the x- and y-directions, F_x and F_y respectively, is

$$F_x = F_{Mx} + F_{Px} + F_{Bx} = \dot{m}(C - C\cos\theta)$$

$$F_y = F_{My} + F_{Py} + F_{By} = \dot{m}C\sin\theta$$

The resultant force on the plate F is given by Equation (10.13) as

$$F = (F_x^2 + F_y^2)^{0.5} = \dot{m}C((1 - \cos\theta)^2 + (\sin\theta)^2)^{0.5}$$

$$\therefore \quad F = \dot{m}C(2 - 2\cos\theta)^{0.5} \tag{10.15}$$

If the resultant force is at an angle θ' to the x-direction, this is given by

$$\tan\theta' = \frac{F_y}{F_x} = \frac{\sin\theta}{1 - \cos\theta} \tag{10.16}$$

As before, this is a simplified analysis for the same reasons discussed for the jet hitting the perpendicular plate. In addition, it is difficult to get the fluid to strike the plate tangentially.

EXAMPLE 10.2

A horizontal jet of water of density 1000 kg/m³ emits from a nozzle of diameter 25 mm with a velocity of 10 m/s into the atmosphere. It strikes a stationary curved plate tangentially which turns it through an angle of 50° in the horizontal plane. The water leaves the plate at 10 m/s. What is the force on the plate (Figure 10.7)?

If the plate is reshaped so that it turns the water through an angle of 150°, what is now the force on the plate?

If the 150° plate is made to move with a velocity of 2 m/s in the same direction as the jet issuing from the nozzle, what is the percentage reduction in the force exerted on the plate?

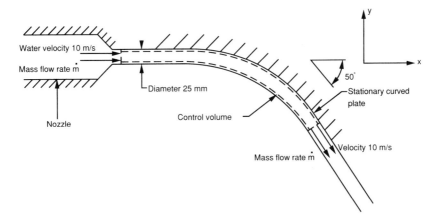

Fig. 10.7 *Example 10.2— Two-dimensional plan view of the control volume for a horizontal jet of water striking a stationary plate curved at 50 degrees in the horizontal plane.*

Solution

The mass flow rate of water $\{\dot{m}\}_w$ leaving the nozzle can be determined from the continuity equation. Applying Equation (8.3) gives

$$\{\dot{m}\}_w = A_{xs}\{\rho C\}_w = 0.25\pi D^2 \{\rho C\}_w = 0.25\pi \times 0.025^2 \times 1000 \times 10 = 4.91 \text{ kg/s}$$

The resultant force F on the plate is given by Equation (10.15) as

$$F = \{\dot{m}C\}_w (2 - 2\cos\theta)^{0.5} = 4.91 \times 10 \times (2 - 2\cos 50)^{0.5} = 41.5 \text{ N}$$

Its angular direction to the x-direction θ' is given by Equation (10.16) as

$$\theta' = \tan^{-1} \frac{\sin 50}{1 - \cos 50} = 65°$$

When the water is turned through a larger angle, there is no need to change the equations because the signs will take care of themselves in the trigonometric functions. All the conditions are the same except that θ is $150°$. Therefore, the resultant force F'' is now, from Equation (10.15):

$$F'' = 4.91 \times 10(2 - 2\cos 150)^{0.5} = 94.9 \text{ N}$$

Its angular direction to the x-direction θ'' is, from Equation (10.16):

$$\theta'' = \tan^{-1} \frac{\sin 150}{1 - \cos 150} = 15°$$

If the plate is moving with a velocity of 2 m/s in the same direction as the jet, it is the relative velocities that must be used. The relative velocity of the water $\{C'\}_w$ is

$$\{C'\}_w = 10 - 2 = 8 \text{ m/s}$$

The relative mass flow rate $\{\dot{m}'\}_w$ becomes

$$\{\dot{m}'\}_w = 0.25 \times 1000\pi \times 0.025^2 \times 8 = 3.93 \text{ kg/s}$$

The resultant force F' is

$$F' = 3.93 \times 8 \times (2 - 2\cos 150)^{0.5} = 60.7 \text{ N}$$

The percentage reduction in the force ($\%F$) when the plate is moving is

$$(\%F) = \frac{F - F'}{F} = \frac{94.9 - 60.7}{94.9} = 36\%$$

The direction of the resultant force remains the same.

It can be seen that the force is increased considerably when the jet is turned back upon itself. In fact, the greatest force theoretically will be obtained when the angle is $180°$, but then the fluid leaving the plate is likely to interfere with the fluid entering. A machine which produces a power output like this is called a Pelton Wheel. The above analysis can also be used to determine the force on blades in rotodynamic machinery such as a rotary turbine or compressor, as shown in Figure 8.20. Here, although the fluid is not open to the atmosphere, the pressure forces can be ignored because they only act in the radial direction, whereas the force is produced in the rotary direction. The shape of the blades causes the pressure to change significantly as the fluid passes through each set of blades, and this affects the value of the velocity of the fluid at exit.

EXAMPLE 10.3

A horizontal jet of water of density 1000 kg/m³ emits from a nozzle of diameter 25 mm with a velocity of 10 m/s into the atmosphere. The jet strikes a flat plate, held at an angle of 45°, at its centre point. The flat plate acts as a flow splitter such that, in a two-dimensional plan view, half the flow goes to one side of the plate and the other half to the other side, in other words both in the horizontal plane, as shown in Figure 10.8. What is the force on the splitter?

If, in fact, the flow is divided such that one-third is turned through an angle of 135° and two-thirds is turned through 45° what is now the force on the splitter?

Solution

The mass flow rate of water $\{\dot{m}\}_w$ leaving the nozzle can be determined from the continuity equation. Applying Equation (8.3) gives

$$\{\dot{m}\}_w = A_{xs}\{\rho C\}_w = 0.25\pi D^2\{\rho C\}_w = 0.25\pi \times 0.025^2 \times 1000 \times 10 = 4.91 \text{ kg/s}$$

As with Examples 10.1 and 10.2, the pressure and body forces are zero. Assume that state 1 is where the water leaves the nozzle and state 2 where the water leaves the flat plate in the two positions. The momentum force in the x-direction F_{Mx}, from Equation (10.1), is

$$F_{Mx} = \{\dot{m}(C_1 - C_2)_x\}_w$$

$$\therefore F_{Mx} = 4.91 \times 10 - (4.91 \times 0.5 \times 10 \cos 45 + (-4.91 \times 0.5 \times 10 \cos 45)) = 49.1 \text{ N}$$

In the y-direction, the flow is split in half. Therefore, the momentum force in the y-direction F_{My}, from Equation (10.2), is

$$F_{My} = \{\dot{m}(C_1 - C_2)_y\}_w$$

$$\therefore F_{My} = 4.91 \times 10 \cos 90 - (0.5 \times 4.91 \times 10 \cos 45 + (-0.5 \times 4.91 \times 10 \cos 45))$$

$$\therefore F_{My} = 0$$

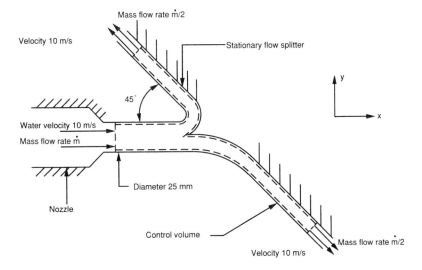

Fig. 10.8 *Two-dimensional plan view of the control volume for a horizontal jet of water striking a flow splitter which divides the water flow in half in the horizontal plane.*

The total force in the x-direction F_x, from Equation (10.10), is

$$F_x = (F_{Mx} + F_{Px} + F_{Bx}) = 49.1 \text{ N}$$

The total force in the y-direction F_y, from Equation (10.11), is

$$F_y = (F_{My} + F_{Py} + F_{By}) = 0$$

The resultant force F, from Equation (10.13), is

$$F = (F_x^2 + F_y^2)^{0.5} = 49.1 \text{ N}$$

Its direction is in the x-direction. The force exerted is the same as when the plate is held perpendicular to the flow, in other words it is independent of the angle of the plate. This is only true when the flow is evenly divided by the splitter. When a third of the flow is turned through 135° and two thirds through 45°, the momentum force in the x-direction F'_{Mx} is, from Equation (10.1):

$$F'_{Mx} = 4.91 \times 10 - (-4.91 \times 0.33 \times 10 \cos 45 + 0.67 \times 4.91 \times 10 \cos 45) = 37.3 \text{ N}$$

The momentum force in the y-direction F'_{My}, from Equation (10.2), is

$$F'_{My} = 0 - (4.91 \times 0.33 \times 10 \cos 45 + (-0.67 \times 4.91 \times 10 \cos 45)) = 11.8 \text{ N}$$

The resultant force F', from Equation (10.13), is

$$F' = (37.3^2 + 11.8^2)^{0.5} = 39.1 \text{ N}$$

Its direction is at an angle θ' to the x-direction of

$$\theta' = \tan^{-1} \frac{11.8}{37.3} = 17.6°$$

The angle at which the force acts depends upon the way the splitter operates, how much water is directed in one way and how much in the other.

10.3.3 Reaction of a jet

When a fluid issues from an opening in the side of a large tank, the force which causes the fluid to be ejected is due to the static pressure of the fluid in the tank. Therefore, there is an equal and opposite force of the fluid on the tank which can be predicted by the SFME. Consider Figure 10.9.

If the tank is large enough, the velocity of the fluid in the tank may be assumed to be approximately zero. If the velocity of the jet is C and its mass flow rate is $\dot{m}$, the force due to momentum in the x-direction F_{Mx} is

$$F_{Mx} = \dot{m}(0 - C) = -\dot{m}C$$

In other words, the momentum force of the jet on the tank is in the negative x-direction. As all the flow is in the x-direction, the momentum forces in the y- and z-directions are zero. If the jet is at approximately atmospheric pressure, it is at the same pressure as the atmosphere

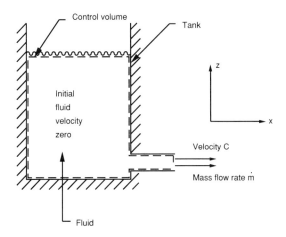

Fig. 10.9 *The control volume for a fluid emitting from a tank.*

surrounding the control volume and pressure forces are not applicable. Body forces due to the jet act vertically downwards and are also not applicable. The resultant force F is

$$F = F_x = F_{Mx} = -\dot{m}C \qquad (10.17)$$

If the tank is moving with a certain velocity, the resultant force is given by the relative velocity between the fluid jet and the tank, which depends upon the direction the tank is moving in.

The reaction of a jet can be used to propel various vessels, for example an aircraft or a ship, although the actual determination of the force is a little more complicated than Equation (10.17).

EXAMPLE 10.4

The performance of a jet engine is being examined on a stationary test bed in a laboratory. Air enters the engine from the atmosphere with a velocity of 10 m/s, a density of 1.25 kg/m³, through an area of 0.6 m². Fuel is burned in the air to produce heat at an air to fuel ratio by mass of 60 : 1. The combustion gases leave the engine and exhaust into the atmosphere with a velocity of 300 m/s (Figure 10.10). What is the force required to hold the engine stationary?

Solution

As the fluid enters and leaves the engine in the x-direction and at atmospheric pressure, Equation (10.17) is applicable, except that the mass flow of fluid changes due to the introduction of the fuel and the air has an initial velocity. If the mass flow rate of air is $\{\dot{m}\}_{air}$ and the mass flow rate of fuel $\{\dot{m}\}_{fuel}$, and the velocity of the fluid at entry and exit is C_1 and C_2 respectively, Equation (10.17) for the total force F becomes

$$F = \{\dot{m}\}_{air} C_1 - (\{\dot{m}\}_{air} + \{\dot{m}\}_{fuel}) C_2$$

The mass flow rate of air at entry $\{\dot{m}\}_{air}$ is given by the continuity equation (Equation (8.3)) as

$$\{\dot{m}\}_{air} = A_{xs} \{\rho C_1\}_{air} = 0.6 \times 1.25 \times 10 = 7.5 \text{ kg/s}$$

THE STEADY FLOW MOMENTUM EQUATION

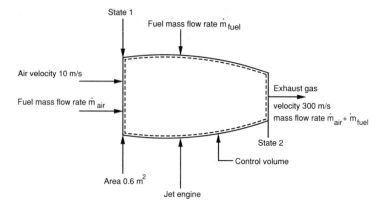

Fig. 10.10 *Example 10.4— the control volume for a jet engine.*

Hence, the mass flow rate of fuel $\{\dot{m}\}_{fuel}$ is

$$\{\dot{m}\}_{fuel} = \frac{\{\dot{m}\}_{air}}{60} = \frac{7.5}{60} = 0.125 \text{ kg/s}$$

The resultant force F is

$$F = 7.5 \times 10 - (7.5 + 0.125) \times 300 = -2212.5 \text{ N}$$

This is the force exerted in the negative x-direction by the fluid on the test stand. Therefore, the force required to hold the engine in place is 2212.5 N in the x-direction.

EXAMPLE 10.5

A garden sprinkler has two outlet nozzles as shown in Figure 10.11. Water of density 1000 kg/m³ enters the base of the sprinkler at the rate of 1.2 kg/s. The outlet nozzles are both of diameter 9 mm. What is the velocity of the water leaving each nozzle when the sprinkler head is held stationary?

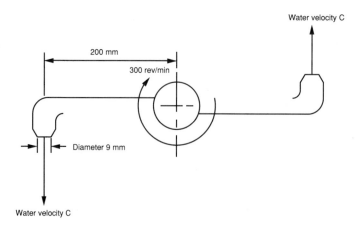

Fig. 10.11 *Example 10.5— the head of a garden sprinkler.*

If the sprinkler head rotates at 300 rev/min in a horizontal plane and the nozzles are on a radius of 200 mm about the sprinkler centre axis, what is now the velocity of the water leaving each nozzle, relative to a stationary reference point?

Determine the torque required to hold the sprinkler stationary and the resisting torque when the sprinkler head is rotating at 300 rev/min.

At what velocity must the sprinkler head rotate for there to be no resisting torque?

Solution

The cross-sectional area A_{xs} of each nozzle of diameter D is given by

$$A_{xs} = 0.25\pi D^2 = 0.25\pi (9 \times 10^{-3})^2 = 6.36 \times 10^{-5} \text{ m}^2$$

As there are two nozzles, the mass flow rate of water leaving each nozzle $\{\dot{m}'\}_w$ is half the total mass flow rate entering the sprinkler $\{\dot{m}\}_w$. Thus:

$$\{\dot{m}'\}_w = 0.5\{\dot{m}\}_w = 0.5 \times 1.2 = 0.6 \text{ kg/s}$$

The velocity of the water leaving each nozzle $\{C\}_w$ when the sprinkler head is stationary is given by the continuity equation (Equation (8.3)) as

$$\{\dot{m}'\}_w = A_{xs}\{\rho C\}_w$$

$$\therefore \{C\}_w = \frac{0.6}{1000 \times 6.36 \times 10^{-5}} = 9.4 \text{ m/s}$$

When the sprinkler head is rotating at 300 rev/min, the nozzles at radius 200 mm are rotating at a velocity C'' given by

$$C'' = \frac{300 \times 2\pi \times 0.2}{60} = 6.3 \text{ m/s}$$

The velocity of the water relative to a stationary reference point $\{C'\}_w$ is the relative velocity between the water and the sprinkler. Thus:

$$\{C'\}_w = \{C\}_w - C'' = 9.4 - 6.3 = 3.1 \text{ m/s}$$

Considering the SFME, the pressure forces can be ignored because the jet of water exits to the atmosphere. The body forces can be ignored too because the jet issues in the horizontal plane. The control volume is somewhat confusing in this example as the sprinkler head rotates, but inlet conditions (state 1) are those at entry to the base of the sprinkler, and outlet conditions (state 2) are those where the water leaves the nozzle, as shown in Figure 10.12.

Considering the tangential direction only, the momentum force F_M due to both nozzles is given by Equation (10.1) as

$$F_M = \{\dot{m}(C_1 - C_2)\}_w = \{(\dot{m})\}_w(\{C_1\}_w - \{C\}_w)$$

$$\therefore F_M = 1.2 C_1 \cos 90 - 2 \times 0.6 \times 9.4 = -11.3 \text{ N}$$

The negative sign shows that the force is in the opposite direction to the water flow.

The torque required to hold the sprinkler head stationary TQ is

$$TQ = F_M \times 0.2 = -11.3 \times 0.2 = -2.3 \text{ N m}$$

When the sprinkler head is rotating, the relative velocity of the water must be used. The resisting torque TQ' becomes

$$TQ' = -2 \times 0.6 \times 3.1 \times 0.2 = -0.74 \text{ N m}$$

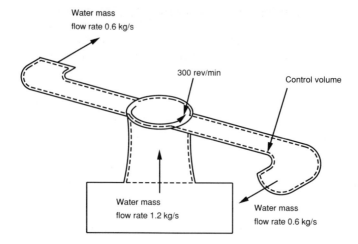

Fig. 10.12 *Example 10.10—the control volume for water entering and leaving a garden sprinkler.*

For the resisting torque to be zero, the water must leave the nozzle at the same velocity as the sprinkler head rotates. If the sprinkler head is rotating at 9.4 m/s, this is equivalent to a rotational velocity N_{rev} of

$$N_{rev} = \frac{9.4}{2\pi \times 0.2} = 7.48 \text{ rev/s}$$

This is the rotational velocity that the sprinkler head would adopt naturally for the given flow rate of 1.2 kg/s of water.

10.3.4 Force when fluid flows over a surface

Whenever a fluid flows over a surface, a boundary layer is formed, as explained in Chapter 9. Outside the boundary layer, the fluid moves with the free stream velocity and is unaffected by the presence of the surface. Inside the boundary layer, a velocity profile in the fluid develops, giving rise to shear stresses between the layers of fluid travelling at different velocities, producing a viscous force which retards the fluid and acts at the surface as a friction force. It is possible to analyse the force from a consideration of the SFME, but not within the scope of this book. Both pressure and momentum forces are applicable and must be integrated over the thickness of the boundary layer.

10.3.5 Force on a solid body in a flowing fluid

Figure 10.2 shows the streamline flow of a fluid over a solid body and Equations (10.1)–(10.13) develop the SFME in order to determine the resultant force generated. In practice, such a simple analysis is not correct because the velocity and pressure at exit from the control volume are not uniform. Indeed, it is the difference in the velocity and pressure profiles of the fluid upstream and downstream of the solid object that determines the force, as discussed in Chapter 9.

10.3.6 Fluid flowing through a straight pipe or duct

The one-dimensional analysis of the SFME is the same for a pipe as for a duct, provided that in the case of the latter, the equivalent diameter is used. In practice, flow in ducts is a little more complicated because of the three-dimensional effects in the corners.

Firstly, consider the simplest case, the steady flow of a fluid through a horizontal pipe of cross-sectional area A_{xs} and diameter D. For a given control volume, the fluid will enter at state 1 with a velocity C_1, pressure p_1 and density ρ_1, and leave at state 2 with a velocity C_2, pressure p_2 and density ρ_2, as shown in Figure 10.13.

As the flow is considered only in the axial direction of the pipe, and all the fluid property values are assumed to be constant across any cross-section, there will only be a contribution to the total force in the direction of fluid flow, call it the x-direction. If the mass flow rate of fluid is $\dot{m}$, the momentum force F_{Mx} is

$$F_{Mx} = \dot{m}(C_1 - C_2)$$

The pressure force F_{Px} is

$$F_{Px} = p_1 A_{xs} + (-p_2 A_{xs}) = A_{xs}(p_1 - p_2)$$

The body force acts vertically downwards only (implying that the pipe must be supported from below) and is not applicable. The total force in the x-direction F_x is:

$$F_x = \dot{m}(C_1 - C_2) + A_{xs}(p_1 - p_2) \qquad (10.18)$$

The total force in the x-direction is also the resultant force F. The only way that this force can act upon the pipe is to overcome the effects of friction at the surface. The force exerted by the fluid on the pipe, by Newton's third law of motion, is equal but opposite to the friction force exerted by the pipe surface on the fluid. The total force in the x-direction must be the friction force opposing the motion of the fluid. For the particular case of the steady flow of an incompressible fluid where the density remains constant, the continuity equation (Equation (8.3)) is

$$\dot{m} = \rho_1 (A_{xs})_1 C_1 = \rho_2 (A_{xs})_2 C_2$$

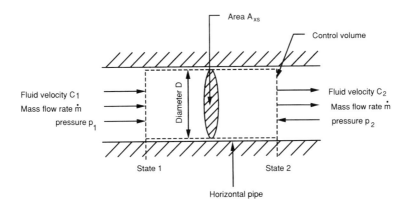

Fig. 10.13 The control volume for the flow of a fluid through a horizontal pipe.

THE STEADY FLOW MOMENTUM EQUATION

If $p_1 = p_2$, and the pipe is of constant diameter so that $(A_{xs})_1 = (A_{xs})_2$, to maintain steady flow, the velocity of the fluid must be constant. The force due to momentum is zero, and in Equation (10.18), the resultant force, and so the friction force, is equal to the pressure force. In other words, for the incompressible flow of a fluid through a horizontal pipeline, steady flow is maintained by providing a pressure force in the direction of flow which balances the friction force opposing the motion of the fluid. The SFME becomes

$$F = F_x = A_{xs}(p_1 - p_2) \qquad (10.19)$$

If the pipe is inclined at some angle θ to the horizontal, but the x-direction is still maintained along the axis of the pipe, the body forces will be included in Equation (10.18) as follows:

$$F_x = \dot{m}(C_1 - C_2) + A_{xs}(p_1 - p_2) \pm mg \sin\theta \qquad (10.20)$$

The sign is plus if the fluid in the pipe goes downwards and minus if it goes upwards.

When the fluid is enclosed within a solid boundary, as with the flow of a fluid in a pipe, it is necessary to consider the effect of the atmosphere on the exterior boundary. With a straight pipe, this is fairly easy to do, but not in devices which change the cross-section of the flow, such as a nozzle. However, the problem can be overcome by using the Gauge Pressures in calculations instead of the absolute pressures. The argument for this is that, if the pipe, for example, were empty except for stationary air, a force would still be exerted on the inside surface. When the pipe is full of a fluid flowing through it, it is the amount by which the force produced by the flowing fluid is greater than the force exerted by the stationary air that is of interest. The force due to the stationary air is, in effect, balanced by the force of the atmosphere on the outside of the pipe.

In practice, with the flow of fluids through straight pipes in which the diameter does not change, it does not make any difference in a calculation to the answer obtained for the resultant force whether absolute static pressures or gauge static pressures are used. This is because the pressure force consists of the difference in the pressure of the fluid at entry and exit multiplied by the cross-sectional area. Hence, for such situations, the absolute pressures will always be used. But if the pipe goes around a bend or brings about a change in the cross-sectional area of the flow, the gauge pressures must be substituted in the SFME. This is one of the few exceptions to the statement in Equation (2.1), that absolute pressures should be used in calculations.

EXAMPLE 10.6

In a factory, a long horizontal pipe of diameter 50 mm is used to transport air. The air enters the pipe at a pressure of 200 kPa, density 2.4 kg/m³ and velocity 2 m/s. At the end of the pipe, the air pressure is 110 kPa and velocity 3.7 m/s. What is the friction force exerted by the pipe opposing the motion of the air (Figure 10.14)?

Elsewhere in the factory, the same diameter pipe is also used to transport water of density 1000 kg/m³. What would be the friction force exerted by the pipe opposing the motion of the water if the water inlet pressure and velocity conditions, and the water pressure drop in the pipe, were the same as for the air? Note that the pressure drop would occur over a much shorter length of the pipe.

Later it is discovered that the pipe carrying the water in fact slopes upwards at an angle of 5°. What is the actual friction force exerted by the pipe opposing the motion of the water if the pipe length required to achieve the pressure drop is 56.25 m?

APPLICATIONS OF THE SFME 287

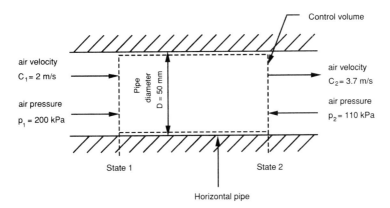

Fig. 10.14 Example 10.6— the control volume for the flow of a fluid in a horizontal pipe.

Solution
Let the entry to the pipe be state 1 and the exit from the pipe state 2.

When the fluid is air, which is compressible, the velocity increases. For steady flow conditions, the mass flow rate of air $\{\dot{m}\}_{air}$ does not change and is given at entry to the pipe by Equation (8.3), the continuity equation, as

$$\{\dot{m}\}_{air} = A_{xs}\{\rho_1 C_1\}_{air} = 0.25\pi \times 0.05^2 \times 2.4 \times 2 = 0.0094 \text{ kg/s}$$

The resultant force on the pipe F' in the x-direction, when the fluid is air, is given by Equation (10.18) as

$$F' = F'_x = \{\dot{m}(C_1 - C_2)\}_{air} + A_{xs}\{p_1 - p_2\}_{air}$$
$$\therefore \quad F' = F'_x = 0.0094(2 - 3.7) + 0.25\pi \times 0.05^2(2 \times 10^5 - 1.1 \times 10^5) = 176.7 \text{ N}$$

As the mass flow rate of the air is fairly small, the momentum force contribution is insignificant. It might seem strange that the velocity of the air increases when friction is present. In fact, the property values of the fluid are determined by the relative magnitude of three forces, those due to momentum, pressure and compressibility, and the relationship between them is complex.

When the fluid is incompressible water, its mass flow rate $\{\dot{m}\}_w$ is given by the continuity equation (Equation (8.3)) as

$$\{\dot{m}\}_w = A_{xs}\{\rho C_1\}_w = 0.25\pi \times 0.05^2 \times 1000 \times 2 = 3.93 \text{ kg/s}$$

If the fluid is incompressible, the velocity will remain constant and the momentum force is zero. Therefore, applying Equation (10.19) gives the resultant force F in the x-direction as

$$F = F_x = A_{xs}\{p_1 - p_2\}_w = 0.25\pi \times 0.05^2(2 \times 10^5 - 1.1 \times 10^5) = 176.7 \text{ N}$$

This must be equal to the friction force opposing the flow. It is the same value as for the airflow because the contribution of the momentum force in the airflow is negligible. However, it occurs over a considerably shorter length of pipe because the density of water is so much greater than that of air.

When the water pipe is inclined at 5° upwards, the body forces affect the resultant force. From Equation (10.20), the resultant force upwards along the axis of the pipe F is given by

$$F = \{\dot{m}(C_1 - C_2)\}_w + A_{xs}\{p_1 - p_2\}_w - \{m\}_w g \sin \theta$$

The mass of water m in a length of 56.25 m of pipe is the product of the cross-sectional area of the pipe A_{xs}, the length of the pipe L and the density of the water $\{\rho\}_w$. Thus:

$$m = A_{xs}L\{\rho\}_w = 0.25\pi \times 0.05^2 \times 56.25 \times 1000 = 110.4 \text{ kg}$$

The friction force F becomes

$$F = 176.7 - 110.4 \times 9.81 \sin 5 = 82.3 \text{ N}$$

The resultant force is reduced considerably because the body forces, in effect the potential energy contribution, are significant when the pipe is so long.

EXAMPLE 10.7

Air, of pressure, velocity and temperature 100 kPa, 50 m/s and 290 K respectively, flows along a horizontal duct of cross-sectional area 0.16 m². The air passes through a uniform wire gauze screen fixed in the duct perpendicular to the flow direction (Figure 10.15). The air pressure falls to 90 kPa after passing through the screen. What is the force exerted by the air on the wire gauze screen if the temperature of the air remains approximately constant?

Solution

If the upstream conditions are at state 1 and the downstream conditions at state 2, the density of the air upstream of the wire gauze screen $\{\rho_1\}_{air}$ is given by the equation of state (Equation (6.1)) as

$$\{\rho_1\}_{air} = \frac{\{p_1\}_{air}}{\{RT_1\}_{air}} = \frac{1 \times 10^5}{287 \times 290} = 1.2 \text{ kg/m}^3$$

Similarly the density of the air $\{\rho_2\}_{air}$ downstream of the gauze is

$$\{\rho_2\}_{air} = \frac{\{p_2\}_{air}}{\{RT_2\}_{air}} = \frac{0.9 \times 10^5}{287 \times 290} = 1.08 \text{ kg/m}^3$$

The steady mass flow rate of air $\{\dot{m}\}_{air}$ is given by the continuity equation applied to the upstream conditions. Equation (8.3) is

$$\{\dot{m}\}_{air} = A_{xs}\{\rho_1 C_1\}_{air} = 0.16 \times 1.2 \times 50 = 9.6 \text{ kg/s}$$

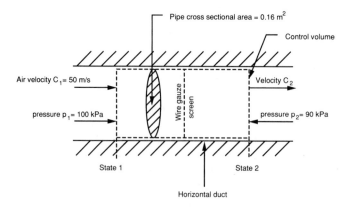

Fig. 10.15 Example 10.7—the control volume for the flow of air through a wire gauze screen in a horizontal duct.

Knowing the mass flow rate of air and applying the continuity equation downstream of the gauze, gives the exit velocity $\{C_2\}_{air}$ as

$$\{C_2\}_{air} = \frac{\{\dot{m}\}_{air}}{\{\rho_2\}_{air} A_{xs}} = \frac{9.6}{1.08 \times 0.16} = 55.6 \text{ m/s}$$

Assume the x-direction is along the axis of the duct and in the direction of motion of the air. Considering the SFME, only forces in the x-direction due to momentum and pressure are relevant. The resultant force in the x-direction F is given by Equation (10.18) as

$$F = F_x = F_{Mx} + F_{Px} = \{\dot{m}(C_1 - C_2)\}_{air} + A_{xs}\{p_1 - p_2\}_{air}$$

$$\therefore \quad F = F_x = 9.6 \times (50 - 55.6) + 0.16 \times 10^5 \times (1 - 0.9) = 1546.2 \text{ N}$$

This is a significant force and care will have to be taken that the gauze screen is strong enough not to shred or pull away from its mounting in the duct.

10.3.7 Fluid flowing in a pipe or duct in which the cross-section changes

It is not possible to consider all the potential situations which could arise in this section, such as flow through contraction joints, expansion joints, nozzles, diffusers, valves and obstructions in general. The basic principles of the SFME have to be applied in each situation. As an example, take the case of the flow of a fluid through a horizontal nozzle positioned in a section of pipework, as shown in Figure 10.16.

Assume that at inlet to the nozzle (state 1), the fluid velocity, fluid pressure and nozzle cross-sectional area are C_1, p_1 and $(A_{xs})_1$ respectively, and that at outlet from the nozzle (state 2) the values are C_2, p_2 and $(A_{xs})_2$ respectively. The mass flow of fluid, which is steady, is $\dot{m}$, and the x-direction is along the axis of the nozzle. As all the flow is in the x-direction, all components of force in the y- and z-directions are not applicable, including body forces, which act vertically downwards. Therefore the momentum force in the x-direction F_{Mx} is

$$F_{Mx} = \dot{m}(C_1 - C_2)$$

The pressure force in the x-direction is

$$F_{Px} = p_1(A_{xs})_1 - p_2(A_{xs})_2$$

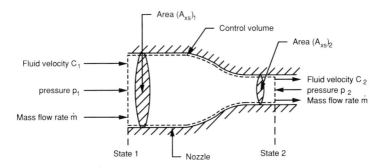

Fig. 10.16 *The control volume for the flow of a fluid through a nozzle in a horizontal pipe.*

THE STEADY FLOW MOMENTUM EQUATION

The resultant force F on the nozzle is

$$F = F_x = \dot{m}(C_1 - C_2) + p_1(A_{xs})_1 - p_2(A_{xs})_2 \qquad (10.21)$$

If the nozzle is at an angle of θ to the horizontal direction but the x-direction is still maintained along the axis of the nozzle, body forces are applicable and the equation becomes

$$F = \dot{m}(C_1 - C_2) + p_1(A_{xs})_1 - p_2(A_{xs})_2 \pm mg \sin \theta \qquad (10.22)$$

The sign is +ve if the fluid in the nozzle goes downwards and −ve if it goes upwards.

If pressure p_2 is not known, it is possible in certain applications to assume that the flow is frictionless and use Bernoulli's equation (Equation (8.20)) to find the change in pressure without too much loss of accuracy in determining the resultant force. In fact, this approach could be extended to produce a measure of the 'efficiency' of the obstruction being examined, if the actual exit pressure is measured and compared to that predicted by Bernoulli's equation, which is for ideal fluid flow.

As discussed in Section 10.3.6 for the flow of fluid through a pipe, it is the gauge pressures which must be used in calculations involving the SFME.

EXAMPLE 10.8

A vertical water nozzle tapers from a diameter of 100 mm at inlet to 50 mm at outlet, in a length of 100 mm. Water, of density 1000 kg/m³, enters the nozzle with a velocity of 5 m/s and a gauge pressure of 300 kPa (Figure 10.17). Assuming frictionless flow in the nozzle, what is the difference in the force required to hold the nozzle in position if the flow of water is downwards compared to when the flow of water is upwards?

Solution

Let the inlet to the nozzle be state 1 and the outlet from the nozzle state 2.

If the inlet and outlet diameters of the nozzle are D_1 and D_2 respectively, the inlet and outlet areas, $(A_{xs})_1$ and $(A_{xs})_2$, are

$$(A_{xs})_1 = 0.25\pi D_1^2 = 0.25\pi \times 0.1^2 = 7.85 \times 10^{-3} \text{ m}^2$$
$$(A_{xs})_2 = 0.25\pi D_2^2 = 0.25\pi \times 0.05^2 = 1.96 \times 10^{-3} \text{ m}^2$$

The mass flow rate of water at entry to the nozzle $\{\dot{m}\}_w$ is given by the continuity equation (Equation (8.3)) as

$$\{\dot{m}\}_w = (A_{xs})_1 \{\rho C_1\}_w = 0.00785 \times 1000 \times 5 = 39.3 \text{ kg/s}$$

As the flow is steady, the mass flow rate of water is constant. The continuity equation can be used to determine the velocity of the water at exit from the nozzle $\{C_2\}_w$, as follows:

$$\{C_2\}_w = \frac{\{\dot{m}\}_w}{\{\rho\}_w (A_{xs})_2} = \frac{39.3}{1000 \times 0.001 \times 96} = 20.1 \text{ m/s}$$

If the flow is frictionless, Bernoulli's equation (Equation (8.20)) can be used to calculate the exit pressure of the water from the nozzle $\{p_2\}_w$, as follows:

$$\{p_1\}_w + 0.5\{\rho C_1^2\}_w + \{\rho\}_w g z_1 = \{p_2\}_w + 0.5\{\rho C_2^2\}_w + \{\rho\}_w g z_2$$

If the flow is downwards:

$$z_1 - z_2 = 0.1 \text{ m}$$

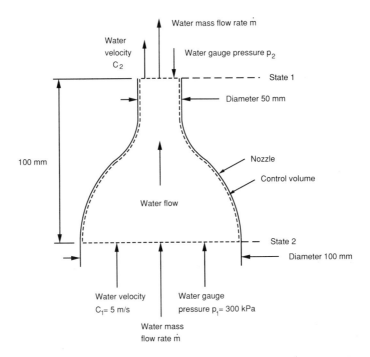

Fig. 10.17 *Example 10.8—the control volume for the flow of a fluid through a nozzle in a vertical pipe.*

Hence:

$$3 \times 10^5 + 0.5 \times 10^3 \times 5^2 + 10^3 g \times 0.1 = \{p_2\}_w + 0.5 \times 10^3 \times 20.1^2$$

$$\therefore \quad \{p_2\}_w = 111.5 \text{ kPa gauge}$$

Note that absolute pressures should be used in Bernoulli's equation, but the atmospheric pressure added to $\{p_1\}_w$ and $\{p_2\}_w$, will cancel out on either side of the equation.

Taking the x-direction down the axis of the nozzle, the resultant force F is given by Equation (10.22) as

$$F = F_x = \{\dot{m}(C_1 - C_2)\}_w + \{p_1\}_w (A_{xs})_1 - \{p_2\}_w (A_{xs})_2 + \{m\}_w g \sin \theta$$

where $\{p_1\}_w$ and $\{p_2\}_w$ are actually gauge pressures. As the height of the cone extended beyond area $(A_{xs})_2$ is 0.1 m, the mass of water in the nozzle $\{m\}_w$ is given by

$$\{m\}_w = \{\rho\}_w (0.33(A_{xs})_1 \times 0.2 - 0.33(A_{xs})_2 \times 0.1)$$

$$\therefore \quad \{m\}_w = 1000 \times (0.33 \times 0.00785 \times 0.2 - 0.33 \times 0.00196 \times 0.1) = 0.45 \text{ kg}$$

$$\therefore \quad F = 39.3 \times (5 - 20.1) + 10^2 \times (3 \times 7.85 - 1.115 \times 1.96) + 0.45 g \sin 90$$

$$\therefore \quad F = 1547.4 \text{ N downwards}$$

If the flow is upwards, the outlet gauge pressure $\{p'_2\}_w$, as predicted by Bernoulli's equation, is slightly different because $(z_1 - z_2)$ is now -0.1 m. Hence, $\{p'_2\}_w$ is given by

$$\{p_1\}_w + 0.5\{\rho C_1^2\}_w + \{\rho\}_w g z_1 = \{p'_2\}_w + 0.5\{\rho C_2^2\}_w + \{\rho\}_w g z_2$$

$$\therefore \quad 3 \times 10^5 + 0.5 \times 10^3 \times 5^2 - 10^3 g \times 0.1 = \{p'_2\}_w + 0.5 \times 10^3 \times 20.1^2$$

$$\therefore \quad \{p'_2\}_w = 109.5 \text{ kPa gauge}$$

The new resultant force F' upwards along the axis of the nozzle, which is now the x-direction, from (Equation (10.22)) is

$$F' = F'_x = \{\dot{m}(C_1 - C_2)\}_w + \{p_1\}_w(A_{xs})_1 - \{p'_2\}_w(A_{xs})_2 - \{m\}_w g \sin \theta$$

Again, $\{p_1\}_w$ and $\{p_2\}_w$ are the gauge pressures.

$$\therefore \quad F' = 39.3 \times (5 - 20.1) + 10^2 \times (3 \times 7.85 - 1.095 \times 1.96) - 0.45 g \sin 90$$

$$\therefore \quad F' = 1542.5 \text{ N upwards}$$

The difference is only 4 N because the pressure force is considerably greater in magnitude than the body force.

10.3.8 Fluid flowing around a bend in a pipe or duct

Consider the steady flow of an incompressible fluid of density ρ and of mass flow rate $\dot{m}$ going around a reducing bend of angle θ, which is an acute angle measured with respect to the x-direction. Assume that the flow is maintained in a horizontal plane, as shown in Figure 10.18, and that state 1 is upstream of the bend and state 2 downstream of the bend. The cross-sectional area of the pipe at inlet is $(A_{xs})_1$ and at outlet is $(A_{xs})_2$. The velocity and pressure of the fluid at entry to the bend are C_1 and p_1 respectively, and at exit from the bend C_2 and p_2 respectively.

The momentum forces in the x- and y-directions, F_{Mx} and F_{My} respectively, are

$$F_{Mx} = \dot{m}(C_1 - C_2 \cos \theta)$$

$$F_{My} = \dot{m}(C_1 \cos 90 - (-C_2 \sin \theta)) = \dot{m} C_2 \sin \theta$$

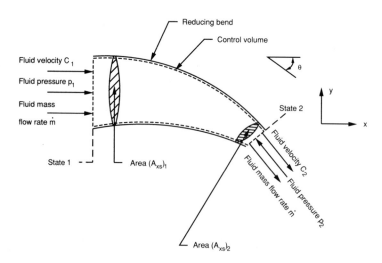

Fig. 10.18 The control volume for the flow of a fluid around a reducing bend in a horizontal pipe.

APPLICATIONS OF THE SFME

The pressure forces in the x- and y-directions, F_{Px} and F_{Py} respectively, are

$$F_{Px} = p_1(A_{xs})_1 + (-p_2(A_{xs})_2 \cos\theta) = p_1(A_{xs})_1 - p_2(A_{xs})_2 \cos\theta$$

$$F_{Py} = p_1(A_{xs})_1 \cos 90 + p_2(A_{xs})_2 \sin\theta = p_2(A_{xs})_2 \sin\theta$$

The body forces in the x- and y-directions are zero because all the weight of the fluid is acting vertically downwards in the z-direction.

The total force in the x-direction F_x is

$$F_x = F_{Mx} = \dot{m}(C_1 - C_2 \cos\theta) + p_1(A_{xs})_1 - p_2(A_{xs})_2 \cos\theta \tag{10.23}$$

The total force in the y-direction F_y is

$$F_y = F_{My} = \dot{m}C_2 \sin\theta + p_2(A_{xs})_2 \sin\theta \tag{10.24}$$

The resultant force F is given by Equation (10.13) as

$$F = (F_x^2 + F_y^2)^{0.5}$$

If the bend is in a vertical plane, the body forces must be included in the equations. As with the nozzle, the pressures used in the calculations with the SFME must be gauge pressures and, if the flow may be assumed frictionless, Bernoulli's equation can be called upon to calculate p_2 knowing p_1 and the velocities of the fluid.

Strictly, to determine the force on any bend in isolation in such a manner, it should be considered to be joined to the rest of the pipe or duct at inlet and outlet with flexible connections which take no force. This will be assumed in the examples that follow.

EXAMPLE 10.9

Oil of density 800 kg/m³ flows through a horizontal 45° reducing bend. The pipe diameter at entry is 100 mm and at exit 75 mm. The oil approaches the bend with a velocity and gauge pressure of 4 m/s and 300 kPa respectively (Figure 10.19). Assuming the flow of the oil around the bend is frictionless, what is the force exerted by the fluid on the bend and at what angle does it act?

Solution

Let the entry to the bend be state 1 and the exit from the bend state 2.

At inlet and outlet to the bend, the cross-sectional areas $(A_{xs})_1$ and $(A_{xs})_2$ are

$$(A_{xs})_1 = 0.25\pi D_1^2 = 0.25\pi \times 0.1^2 = 0.0079 \text{ m}^2$$

$$(A_{xs})_2 = 0.25\pi D_2^2 = 0.25\pi \times 0.075^2 = 0.0044 \text{ m}^2$$

From the continuity equation (Equation (8.3)) applied at entry, the steady mass flow rate of oil $\{\dot{m}\}_{oil}$ is

$$\{\dot{m}\}_{oil} = (A_{xs})_1 \{\rho C_1\}_{oil} = 0.0079 \times 800 \times 4 = 25.3 \text{ kg/s}$$

Hence, the velocity of the oil at exit from the bend is

$$\{C_2\}_{oil} = \frac{\{\dot{m}\}_{oil}}{\{\rho\}_{oil}(A_{xs})_2} = \frac{25.3}{800 \times 0.0044} = 7.2 \text{ m/s}$$

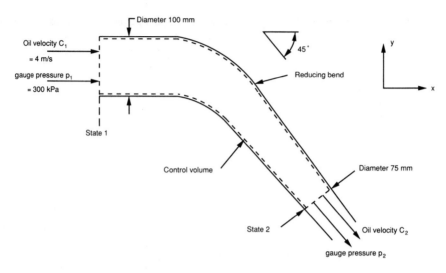

Fig. 10.19 *Example 10.9— the control volume for the flow of oil through a 45 degree reducing bend in a horizontal pipe.*

From Bernoulli's equation, the outlet pressure of the oil from the bend $\{p_2\}_{oil}$ can be determined. Equation (8.20) is

$$\{p_1\}_{oil} + 0.5\{\rho C_1^2\}_{oil} + \{\rho\}_{oil}gz_1 = \{p_2\}_{oil} + 0.5\{\rho C_2^2\}_{oil} + \{\rho\}_{oil}gz_2$$
$$\therefore \quad 3 \times 10^5 + 0.5 \times 800 \times 4^2 = \{p_2\}_{oil} + 0.5 \times 800 \times 7.2^2$$
$$\therefore \quad \{p_2\}_{oil} = 286 \text{ kPa gauge}$$

Note that, as with the nozzle, absolute pressures should be used in Bernoulli's equation, but the atmospheric pressure added to $\{p_1\}_{oil}$ and $\{p_2\}_{oil}$ will cancel out on either side of the equation.

The total force in the x-direction F_x is given by Equation (10.23) as

$$F_x = \{\dot{m}(C_1 - C_2 \cos 45)\}_{oil} + \{p_1\}_{oil}(A_{xs})_1 - \{p_2\}_{oil}(A_{xs})_2 \cos 45$$

where $\{p_1\}_{oil}$ and $\{p_2\}_{oil}$ are actually the gauge pressures.

$$\therefore \quad F_x = 25.3(4 - 7.2 \cos 45) + 10^3 \times (3 \times 0.79 - 2.86 \times 0.44 \cos 45) = 1452.6 \text{ N}$$

The total force in the y-direction F_y is given by Equation (10.24) as

$$F_y = \{\dot{m}C_2\}_{oil} \sin 45 + \{p_2\}_{oil}(A_{xs})_2 \sin 45$$
$$\therefore \quad F_y = 25.3 \times 7.2 \sin 45 + 2.86 \times 10^5 \times 0.0044 \sin 45 = 1018.6 \text{ N}$$

The resultant force F is given by Equation (10.13) as

$$F = (F_x^2 + F_y^2)^{0.5} = (1452.6^2 + 1018.6^2)^{0.5} = 1774.1 \text{ N}$$

If its direction with respect to the x-direction is at an angle of θ', this is given by

$$\theta' = \tan^{-1} \frac{1018.6}{1452.6} = 35.1°$$

APPLICATIONS OF THE SFME   295

The force produced by a fluid flowing through a reducing bend can be substantial and must be allowed for, either in the strength of the flanges that locate the bend in the pipe, or by providing solid foundations such as concrete anchorages to hold the pipe in position. These are in addition to the supports required to hold the bend and fluid upwards.

The force produced by a fluid flowing through a bend of constant diameter is also substantial. However, it can only be calculated if the fluid inlet and outlet pressures to the bend are measured. If the outlet pressure is determined using Bernoulli's equation, it is the same as the inlet pressure in a constant area bend because the velocity does not change in an incompressible fluid. The reducing bend was considered for the example as it allows the use of Bernoulli's equation to be demonstrated.

EXAMPLE 10.10

Air flows around a 90° reducing bend in a horizontal wind tunnel. The air approaches the bend with an absolute pressure of 112 kPa, velocity of 20 m/s and a density of 1.2 kg/m³. Upstream of the bend, the wind tunnel has a cross-section measuring 1.2 m wide by 1.2 m deep. Downstream of the bend, the depth is the same but the width is only 0.6 m. There are 10 guide vanes fitted in a cascade in the bend which help to direct the flow (Figure 10.20). Assuming that the air behaves as an incompressible fluid, and that friction effects in the bend may be ignored, what is the force on each of the guide vanes in the cascade? Take the atmospheric pressure as 108 kPa.

Solution
Let the inlet to the bend be state 1 and the outlet from the bend state 2.

The cross-sectional areas of the wind tunnel upstream and downstream of the bend, $(A_{xs})_1$ and $(A_{xs})_2$ respectively, are

$$(A_{xs})_1 = 1.2 \times 1.2 = 1.44 \text{ m}^2$$

$$(A_{xs})_2 = 1.2 \times 0.6 = 0.72 \text{ m}^2$$

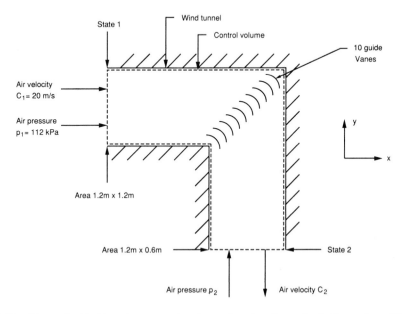

Fig. 10.20 *Example 10.10—the control volume for the flow of air through a 90-degree reducing bend in a horizontal wind tunnel.*

The steady mass flow rate of air $\{\dot{m}\}_{air}$ can be determined from the continuity equation applied to the upstream conditions. Equation (8.3) gives

$$\{\dot{m}\}_{air} = (A_{xs})_1\{\rho C_1\}_{air} = 1.44 \times 1.2 \times 20 = 34.6 \text{ kg/s}$$

The velocity of air at outlet from the bend $\{C_2\}_{air}$ can also be calculated from the continuity equation. Thus:

$$\{C_2\}_{air} = \frac{\{\dot{m}\}_{air}}{\{\rho\}_{air}(A_{xs})_2} = \frac{34.6}{1.2 \times 0.72} = 40 \text{ m/s}$$

The pressure at outlet from the bend $\{p_2\}_{air}$, assuming negligible friction effects, can be determined from Bernoulli's equation. Equation (8.20) gives

$$\{p_1\}_{air} + 0.5\{\rho C_1^2\}_{air} = \{p_2\}_{air} + 0.5\{\rho C_2^2\}_{air}$$

$$\therefore \quad 1.12 \times 10^5 + 0.5 \times 1.2 \times 20^2 = \{p_2\}_{air} + 0.5 \times 1.2 \times 40^2$$

$$\therefore \quad \{p_2\}_{air} = 111.3 \text{ kPa absolute}$$

For the SFME, gauge pressures must be used. Thus:

$$\{p_1\}_{air} = 112 - 108 = 4 \text{ kPa gauge}$$

$$\{p_2\}_{air} = 111.3 - 108 = 3.3 \text{ kPa gauge}$$

The total force in the x-direction F_x is given by Equation (10.23) as

$$F_x = \{\dot{m}(C_1 - C_2 \cos 90)\}_{air} + \{p_1\}_{air}(A_{xs})_1 - \{p_2\}_{air}(A_{xs})_2 \cos 90$$

$$\therefore \quad F_x = 34.6 \times 20 + 0.04 \times 10^5 \times 1.44 = 6452 \text{ N}$$

The total force in the y-direction F_y is given by Equation (10.24) as

$$F_y = \{\dot{m}C_2\}_{air} \sin 90 + \{p_2\}_{air}(A_{xs})_2 \sin 90$$

$$\therefore \quad F_y = 34.6 \times 40 + 0.033 \times 10^5 \times 0.72 = 3760 \text{ N}$$

The resultant force F is given by Equation (10.13) as

$$F = (F_x^2 + F_y^2)^{0.5} = (6452^2 + 3760^2)^{0.5} = 7468 \text{ N}$$

If there are 10 guide vanes, the force on each vane F' is

$$F' = \frac{7468}{10} = 747 \text{ N}$$

The actual force will be somewhat larger because ideal fluid flow has been assumed between entry and exit to the bend in order to calculate the exit pressure of the air. However, guide vanes are essential in right-angle bends if a smooth airflow is required and their extra capital cost is balanced by the reduction in pressure drop due to friction and turbulence effects.

EXAMPLE 10.11

Water of density 1000 kg/m^3 flows through a 150 mm diameter horizontal pipe at the steady mass flow rate of 60 kg/s. A junction is fitted into the pipe which divides the water flow into two. Upstream of the junction, the water gauge pressure is 110 kPa gauge. After the junction, 20 kg/s of the water flows through one section of the pipe which has a diameter of 75 mm and is at an angle of 45 degrees to the original flow direction, and 40 kg/s flows through the other section of pipe of diameter 100 mm which is at an angle of 30 degrees to the original flow direction (Figure 10.21). Assuming that friction effects in the junction are negligible, what is the force required to hold the pipe in place?

Solution
Let the inlet to the junction be state 1, the outlet from the 75 mm diameter pipe be state 2 and the outlet from the 100 mm diameter pipe be state 3.
The cross-sectional areas of the three pipes are

$$(A_{xs})_1 = 0.25\pi D_1^2 = 0.25\pi \times 0.15^2 = 0.018 \text{ m}^2$$
$$(A_{xs})_2 = 0.25\pi D_2^2 = 0.25\pi \times 0.075^2 = 0.0044 \text{ m}^2$$
$$(A_{xs})_3 = 0.25\pi D_3^2 = 0.25\pi \times 0.1^2 = 0.0079 \text{ m}^2$$

The velocity of water in each pipe is given by the continuity equation (Equation (8.3)) as

$$\{C_1\}_w = \frac{\{\dot{m}_1\}_w}{\{\rho\}_w (A_{xs})_1} = \frac{60}{1000 \times 0.018} = 3.3 \text{ m/s}$$

$$\{C_2\}_w = \frac{\{\dot{m}_2\}_w}{\{\rho\}_w (A_{xs})_2} = \frac{20}{1000 \times 0.0044} = 4.5 \text{ m/s}$$

$$\{C_3\}_w = \frac{\{\dot{m}_3\}_w}{\{\rho\}_w (A_{xs})_3} = \frac{40}{1000 \times 0.0079} = 5.1 \text{ m/s}$$

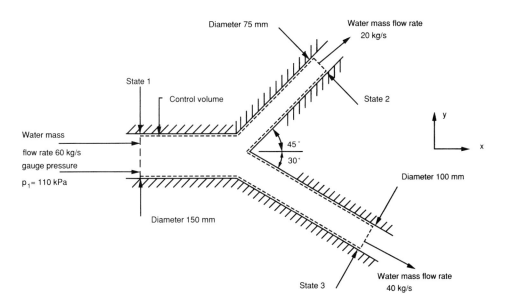

Fig. 10.21 *Example 10.11 — the control volume for the flow of water through a horizontal divider.*

Ignoring friction effects, Bernoulli's equation can be used to determine the pressure in each of the outlet pipes, $\{p_2\}_w$ and $\{p_3\}_w$. Equation (8.20) for a horizontal system gives

$$\{p_1\}_w + 0.5\{\rho C_1^2\}_w = \{p_2\}_w + 0.5\{\rho C_2^2\}_w = \{p_3\}_w + 0.5\{\rho C_3^2\}_w$$

$$\therefore \quad 1.1 \times 10^5 + 500 \times 3.3^2 = \{p_2\}_w + 500 \times 4.5^2 = \{p_3\}_w + 500 \times 5.1^2$$

$$\therefore \quad \{p_2\}_w = 105 \text{ kPa gauge} \quad \text{and} \quad \{p_3\}_w = 102 \text{ kPa gauge}$$

As with the nozzle, absolute pressures should be used in Bernoulli's equation, but the atmospheric pressure added to $\{p_1\}_w$, $\{p_2\}_w$ and $\{p_3\}_w$ will cancel out in the equations.

Taking the x-direction as being in the original direction of motion of the water, the momentum force F_{Mx} in the x-direction, from Equation (10.1), is

$$F_{Mx} = \{\dot{m}(C_1 - C_2)_x\}_w = 60 \times 3.3 - 20 \times 4.5 \cos 45 - 40 \times 5.1 \cos 30 = -42.3 \text{ N}$$

The pressure force in the x-direction F_{Px}, from Equation (10.4) (where all the pressures are gauge values), is

$$F_{Px} = \sum(\{p\}_w A_{xs})_x$$

$$\therefore \quad F_{Px} = \{p_1\}_w (A_{xs})_1 - \{p_2\}_w (A_{xs})_2 \cos 45 - \{p_3\}_w (A_{xs})_3 \cos 30$$

$$\therefore \quad F_{Px} = 10^5 \times (1.1 \times 0.018 - 1.05 \times 0.0044 \cos 45 - 1.03 \times 0.0079 \cos 30)$$

$$\therefore \quad F_{Px} = 955.5 \text{ N}$$

As the junction is in the horizontal plane, the body forces in the x-direction can be ignored. The net force in the x-direction F_x, from Equation (10.10), is

$$F_x = F_{Mx} + F_{Px} = -42.3 + 955.5 = 913.2 \text{ N}$$

The momentum force in the y-direction F_{My}, from Equation (10.2), is

$$F_{My} = \{\dot{m}(C_1 - C_2)_y\}_w$$

$$\therefore \quad F_{My} = 60 \times 3.3 \cos 90 - 20 \times 4.5 \sin 45 - (-40 \times 5.1 \sin 30)$$

$$\therefore \quad F_{My} = 38.4 \text{ N}$$

The pressure force in the y-direction F_{Py}, from Equation (10.5) (again, all the pressures are gauge values), is

$$F_{Py} = \sum(\{p\}_w A_{xs})_y$$

$$\therefore \quad F_{Py} = \{p_1\}_w (A_{xs})_1 \cos 90 - \{p_2\}_w (A_{xs})_2 \sin 45 + \{p_3\}_w (A_{xs})_3 \sin 30$$

$$\therefore \quad F_{Py} = 0 - 1.05 \times 10^5 \times 0.0044 \sin 45 + 1.03 \times 10^5 \times 0.0079 \sin 30$$

$$\therefore \quad F_{Py} = 76.2 \text{ N}$$

As the junction is in the horizontal plane, the body forces in the y-direction can be ignored. The net force in the y-direction F_y, from Equation (10.11), is

$$F_y = F_{My} + F_{Py} = 38.4 + 76.2 = 114.6 \text{ N}$$

The resultant force on the bend F, from Equation (10.13), is

$$F = (F_x^2 + F_y^2)^{0.5} = (913.2^2 + 114.6^2)^{0.5} = 920.4 \text{ N}$$

The direction of the resultant force is at an angle to the x-direction θ' of

$$\theta' = \tan^{-1} \frac{114.6}{913.2} = 7.15°$$

The angle of the resultant force depends upon the way the pipes after the junction diverge, and allowance must be made not just for the magnitude of the force but also its direction.

10.4 APPLICATION IN THERMAL SYSTEMS

It is now possible, to a first approximation, to estimate the forces that arise due to the motion of a fluid in a number of applications. Those due to the external flow of a fluid over a body, or through the blading of a rotodynamic machine, require a deeper analysis, beyond the scope of this book. But those that occur as a consequence of a fluid flowing through a duct, and which are of major importance in a thermal system when the fluid is transferred from component to component, can be determined with a reasonable degree of accuracy.

11 The Steady Flow Energy Equation Applied to Pipe Flow

When a fluid is being transported through a duct, it is subject to forces which affect its motion, as described in Chapter 10. The forces cause a change in the fluid properties, in particular the fluid pressure. Use of the Steady Flow Energy Equation (SFEE) allows the fluid properties to be determined at any cross-section in the duct.

The SFEE for a fluid entering and leaving a generalised thermal system has been derived in Chapter 8. When applied between the fluid inlet and outlet positions, state 1 and state 2 respectively, where all the relevant fluid property values are identified, it may be written in the form of Equation (8.5) as follows:

$$\dot{Q}_{12} + \dot{W}_{12} + \dot{m}(u_1 + \frac{p_1}{\rho_1} + 0.5C_1^2 + gz_1) = \dot{m}(u_2 + \frac{p_2}{\rho_2} + 0.5C_2^2 + gz_2)$$

The SFEE is equally valid when the generalised thermal system is a pipe or duct through which a fluid is flowing. However, it is preferable to modify the equation for this application as follows.

Whenever a fluid flows over a surface, friction effects act to slow it down, as explained in Chapters 2 and 9. In the steady flow of a fluid through a pipe, the friction forces acting on the inside pipe walls will tend to retard the motion of the fluid. If the mass flow rate of the fluid is to be maintained constant, there must be an equal and opposite force in the direction of motion of the fluid which will counteract the effects of the friction force. This is provided by the fluid pressure in that the pressure of the fluid falls as it passes through the pipe, the force due to the pressure drop balancing the force due to friction.

The main effect of the friction force opposing the motion is to increase the internal energy of the fluid. This raises the fluid temperature, which results in some heat transfer from the fluid to the surroundings. Useful energy is 'lost' from the fluid. It is convenient to group the terms in the SFEE related to the internal energy and the heat transfer together and to refer to them as a friction energy E_F, defined as

$$E_F = (\dot{m}u_2 - \dot{m}u_1 - \dot{Q}_{12})$$

The minus sign before the heat transfer term is because, when it is a heat output from the system which must be negative when a numerical value is ascribed to it, in accordance with sign rule B, the heat output increases the energy 'lost' due to friction.

Substituting for the friction energy, the SFEE becomes

$$\frac{p_1}{\rho_1 g} + \frac{0.5C_1^2}{g} + z_1 = \frac{p_2}{\rho_2 g} + \frac{0.5C_2^2}{g} + z_2 + \frac{E_F}{\dot{m}g} - \frac{\dot{W}_{12}}{\dot{m}g} \qquad (11.1)$$

THE STEADY FLOW ENERGY EQUATION APPLIED TO PIPE FLOW

The friction energy written as above in Equation (11.1) has the units of head. It is given the symbol Z_F and called the Head Lost to Friction because it represents the useful energy 'lost' by the fluid to the surroundings due to the effects of friction. The SFEE becomes

$$\frac{p_1}{\rho_1 g} + \frac{0.5 c_1^2}{g} + z_1 = \frac{p_2}{\rho_2 g} + \frac{0.5 c_2^2}{g} + z_2 + Z_F - \frac{\dot{W}_{12}}{\dot{m} g} \quad (11.2)$$

The SFEE of Equation (11.2) is the most suitable when the application is the flow of a fluid through a pipe. For a solution, it requires a knowledge of the term Z_F. This can be achieved for laminar and turbulent flow.

Note that in an ideal fluid, there is no friction and Z_F is zero. If there is no work transfer in the section of pipe under consideration, the equation reduces to the Bernoulli equation (Equation (8.20)).

EXAMPLE 11.1

A tank of water of density 1000 kg/m³ sits on the roof of a building 85 m high. A 50 mm diameter pipe connects this tank to another tank situated on the ground floor. The lower tank is used for watering the lawns and is normally kept filled by rain-water from the gutters. However, when there has not been sufficient rainfall, the lower tank is filled with water from the tank on the roof (Figure 11.1). What is the head lost to friction when the water flows from the tank on the roof to the tank on the ground floor under steady flow conditions?

When there has been excessive rainfall, it is necessary to pump the water back from the ground floor tank into the roof tank, and a pump is placed in the pipeline to do this. If the water is pumped back up at a velocity of 0.6 m/s such that the head lost to friction is the same as when the water flowed downwards of its own accord, what is the power input required at the pump if it has an energy transfer efficiency of 70%?

Solution

When the water is flowing downwards, let the SFEE be applied between the surface of the water in the roof tank (state 1), and the surface of the water in the ground floor tank (state 2). Considering Equation (11.2):

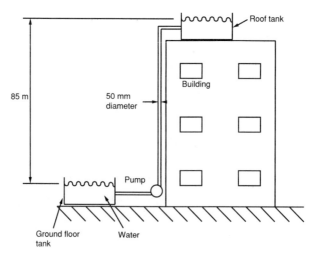

Fig. 11.1 Example 11.1—two water tanks in a building connected by a pipeline containing a pump.

$\{p_1\}_w = \{p_2\}_w =$ atmospheric pressure

$\{C_1\}_w = \{C_2\}_w = 0$ (because the water level in each tank will only fall or rise slowly and so the velocity there can be assumed negligible)

$z_1 - z_2 = 85$ m

$\dot{W}_{12} = 0$ (because the pump is not operated when the water is flowing downwards)

Hence Equation (11.2) becomes

$$Z_F = 85 \text{ m}$$

The head lost to friction is the potential energy of the water.

When the flow is reversed, state 1 is now the surface water level of the ground floor tank, and state 2 is the surface water level of the roof tank. The mass flow rate of water $\{\dot{m}\}_w$ can be determined from the continuity equation, (Equation (8.3)) given a water velocity $\{C\}_w$ of 0.6 m/s, as follows:

$$\{\dot{m}\}_w = A_{xs}\{\rho C\}_w = 0.25\pi D^2 \{\rho C\}_w = 0.25\pi \times 0.05^2 \times 1000 \times 0.6 = 1.18 \text{ kg/s}$$

Considering Equation (11.2):

$$\{p_1\}_w = \{p_2\}_w = \text{atmospheric pressure}$$

$$\{C_1\}_w = \{C_2\}_w = 0 \text{ (for the same reasons as before)}$$

$$z_1 - z_2 = -85 \text{ m}$$

$$Z_F = 85 \text{ m}$$

$$\{\dot{m}\}_w = 1.18 \text{ kg/s}$$

Substituting in Equation (11.2) gives:

$$-85 = 85 - \frac{\dot{W}_{12}}{1.18g} \quad \therefore \quad \dot{W}_{12} = +1967.9 \text{ W}$$

At an energy transfer efficiency of 70%, the actual power input required at the pump $(\dot{W}_{12})_{act}$ is

$$(\dot{W}_{12})_{act} = \frac{1967.9}{0.7} = +2.8 \text{ kW}$$

The positive sign is because it is a pump and, therefore, a power input to the system. For Z_F to be the same when the water flows downwards of its own accord as when it is pumped upwards from one tank to the other, the mass flow rate and velocity of the water in the pipe must be the same, which is a particular case.

11.1 STEADY LAMINAR FLOW OF FLUID IN A STRAIGHT PIPE

In the simplest case, let an incompressible fluid flow under steady laminar conditions through a horizontal straight pipe. Consider a small element of the fluid about the axis of the pipe, as shown in Figure 11.2.

The element is of thickness $2r$ about the axis, and of length dx. There is a pressure p acting on the upstream face, and a pressure $(p + dp)$ on the downstream face. The pressure

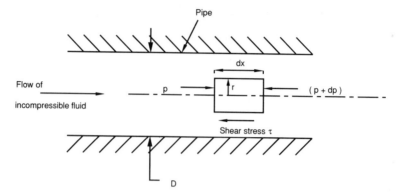

Fig. 11.2 *The steady laminar flow of an incompressible fluid through a horizontal pipe.*

is different because it falls in the direction of motion in order to provide a force to counter the friction force opposing the motion. Assuming the flow in the pipe is fully developed, there is a fluid velocity profile across the pipe (of diameter D). Each layer of fluid is moving at a different velocity and there are shear stresses between the layers. Thus, in Figure 11.2, there is a shear stress τ on the edge of the element opposing the flow direction.

For the steady mass flow ($\dot{m} = $ const.) of an incompressible fluid ($\rho = $ const) through a straight pipe ($A_{xs} = $ const.), the continuity equation predicts that the velocity of the fluid will also remain constant. In Figure 11.2, the velocity of any layer does not change as the fluid flows through the pipe, but the velocity of each layer is different. In other words there is a velocity profile across the pipe diameter. When applying the SFME to the element of fluid in Figure 11.2 in order to find the total force in the x-direction, the momentum force in the direction of flow is zero because the velocity does not change along the axis of the pipe. The body forces act vertically downwards and do not contribute to the total x-direction force. Therefore, the SFME applied to the element of fluid is

$$F_x = p\pi r^2 - (p + dp)\pi r^2$$

But as discussed in Section 10.3.6, the total force in the x-direction can only be the contribution due to the viscous force opposing the motion of the fluid. Hence, utilising Equation (2.12) for the viscous force, the SFME becomes

$$F_x = \tau 2\pi r \, dx = p\pi r^2 - (p + dp)\pi r^2$$

$$\therefore \quad \tau = -0.5r \frac{dp}{dx} \tag{11.3}$$

From Equation (2.13), the shear stress τ in laminar flow is given by

$$\tau = \mu \frac{dC}{dz}$$

In the radial direction, as the velocity is decreasing away from the centre line of the pipe, the equation is

$$\tau = -\mu \frac{dC}{dr} \tag{11.4}$$

Substituting Equation (11.4) into (11.3) gives

$$\mu \frac{dC}{dr} = 0.5r \frac{dp}{dx}$$

Given that the pressure varies only in the x-direction and not over the cross-section of the pipe, this equation can be integrated to give

$$C = \frac{0.25r^2}{\mu} \frac{dp}{dx} + A$$

The constant of integration A can be determined from the boundary conditions that, at the surface of the pipe, the velocity of the fluid is zero. For a pipe of diameter D, at $r = 0.5D$, $C = 0$:

$$\therefore \quad A = \frac{-D^2}{16\mu} \frac{dp}{dx}$$

Hence, the velocity variation across the pipe is

$$C = \frac{-0.25(0.25D^2 - r^2)}{\mu} \frac{dp}{dx} \quad (11.5)$$

The maximum velocity of the fluid C_{max} is at the centre of the pipe where $r = 0$, given by

$$C_{max} = \frac{-0.25^2 D^2}{\mu} \frac{dp}{dx}$$

Equation (11.5) shows that the velocity variation across the pipe is parabolic with the maximum velocity at the middle. This velocity distribution can be confirmed by measurements of the total and static pressure over a cross-section.

The volumetric rate of flow of fluid through the pipe $\dot{V}$ can be determined by considering an elemental ring of the fluid at radius r and of thickness dr, at any cross-section, as shown in Figure 11.3.

The volumetric flow rate $\dot{V}$ is given by the continuity equation as the velocity of the fluid multiplied by the cross-sectional area of flow. As the velocity varies across the pipe diameter, it is necessary to consider the flow in the elemental ring, and to integrate it over the cross-section. Thus:

$$\dot{V} = \int_0^{D/2} C 2\pi r \, dr$$

Substituting for C from Equation (11.5) and integrating gives

$$\dot{V} = \frac{-\pi D^4}{128\mu} \frac{dp}{dx}$$

The mean velocity of the fluid C_{mean} is the volumetric flow rate divided by the cross-sectional area of flow. Thus:

$$C_{mean} = \frac{-D^2}{32\mu} \frac{dp}{dx} = 0.5 C_{max} \quad (11.6)$$

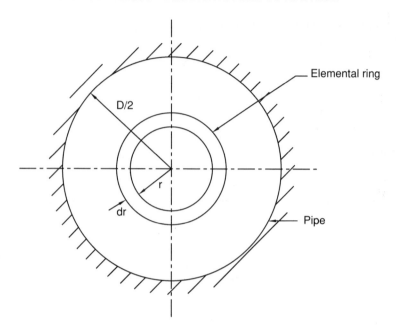

Fig. 11.3 *Elemental ring of fluid at any cross-section in a pipe through which the incompressible fluid is flowing under laminar conditions.*

The mean velocity is half the maximum velocity for the fully developed steady laminar flow of an incompressible fluid through a straight pipe. Equation (11.6) can be rearranged for a pipe of length L as follows:

$$dp = \frac{-32\mu C_{\text{mean}} L}{D^2}$$

$$\therefore \quad \frac{-dp}{\rho g} = \frac{64 L C_{\text{mean}}^2}{2g D Re} \tag{11.7}$$

The negative sign indicates that the pressure drops in the direction of motion. But as discussed in Section 10.3.6, the SFME for the flow of an incompressible fluid through a straight pipe equates the force due to the pressure drop in the pipe with the friction force opposing the motion of the fluid. In other words, the pressure of the fluid falls just sufficiently to maintain steady flow and constant velocity conditions. Therefore, the energy 'lost' to friction, the head lost to friction term Z_F in Equation (11.2), is equal to the pressure drop in the pipe expressed as a head. Equation (11.7) becomes

$$Z_F = \frac{-dp}{\rho g} = \frac{64 L C_{\text{mean}}^2}{2g D Re}$$

In practice, the mean subscript on the velocity term is usually ignored and the equation is written:

$$Z_F = \frac{-dp}{\rho g} = \frac{4 f L C^2}{2g D} \tag{11.8}$$

Here f is the friction factor discussed in Chapter 9. For the case of the fully developed steady laminar flow of an incompressible fluid through a straight pipe, the friction factor is given by

$$f = \frac{16}{Re} \qquad (11.9)$$

Such a result, and the validity of Equation (11.8), can be verified by experiment.

11.2 STEADY TURBULENT FLOW OF FLUID IN A STRAIGHT PIPE

When the fluid is again incompressible but turbulent, a theoretical analysis is not possible because Newton's law relating the shear stress to the velocity gradient cannot be used. Instead, recourse must be made to experiments to determine an empirical relationship. However, it is a reasonable assumption that, if Equation (11.8) holds true for laminar flow, a similar equation could be valid for turbulent flow, and Darcy showed this to be so. Equation (11.8) is often referred to as Darcy's equation. With turbulent flow though, the friction factor is dependent upon not only the Reynolds number but also the surface roughness of the pipe.

Note that, in some texts, the formula may be written:

$$Z_F = \frac{-dp}{\rho g} = \frac{fLC^2}{2gD}$$

In other words, the friction factor in some texts is four times the friction factor in this text. Care needs to be taken to avoid any confusion.

11.3 CONDITIONS AT ENTRY TO A PIPE

The equation for the head lost to friction Z_F only applies to that part of the flow which is fully developed, as described in Section 9.8.2. In the entry length, the velocity profile is changing and Equation (11.8) is not strictly valid. In most applications, the entry length is a small proportion of the total length of pipe in which the fully developed conditions prevail, and fully developed conditions can be assumed with little loss of accuracy. But care must be taken in those circumstances when this is not the case.

11.4 VARIATION OF FRICTION FACTOR

The friction factor for the fully developed steady flow of an incompressible fluid through a straight pipe can be plotted against the Reynolds number of the flow. For turbulent flow, an additional factor must be included to account for the surface roughness of the pipe. Such a set of curves was produced by Nikuradse for pipe which was artificially roughened by sand of known diameter, glued on to the inside of the pipe. The roughness was quantified by the ratio of the average height of the bumps on the surface k to the pipe diameter D. Nikuradse adopted this approach because it was easy to determine the value of k. Unfortunately, commercially available pipe does not have a uniform roughness and so, of more use, is the Moody diagram, after Moody who derived a similar set of curves but based

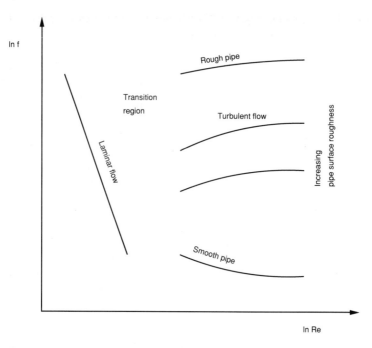

Fig. 11.4 Graph of ln(f) against ln(Re), based on the Moody diagram, for the laminar and turbulent flow of incompressible fluid in pipes of different roughness.

upon experiments carried out with ordinary commercial pipe. A diagram of a typical set of curves is shown in Figure 11.4.

11.5 EMPIRICAL FORMULA FOR FRICTION FACTOR

Figure 11.4 is extremely useful when sizing up pipes to transport a fluid from one place to another. Exact copies of the Moody diagram are available from pipe manufacturers, or can be purchased from suitable outlets. The laminar and turbulent regions can be distinguished clearly. Problems can arise when the friction factor of a pipe is required, but either the diameter or the velocity of flow is not known. In this case, it is necessary to assume a value, and carry out an iteration to achieve the correct answer.

In certain circumstances, it is easier to use an empirical formula. For laminar flow, there is only one formula, that of Equation (11.8). For turbulent flow, there are many possible formulae. One common equation for smooth pipes between a Reynolds number of 3000 and 10^5, is that proposed by Blasius, which is

$$f = \frac{0.079}{Re^{0.25}} \qquad (11.10)$$

Only this equation will be utilised further here, but a reference book on pipe flow will reveal many more for all the possible conditions of turbulent flow.

11.6 COMPRESSIBLE FLOW THROUGH A STRAIGHT PIPE

If the fluid flowing through the pipe is compressible, the Darcy relationship may only be applied when the variation of density is small. Even with a fluid such as air or natural gas, if the pipe is very long, the assumptions for an incompressible flow are not valid. When the density varies, the analysis is more complicated and is beyond the scope of this book.

EXAMPLE 11.2

Oil of density 800 kg/m³ and dynamic viscosity 0.3 kg/m s, is recirculated through a pipework system by a pump at the rate of 18 kg/s. The pipe is 25 m long and of diameter 75 mm (Figure 11.5). What is the power required to move the oil around the circuit if the pump has an energy transfer efficiency of 65%?

What is the maximum velocity of the oil in the pipe and the pressure drop of the oil per unit length of pipe?

Solution

The mean velocity of the oil in the pipe $\{C\}_{oil}$ can be determined from the continuity equation (Equation (8.3)) as follows:

$$\{\dot{m}\}_{oil} = A_{xs}\{\rho C\}_{oil}$$

$$\therefore \{C\}_{oil} = \frac{\{\dot{m}\}_{oil}}{0.25\pi D^2 \{\rho\}_{oil}} = \frac{18}{0.25\pi \times 0.075^2 \times 800} = 5.09 \text{ m/s}$$

The Reynolds number of the oil flow in the pipe Re is given by Equation (2.15) as

$$Re = \frac{\{\rho C\}_{oil} D}{\{\mu\}_{oil}} = \frac{800 \times 5.09 \times 0.075}{0.3} = 1018$$

Hence, the flow in the pipe is laminar, as $Re < 2000$, and the friction factor of the pipe f is given by Equation (11.9) as

$$f = \frac{16}{Re} = \frac{16}{1018} = 0.0157$$

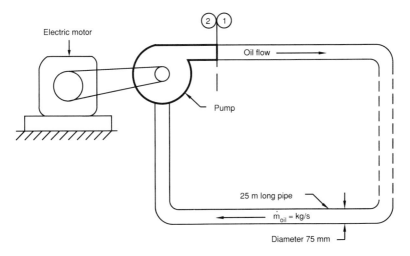

Fig. 11.5 Example 11.2—an oil flow through a pipe circuit containing a pump.

The head lost to friction in the pipe Z_F is given by Equation (11.8) as

$$Z_F = \frac{4fL\{C^2\}_{oil}}{2gD} = \frac{4 \times 0.0157 \times 25 \times 5.09^2}{2g \times 0.075} = 27.64 \text{ m}$$

Apply the SFEE from the outlet of the pump (state 1), around the circuit and back to the outlet of the pump (call it state 2 although it is the same place as state 1). Considering Equation (11.2):

$$\{p_1\}_{oil} = \{p_2\}_{oil}$$

$$\{C_1\}_{oil} = \{C_2\}_{oil}$$

$$z_1 = z_2$$

$$Z_F = 27.64 \text{ m}$$

Substituting in Equation (11.2) gives

$$0 = 27.64 - \frac{\dot{W}_{12}}{18g}$$

$$\therefore \quad \dot{W}_{12} = 27.64 \times 18 \times 9.81 = +4880.7 \text{ W}$$

At an energy transfer efficiency of 65%, the actual power input required to pump the oil $(\dot{W}_{12})_{act}$ is

$$(\dot{W}_{12})_{act} = \frac{4880.7}{0.65} = +7508.7 \text{ W}$$

The positive sign is because it is a work input to the system. The maximum velocity $\{C_{max}\}_{oil}$, because the flow is laminar, is given by Equation (11.6) as

$$\{C_{max}\}_{oil} = \{2C\}_{oil} = 2 \times 5.09 = 10.18 \text{ m/s}$$

The pressure drop of the oil per unit length of pipe $\{dp\}_{oil}/dx$ is also given by Equation (11.6) as:

$$\{C_{mean}\}_{oil} = \frac{-D^2 \{dp\}_{oil}}{32\{\mu\}_{oil} \, dx}$$

$$\therefore \quad 5.09 = \frac{-0.075^2 \, \{dp\}_{oil}}{32 \times 0.3 \, dx}$$

$$\therefore \quad \frac{\{dp\}_{oil}}{dx} = -8686.9 \text{ Pa per metre length}$$

This could also have been determined by applying the SFEE between the outlet of the pump (state 1), around the circuit to the inlet of the pump (state 3). Considering Equation (11.2):

$\{C_1\}_{oil} = \{C_3\}_{oil}$ (because the oil flows at constant velocity)

$z_1 = z_3$ (ignoring any difference in height of the pump inlet and outlet)

$Z_F = 27.64$ m

$\dot{W}_{13} = 0$ (because the pump is no longer within the inlet and outlet positions of the SFEE)

Substituting in Equation (11.2) gives

$$\frac{\{p_1\}_{oil}}{800g} = \frac{\{p_3\}_{oil}}{800g} + 27.64$$

$$\therefore \{p_3 - p_1\}_{oil} = -27.64 \times 800g = -216\,918.7 \text{ Pa}$$

If the pipe is 25 m long, the pressure drop per unit length of pipe is

$$\{p_3 - p_1\}_{oil} = \frac{-216\,918.7}{25} = -8676.7 \text{ Pa per metre length}$$

The difference is in the rounding up of numbers. As the pump recirculates the oil around the system, all the work input to the pump is required to overcome friction effects opposing the motion of the oil, there is no net contribution to or from the potential energy.

EXAMPLE 11.3

An oil is used to lubricate a hydrostatic thrust bearing. The oil has a density of 800 kg/m³ and a dynamic viscosity of 0.035 kg/m s. The oil is pumped to the bearing through a 10 mm diameter horizontal pipe of length 3 m at the rate of 0.16 kg/s. It enters the bearing with a pressure of 550 kPa (Figure 11.6). What is the pressure of the oil at outlet from the pump?

After some time, the piece of equipment in which the bearing is situated is redesigned so that the bearing ends up in a position 2 m above the pump. What is now the pressure of the oil at outlet from the pump?

Solution

The mean velocity of the oil in the pipe $\{C\}_{oil}$ can be determined from the continuity equation (Equation (8.3)) as follows:

$$\{\dot{m}\}_{oil} = A_{xs}\{\rho C\}_{oil}$$

$$\therefore \{C\}_{oil} = \frac{\{\dot{m}\}_{oil}}{0.25\pi D^2\{\rho\}_{oil}} = \frac{0.16}{0.25\pi \times 0.01^2 \times 800} = 2.6 \text{ m/s}$$

The Reynolds number of the oil flow in the pipe Re is given by Equation (2.15) as

$$Re = \frac{\{\rho C\}_{oil} D}{\{\mu\}_{oil}} = \frac{800 \times 2.6 \times 0.01}{0.035} = 594.3$$

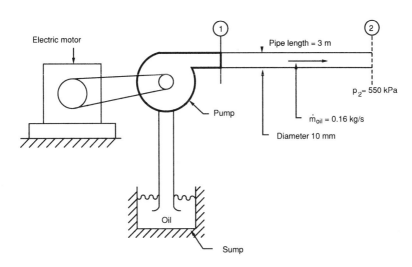

Fig. 11.6 *Example 11.3—oil being pumped to a hydrostatic thrust bearing.*

Hence, the flow in the pipe is laminar, as $Re < 2000$, and the friction factor of the pipe f is given by Equation (11.9) as

$$f = \frac{16}{Re} = \frac{16}{594.3} = 0.027$$

The head lost to friction in the pipe is given by Equation (11.8) as

$$Z_F = \frac{4fL\{C^2\}_{oil}}{2gD} = \frac{4 \times 0.027 \times 3 \times 2.6^2}{2g \times 0.01} = 11.2 \text{ m}$$

Apply the SFEE between the outlet of the pump, and hence the inlet to the pipe (state 1), and the outlet of the pipe (state 2). Considering Equation (11.2):

$\{C_1\}_{oil} = \{C_2\}_{oil}$ (because the oil flows at constant velocity in the pipe)

$z_1 = z_2$ (because the pipe is horizontal)

$\{p_2\}_{oil} = 5.5 \times 10^5$ Pa

$\{\rho_1\}_{oil} = \{\rho_2\}_{oil} = 800$ kg/m^3 (because the oil is incompressible)

$Z_F = 11.2$ m

$\dot{W}_{12} = 0$ (because the pump is not within the inlet and outlet of the pipe)

Substituting in Equation (11.2) gives:

$$\frac{\{p_1\}_{oil}}{800g} = \frac{5.5 \times 10^5}{800g} + 11.2$$

$$\therefore \{p_1\}_{oil} = 638 \text{ kPa}$$

If the bearing is 2 m above the pump, then

$$z_2 - z_1 = 2 \text{ m}$$

Substituting in Equation (11.2) gives

$$\frac{\{p_1\}_{oil}}{800g} = \frac{5.5 \times 10^5}{800g} + 2 + 11.2$$

$$\therefore \{p_1\}_{oil} = 654 \text{ kPa}$$

When the bearing is in a position above the pump, most of the energy put into the fluid by the pump goes towards overcoming friction in the pipe, not in overcoming the potential energy due to the height difference.

EXAMPLE 11.4

Oil is pumped at the rate of 2400 kg/s from storage tanks at an ocean terminal to storage tanks situated in a refinery 150 km inland, through a 1.25 m diameter pipe. The refinery is at an elevation of 50 m above the ocean terminal. The oil has a density of 800 kg/m^3 and a dynamic viscosity of 0.25 kg/m s (Figure 11.7). What is the power input required to drive the pump if it has an energy transfer efficiency of 75%?

Due to a mistake by the contractor, only 75 km of the 1.25 m diameter pipe was installed, the other 75 km of the pipe was 1.0 m in diameter. What is the percentage increase in the power input required to drive the pump?

If the pump is driven by a 25% efficient gas turbine using natural gas at $0.03/kWh, what is the extra annual running cost of the gas turbine caused by the contractor's mistake? Assume that oil is pumped for 8000 hours per year.

COMPRESSIBLE FLOW THROUGH A STRAIGHT PIPE

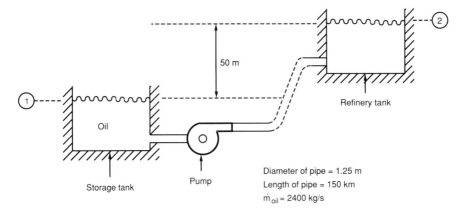

Fig. 11.7 *Example 11.4—oil being pumped from storage tanks at an ocean terminal to storage tanks at a refinery 150 km inland.*

Solution

In the first case, the mean velocity of the oil in the pipe $\{C\}_{\text{oil}}$ can be determined from the continuity equation (Equation (8.3)) as follows:

$$\{\dot{m}\}_{\text{oil}} = A_{\text{xs}}\{\rho C\}_{\text{oil}}$$

$$\therefore \{C\}_{\text{oil}} = \frac{\{\dot{m}\}_{\text{oil}}}{0.25\pi D^2 \{\rho\}_{\text{oil}}} = \frac{2400}{0.25\pi \times 1.25^2 \times 800} = 2.44 \text{ m/s}$$

The Reynolds number of the flow in the pipe Re is given by Equation (2.15) as

$$Re = \frac{\{\rho C\}_{\text{oil}} D}{\{\mu\}_{\text{oil}}} = \frac{800 \times 2.44 \times 1.25}{0.25} = 9760$$

Hence, the flow in the pipe is turbulent, as $Re > 3500$. The friction factor of the pipe can be determined from the Moody diagram or, assuming the pipe is smooth, from the Blasius relationship of Equation (11.10). No information about the pipe roughness is known, so the latter approach is adopted, as follows:

$$f = \frac{0.079}{Re^{0.25}} = \frac{0.079}{9760^{0.25}} = 0.007\,95$$

The head lost to friction in the pipe is given by Equation (11.8) as

$$Z_F = \frac{4fL\{C^2\}_{\text{oil}}}{2gD} = \frac{4 \times 0.00795 \times 150 \times 10^3 \times 2.44^2}{2g \times 1.25} = 1157.9 \text{ m}$$

Apply the SFEE between the surface level of the oil in the storage tanks at the ocean terminal (state 1) and the surface level of the oil in the storage tanks at the refinery (state 2). Considering Equation (11.2):

$\{p_1\}_{\text{oil}} = \{p_2\}_{\text{oil}} =$ atmospheric pressure

$\{C_1\}_{\text{oil}} = \{C_2\}_{\text{oil}} = 0$ (because the levels of the oil in the tanks will only slowly fall or rise)

$z_1 - z_2 = -50$ m

$Z_F = 1157.9$ m

Substituting in Equation (11.2) gives

$$-50 = 1157.9 - \frac{\dot{W}_{12}}{2400g}$$

$$\therefore \quad \dot{W}_{12} = 2400g \times 1207.9 = 28.44 \text{ MW}$$

At an energy transfer efficiency of 75%, the actual power input required $(\dot{W}_{12})_{act}$ is

$$(\dot{W}_{12})_{act} = \frac{28.44}{0.75} = 37.92 \text{ MW}$$

As the gas turbine is only 25% efficient, the heat input $\dot{Q}_i$ from the combustion of the natural gas fuel required to produce a power output of 37.92 MW is

$$\dot{Q}_i = \frac{37.92}{0.25} = 151.7 \text{ MW}$$

The annual running cost $\$_{ARC}$ for this pump is given by Equation (4.8) as

$$\$_{ARC} = 151.7 \times 10^3 \times 8000 \times 0.03 = \$36.4 \text{million per year}$$

In the second case, due to the change in the diameter of the pipe, the velocity, Reynolds number, friction factor and head lost to friction of the oil in the pipe will change from one section of the pipe to the other. Assuming the first 75 km of the pipe is at 1.25 m diameter, the values will be as above except that the head lost to friction will be half the original value. Thus, the head lost to friction in the 1.25 m diameter length of the pipe Z'_F is

$$Z'_F = 1157.9 \times 0.5 = 578.95 \text{ m}$$

In the 1.0 m diameter pipe of cross-sectional area A''_{xs} and diameter D'', the mean velocity of the oil $\{C''\}_{oil}$, from the continuity equation (Equation (8.3)) is

$$\{\dot{m}\}_{oil} = A''_{xs}\{\rho C''\}_{oil}$$

$$\therefore \quad \{C''\}_{oil} = \frac{\{\dot{m}_{oil}\}}{0.25\pi D''^2\{\rho\}_{oil}} = \frac{2400}{0.25\pi \times 1^2 \times 800} = 3.82 \text{ m/s}$$

The Reynolds number of the flow in the 1.0 m diameter pipe Re'' is given by Equation (2.15) as

$$Re'' = \frac{\{\rho C''\}_{oil} D''}{\{\mu\}_{oil}} = \frac{800 \times 3.82 \times 1}{0.25} = 12\,224$$

Hence, the flow in the pipe is turbulent, as $Re'' > 3500$. Using the Blasius formula of Equation (11.10) to find the friction factor in the 1.0 m diameter pipe f'' gives

$$f'' = \frac{0.079}{Re''^{0.25}} = \frac{0.079}{12\,224^{0.25}} = 0.0075$$

The head lost to friction in the 1.0 m diameter pipe Z''_F of length L'' is given by Equation (11.8) as

$$Z''_F = \frac{4f''L''\{C''^2\}_{oil}}{2gD''} = \frac{4 \times 0.0075 \times 75 \times 10^3 \times 3.82^2}{2g \times 1} = 1673.4 \text{ m}$$

Apply the SFEE between the surface level of the oil in the storage tanks at the ocean terminal and the surface level of the oil in the storage tanks at the refinery again. Considering Equation (11.2):

$\{p_1\}_{oil} = \{p_2\}_{oil}$ = atmospheric pressure

$\{C_1\}_{oil} = \{C_2\}_{oil} = 0$ (because the levels of the oil in the tanks will only fall or rise slowly)

$z_1 - z_2 = -50$ m

$Z_F = Z'_F + Z''_F = 578.95 + 1673.4 = 2252.35$ m

Substituting in Equation (11.2) gives

$$-50 = 2252.35 - \frac{\dot{W}_{12}}{2400g}$$

$$\therefore \dot{W}_{12} = 2400g \times 2302.35 = 54.21 \text{ MW}$$

At an energy transfer efficiency of 75%, the actual power input required $(\dot{W}_{12})_{act}$ is

$$(\dot{W}_{12})_{act} = \frac{54.21}{0.75} = 72.28 \text{ MW}$$

As the gas turbine is only 25% efficient, the heat input $\dot{Q}_i$ from the combustion of the natural gas fuel required to produce a power output of 72.28 MW is

$$\dot{Q}_i = \frac{72.28}{0.25} = 289.1 \text{ MW}$$

The annual running cost for this pump $\$'_{ARC}$ is given by Equation (4.8) as

$$\$'_{ARC} = 289.1 \times 10^3 \times 8000 \times 0.03 = \$69.4 \text{ million per year}$$

The percentage increase in the annual running cost due to the contractor's mistake $(\%\$_{ARC})$ is

$$(\%\$_{ARC}) = \frac{69.4 - 36.4}{36.4} = 90.7\%$$

Note that, in the 1.0 m diameter pipe, the velocity of the oil is greater. In the equation for Z_F, the velocity term is squared and this has a considerable effect upon the amount of energy required to overcome friction. In addition, the smaller the diameter pipe, the larger is the head lost term. As a result, the Z_F for the 1.0 m diameter pipe is nearly three times the Z_F of the 1.25 m diameter pipe, although they are the same length and transport the same mass flow rate of oil.

In practice, a pump sized for a 1.25 m diameter pipe will not be able to achieve the oil flow rate anticipated if half of the pipeline is pipe of diameter only 1.0 m.

EXAMPLE 11.5

A farmer wants to pump water of density 1000 kg/m³ and dynamic viscosity 10^{-3} kg/m s from a well into a storage reservoir. He places a pump on the ground next to the well, and couples an 8 m long suction pipe of diameter 100 mm from the pump and into the well. The well water level is 5 m below the pump axis. On the discharge side, he couples a 25 m long delivery pipe of diameter 125 mm from the pump outlet and into the storage reservoir. The water level in the reservoir is at a height of 15 m above the pump axis (Figure 11.8). If both the suction and

316 THE STEADY FLOW ENERGY EQUATION APPLIED TO PIPE FLOW

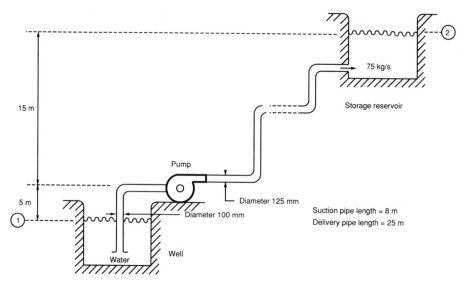

Fig. 11.8 *Example 11.5—water being pumped from a well into a storage reservoir.*

delivery pipes can be assumed smooth and for the Blasius formula for friction factor to prevail, what is the power input required at the pump if the water flow rate is 75 kg/s and the pump has an energy transfer efficiency of 65%?

Solution

The mean velocity of the water in the suction pipe $\{C''\}_w$ of cross-sectional area A''_{xs} and diameter D'' can be determined from the continuity equation (Equation (8.3)) as follows:

$$\{\dot{m}\}_w = A''_{xs}\{\rho C''\}_w$$

$$\therefore \{C''\}_w = \frac{\{\dot{m}\}_w}{0.25\pi D''^2 \{\rho\}_w} = \frac{75}{0.25\pi \times 0.1^2 \times 1000} = 9.55 \text{ m/s}$$

The Reynolds number of the flow in the suction pipe Re'' is given by Equation (2.15) as

$$Re'' = \frac{\{\rho C''\}_w D''}{\{\mu\}_w} = \frac{1000 \times 9.55 \times 0.1}{10^{-3}} = 955\,000$$

Hence, the flow in the suction pipe is turbulent, as $Re'' > 3500$. The friction factor of the suction pipe f'' can be determined from the Blasius relationship of Equation (11.10), as follows:

$$f'' = \frac{0.079}{Re''^{0.25}} = \frac{0.079}{955\,000^{0.25}} = 0.0025$$

The head lost to friction in the suction pipe Z''_F of length L'' is given by Equation (11.8) as

$$Z''_F = \frac{4f''L''\{C''^2\}_w}{2gD''} = \frac{4 \times 0.0025 \times 8 \times 9.55^2}{2g \times 0.1} = 3.72 \text{ m}$$

The mean velocity of the water in the delivery pipe $\{C'\}_w$ of cross-sectional area A'_{xs} and diameter D' can be determined from the continuity equation (Equation (8.3)) as follows:

$$\{\dot{m}\}_w = A'_{xs}\{\rho C'\}_w$$

$$\therefore \{C'\}_w = \frac{\{\dot{m}\}_w}{0.25\,\pi D'^2 \{\rho\}_w} = \frac{75}{0.25\pi \times 0.125^2 \times 1000} = 6.11 \text{ m/s}$$

The Reynolds number of the flow in the delivery pipe Re' is given by Equation (2.15) as

$$Re' = \frac{\{\rho C'\}_w D'}{\{\mu\}_w} = \frac{1000 \times 6.11 \times 0.125}{10^{-3}} = 763\,750$$

Hence, the flow in the delivery pipe is turbulent, as $Re' > 3500$. The friction factor of the delivery pipe f' can be determined from the Blasius relationship of Equation (11.10), as follows:

$$f' = \frac{0.079}{Re'^{0.25}} = \frac{0.079}{763\,750^{0.25}} = 0.002\,67$$

The head lost to friction in the delivery pipe Z'_F of length L' is given by Equation (11.8) as

$$Z'_F = \frac{4f'L'\{C'^2\}_w}{2gD'} = \frac{4 \times 0.00267 \times 25 \times 6.11^2}{2g \times 0.125} = 4.06 \text{ m}$$

Apply the SFEE between the surface level of the water in the well (state 1) and the surface level of the water in the storage reservoir (state 2). Considering Equation (11.2):

$\{p_1\}_w = \{p_2\}_w$ = atmospheric pressure

$\{C_1\}_w = \{C_2\}_w$ = 0 (because the level of the water in the well and the storage reservoir will only slowly fall or rise)

$z_1 - z_2 = -(5 + 15) = -20$ m

$Z_F = Z''_F + Z'_F = 3.72 + 4.06 = 7.78$ m

Substituting in Equation (11.2) gives

$$-20 = 7.78 - \frac{\dot{W}_{12}}{75g}$$

$$\therefore \dot{W}_{12} = 75g \times 27.78 = 20.4 \text{ kW}$$

At an energy transfer efficiency of 65%, the actual power input required $(\dot{W}_{12})_{act}$ is

$$(\dot{W}_{12})_{act} = \frac{20.4}{0.65} = 31.4 \text{ kW}$$

Note that the head lost to friction in the delivery pipe is approximately the same as in the suction pipe even though it is three times as long. This is because the diameter of the delivery pipe is larger than the suction pipe. Therefore, the velocity of water in the delivery pipe is less than the velocity of water in the suction pipe, and the velocity term is the dominant one in the equation for the head lost.

A suction pipe and pump in such an installation must be full of the liquid, water in this case, before the pump is switched on, otherwise the pump will turn with air only in the blading and not be effective. This is known as priming the pump.

11.7 OTHER 'HEAD LOSSES' IN PIPES

As well as the energy in the fluid used to overcome the effects of friction when it flows through a pipe, the head lost to friction term Z_F, there are other head losses which are grouped under the general heading of Minor Losses. These occur whenever the fluid encounters a disturbance to its flow such as a bend in the pipe, a contraction in the pipe diameter, an expansion in the pipe diameter, at the entrance and exit of the pipe and at any fitting in the pipe such as a valve. In a short length of pipe, the minor losses may be greater than Z_F, but in longer lengths of pipe, the head lost to friction generally predominates.

Some of the minor loss situations can be examined analytically. However, most cannot, and the easiest approach is to quantify each minor head loss Z_{FM} as a number of velocity heads as follows:

$$Z_{FM} = \frac{k_{FM} C^2}{2g}$$

Here k_{FM} is dimensionless and is called the Loss Coefficient. It will have a value dependent upon what is causing the disturbance to the flow. It is not always possible to give an exact value for k_{FM} because it can vary in different circumstances, but Table 11.1 provides a

Table 11.1 Representative values for the loss coefficient k_{FM} for pipe flow.

Type of disturbance to flow	k_{FM}
Sudden enlargement (based upon $C_{DN}^2/2g$)	$((A_{xs})_{DN}/(A_{xs})_U) - 1)^2$
Sudden enlargement (based upon $C_U^2/2g$)	$(1 - (A_{xs})_U/(A_{xs})_{DN}))^2$
Gradual enlargement	1 (inc. angle 50°) 0.7 (inc. angle 30°) 0.15 (inc. angle 10°)
Exit from pipe to large reservoir (based upon $C_U^2/2g$)	1
Sudden contraction (based upon $C_{DN}^2/2g$)	0.5 ($D_{DN}/D_U = 0$) 0.45 ($D_{DN}/D_U = 0.2$) 0.38 ($D_{DN}/D_U = 0.4$) 0.28 ($D_{DN}/D_U = 0.6$) 0.14 ($D_{DN}/D_U = 0.8$) 0 ($D_{DN}/D_U = 1.0$)
Gradual contraction	0
Entry from large reservoir to pipe (based upon $C_{DN}^2/2g$)	0.5 (sharp entry) 0 (radiused entry)
Small radius 90° bend (based upon $C_U^2/2g$)	1.1
Large radius 90° bend (based upon $C_U^2/2g$)	0.6
90° bend with cascades (based upon $C_U^2/2g$)	0.2

Table 11.2 *Representative values for the loss coefficient k_{FM} for pipe fittings.*

Pipe fitting	k_{FM}
90-degree elbow (based upon $C_U^2/2g$)	0.9
45-degree elbow (based upon $C_U^2/2g$)	0.4
Main inlet to T-junction (based upon $C_U^2/2g$)	0.4
Side outlet from T-junction	1.8
Globe valve (based upon $C_U^2/2g$)	10 (wide open) 11 (75% open) 12.5 (50% open)
Gate valve (based upon $C_U^2/2g$)	0.2 (Wide open) 1 (75% open) 5 (50% open) 24 (25% open)
Foot valve (based upon $C_U^2/2g$)	1.5 (hinged) 10 (lift)
Check valve (based upon $C_U^2/2g$)	2.5 (hinged) 4 (ball) 15 (lift)

summary of suitable values that are appropriate to most applications for flow in pipes and Table 11.2 for pipe fittings. They are based upon using either the upstream or the downstream velocity in Z_{FM}. The upstream conditions have a subscript U and the downstream conditions have a subscript DN.

The SFEE, including minor losses, becomes

$$\frac{p_1}{\rho_1 g} + \frac{0.5 C_1^2}{g} + z_1 = \frac{p_2}{\rho_2 g} + \frac{0.5 C_2^2}{g} + z_2 + Z_F + Z_{FM} - \frac{\dot{W}_{12}}{\dot{m} g} \quad (11.11)$$

EXAMPLE 11.6

Water is pumped from one reservoir to another through a vertical height of 45 m. Just after the entrance to the pipe from the lower reservoir, and before the pump, there is a fully open globe valve fitted. The pump suction pipe is 5 m long, 50 mm in diameter and has a friction factor of 0.008. The delivery pipe is 120 m long, of diameter 75 mm, has a friction factor of 0.01, and contains two 90° elbows (Figure 11.9). What is the power required to pump the water, of density 1000 kg/m³, at the rate of 32 kg/s, if the pump has an energy transfer efficiency of 70%? Account for minor losses.

Solution

The mean velocity of the water in the suction pipe $\{C''\}_w$ of cross-sectional area A''_{xs} and diameter D'' can be determined from the continuity equation (Equation (8.3)) as follows:

$$\{\dot{m}\}_w = A''_{xs} \{\rho C''\}_w$$

320 THE STEADY FLOW ENERGY EQUATION APPLIED TO PIPE FLOW

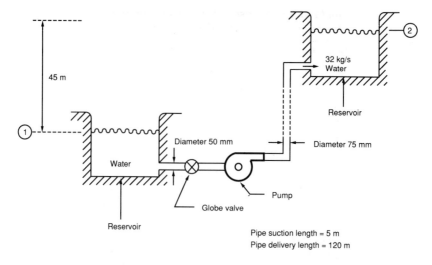

Fig. 11.9 Example 11.6—water being pumped from one reservoir to another.

$$\therefore \{C''\}_w = \frac{\{\dot{m}\}_w}{0.25\pi\, D''^2 \{\rho\}_w} = \frac{32}{0.25\pi \times 0.05^2 \times 1000} = 16.3 \text{ m/s}$$

The head lost to friction in the suction pipe Z''_F of length L'' and friction factor f'' is given by Equation (11.8) as

$$Z''_F = \frac{4f''L''\{C''^2\}_w}{2g\,D''} = \frac{4 \times 0.008 \times 5 \times 16.3^2}{2g \times 0.05} = 43.33 \text{ m}$$

The mean velocity of the water in the delivery pipe $\{C'\}_w$ of cross-sectional area A'_{xs} and diameter D' can be determined from the continuity equation (Equation (8.3)) as follows:

$$\{\dot{m}\}_w = A'_{xs} \{\rho C'\}_w$$

$$\therefore \{C'\}_w = \frac{\{\dot{m}\}_w}{0.25\pi\, D'^2 \{\rho\}_w} = \frac{32}{0.25\pi \times 0.075^2 \times 1000} = 7.24 \text{ m/s}$$

The head lost to friction in the delivery pipe Z'_F of length L' and friction factor f' is given by Equation (11.8) as

$$Z'_F = \frac{4f'L'\{C'^2\}_w}{2g\,D'} = \frac{4 \times 0.01 \times 120 \times 7.24^2}{2g \times 0.075} = 171 \text{ m}$$

The total head lost to friction Z_F in the suction and delivery pipes is

$$Z_F = Z''_F + Z'_F = 43.33 + 171 = 214.33 \text{ m}$$

The minor losses Z_{FM} include the entrance (assumed to be sharp), the globe valve, the two 90° elbows and the exit loss. Using Table 11.1 with the appropriate upstream and downstream water velocities, the minor losses are

$$Z_{FM} = \frac{0.5\{C''^2\}_w}{2g} + \frac{10\{C''^2\}_w}{2g} + \frac{2 \times 0.9\{C'^2\}_w}{2g} + \frac{\{C'^2\}_w}{2g}$$

$$\therefore Z_{FM} = 0.54 \times 16.3^2 + 0.14 \times 7.24^2 = 150.8 \text{ m}$$

Apply the SFEE between the surface level of the water in the lower reservoir (state 1) and the surface level of the water in the upper reservoir (state 2). Considering Equation (11.11):

$\{p_1\}_w = \{p_2\}_w$ = atmospheric pressure

$\{C_1\}_w = \{C_2\}_w = 0$ (because the level of the water in the storage reservoirs will only fall or rise slowly)

$z_1 - z_2 = -45$ m

$Z_F + Z_{FM} = 214.33 + 150.8 = 365.13$ m

Substituting in Equation (11.11) gives

$$-45 = 365.13 - \frac{\dot{W}_{12}}{32g}$$

$$\therefore \dot{W}_{12} = 32g \times 410.13 = 128.8 \text{ kW}$$

At an energy transfer efficiency of 70%, the actual power input required $(\dot{W}_{12})_{act}$ is

$$(\dot{W}_{12})_{act} = \frac{128.8}{0.7} = 184 \text{ kW}$$

The minor losses represent 41% of the head required to overcome friction forces and 37% of the power input required at the pump, a not-insignificant proportion.

11.8 EQUIVALENT LENGTH

The minor losses incurred in fittings are sometimes expressed as an Equivalent Length L_{EQ} of straight pipe of the same diameter as that which includes the fitting and which will produce the same head loss as the fitting. In other words, equating Z_F with Z_{FM} gives

$$\frac{4fL_{EQ}C^2}{D\ 2g} = \frac{k_{FM}C^2}{2g}$$

$$\therefore L_{EQ} = \frac{k_{FM}D}{4f} \quad (11.12)$$

L_{EQ} is normally calculated as a number of diameters.

EXAMPLE 11.7

A reservoir, situated outside a town, is to be used to supply water of density 1000 kg/m³ at the local fire station. The surface level of the water in the reservoir is 155 m above the outlet hydrant in the station. It is estimated that the distance between the reservoir and the hydrant is 25 km, and that any pipe fitted will have to have 5 large radius 90° bends, 3 small radius 90° bends, 12 gate valves and 1 globe valve. It is decided to use a pipe with a friction factor of 0.01. It is stipulated by the local authorities that the pressure at the hydrant must be 200 kPa gauge when the flow rate of water is 4 kg/s (Figure 11.10). What diameter pipe should be specified, accounting for minor losses and assuming that all the valves are fully open? Express the minor losses as an equivalent length of the pipe.

Solution
The continuity equation relates the velocity of water in the pipe $\{C\}_w$ with the pipe diameter D. Equation (8.3) is

$$\{\dot{m}\}_w = A_{xs}\{\rho C\}_w$$

322 THE STEADY FLOW ENERGY EQUATION APPLIED TO PIPE FLOW

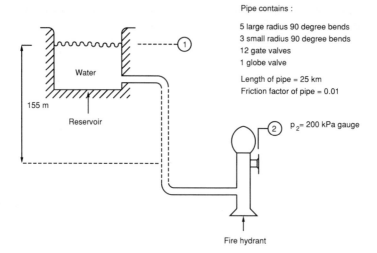

Fig. 11.10 Example 11.7—water flowing from a reservoir to a hydrant in a fire station.

$$\therefore \quad 4 = 0.25\pi D^2 \times 1000 \{C\}_w$$

$$\therefore \quad \{C\}_w = \frac{5.1 \times 10^{-3}}{D^2}$$

The head lost to friction in the pipe Z_F is given by Equation (11.8) as

$$Z_F = \frac{4fL\{C^2\}_w}{2gD} = \frac{4 \times 0.01 \times 25 \times 10^3 \times 5.1^2 \times 10^{-6}}{2gD^5} = \frac{1.33 \times 10^{-3}}{D^5}$$

The minor losses Z_{FM} include the entrance to the pipe (assumed to be sharp), the 5 large radius bends, the 3 small radius bends, the 12 gate valves, and the globe valve. Using Table 11.1 with the upstream velocity being $\{C\}_w$, the minor losses are

$$Z_{FM} = (0.5 + 5 \times 0.6 + 3 \times 1.1 + 12 \times 0.2 + 1 \times 10)\frac{\{C\}_w^2}{2g}$$

$$\therefore \quad Z_{FM} = \frac{19.2\{C\}_w^2}{2g} = \frac{25.45 \times 10^{-6}}{D^4}$$

Note that the exit loss is not included because it appears in the SFEE.

Apply the SFEE between the surface level of the water in the reservoir (state 1) and the outlet of water at the hydrant (state 2). Considering Equation (11.11):

$$\{p_1\}_w = \text{atmospheric pressure}$$

$$\{C_1\}_w = 0 \text{ (because the level of the water in the reservoir will fall only slowly)}$$

$$z_1 - z_2 = 155 \text{ m}$$

$$\{p_2\}_w = 2 \times 10^5 \text{ Pa} + \text{atmospheric pressure}$$

$$\{C_2\}_w = \{C\}_w = \frac{5.1 \times 10^{-3}}{D^2}$$

$$(Z_F + Z_{FM}) = \frac{1.33 \times 10^{-3}}{D^5} + \frac{25.45 \times 10^{-6}}{D^4}$$

$$\dot{W}_{12} = 0 \text{ (because there is no pump)}$$

Substituting in Equation (11.11) gives

$$155 = \frac{2 \times 10^5}{10^3 g} + \frac{(5.1 \times 10^{-3})^2}{2g D^4} + \frac{1.33 \times 10^{-3}}{D^5} + \frac{25.45 \times 10^{-6}}{D^4}$$

A trial and error solution gives the approximate answer of $D = 100$ mm. The equivalent length of pipe L_{EQ} corresponding to the minor losses is given by Equation (11.12) as

$$L_{EQ} = \frac{k_{FM} D}{4f} = \frac{19.2 \times 0.1}{4 \times 0.01} = 48 \text{ m of pipe}$$

Almost any analysis in which the pipe diameter is an unknown requires a trial and error solution, which can be somewhat time-consuming.

11.9 COMBINATIONS OF PIPES

11.9.1 Pipes in series

If pipes are connected end to end, the equation of continuity must apply and the mass flow rate of fluid passing through one pipe must be the same as the mass flow rate of fluid passing through the other pipe. The total head lost will be the sum of the head lost to friction in each pipe plus losses due to any fittings in either pipe.

EXAMPLE 11.8

Two water reservoirs are connected by a 240 m long reasonably straight pipeline. There is a difference in height of 30 m between the water levels in each reservoir. The friction factor of the pipe is 0.007 and its diameter is 100 mm. What will be the flow rate of the water of density 1000 kg/m³ through the pipe, accounting for minor losses? There is a fully open gate valve situated in the pipeline near the entrance to the pipe (Figure 11.11).

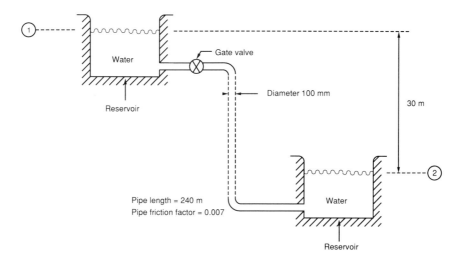

Fig. 11.11 Example 11.8—water flowing from one reservoir to another.

If the diameter of the pipe was 125 mm, but with the same friction factor, what would be the percentage increase in the flow rate of the water?

If the first half of the pipe length was of diameter 100 mm, and the second half of diameter 125 mm, what would now be the flow rate of the water, assuming the friction factor for both pipes is again 0.007?

Solution

For the 100 mm diameter pipe, the head lost to friction in the pipe Z_F is given by Equation (11.8) as

$$Z_F = \frac{4fL\{C^2\}_w}{2gD} = \frac{4 \times 0.007 \times 240\{C\}_w^2}{2g \times 0.1} = 3.43\{C\}_w^2$$

The minor losses Z_{FM} are due to the entrance to the pipe (assumed to be sharp), the gate valve and the exit from the pipe. From Table 11.1 with an upstream velocity of $\{C\}_w$, the minor losses are

$$Z_{FM} = (0.5 + 0.2 + 1.0)\frac{\{C\}_w^2}{2g} = 0.087\{C\}_w^2$$

Apply the SFEE between the water surface level in the upper reservoir (state 1) and the water surface level in the lower reservoir (state 2). Considering Equation (11.11):

$\{p_1\}_w = \{p_2\}_w =$ atmospheric pressure

$\{C_1\}_w = \{C_2\}_w = 0$ (because the level of the water in the reservoirs will only slowly fall or rise)

$z_1 - z_2 = 30$ m

$Z_F + Z_{FM} = (3.43 + 0.087)\{C\}_w^2 = 3.517\{C\}_w^2$

$\dot{W}_{12} = 0$ (because there is no pump)

Substituting in Equation (11.11) gives

$$30 = 3.517\{C\}_w^2 \quad \therefore \quad \{C\}_w = 2.92 \text{ m/s}$$

The mass flow rate of water $\{\dot{m}\}_w$ is given by the continuity equation (Equation (8.3)) as:

$$\{\dot{m}\}_w = A_{xs}\{\rho C\}_w = 0.25\pi \times 0.1^2 \times 1000 \times 2.92 = 22.9 \text{ kg/s}$$

When the pipe is of diameter 125 mm, the water will flow with an increased velocity. The only change to the calculation required to achieve this is in the term for the head lost to friction, which contains the diameter of the pipe, and becomes

$$Z_F = \frac{4fL\{C^2\}_w}{2gD} = \frac{4 \times 0.007 \times 240\{C\}_w^2}{2g \times 0.125} = 2.74\{C\}_w^2$$

Substituting in Equation (11.11) gives

$$30 = (2.74 + 0.087)\{C\}_w^2$$

$$\therefore \quad \{C\}_w = 3.26 \text{ m/s}$$

The mass flow rate with the bigger pipe is

$$\{\dot{m}\}_w = A_{xs}\{\rho C\}_w = 0.25\pi \times 0.125^2 \times 1000 \times 3.26 = 40 \text{ kg/s}$$

COMBINATIONS OF PIPES

The percentage increase in the mass flow rate of water $\{\%\dot{m}\}_w$ is

$$\{\%\dot{m}\}_w = \frac{40 - 22.9}{22.9} = 74.7\%$$

When the first half of the pipe is 100 mm in diameter (superscript ′) and the second half is 125 mm in diameter (superscript ″), the mass flow rate of water $\{\dot{m}\}_w$ in each pipe must be the same. From the continuity equation (Equation (8.3)):

$$\{\dot{m}\}_w = A'_{xs}\{\rho C'\}_w = A''_{xs}\{\rho C''\}_w$$

$$\therefore \{\dot{m}\}_w = 0.25\pi \times 0.1^2 \times 1000\{C'\}_w = 0.25\pi \times 0.125^2 \times 1000\{C''\}_w$$

$$\therefore \{\dot{m}\}_w = 7.85\{C'\}_w = 12.27\{C''\}_w$$

For the first half of the pipe, the head lost to friction Z'_F will be

$$Z'_F = \frac{4f'L'\{C'^2\}_w}{2gD'} = \frac{4 \times 0.007 \times 120\{C'\}^2_w}{2g \times 0.1} = 1.71\{C'\}^2_w$$

$$\therefore Z'_F = 0.028\{\dot{m}\}^2_w$$

In the second half of the pipe, the head lost to friction Z''_F will be

$$Z''_F = \frac{4f''L''\{C''^2\}_w}{2gD''} = \frac{4 \times 0.007 \times 120\{C''\}^2_w}{2g \times 0.125} = 1.37\{C''\}^2_w$$

$$\therefore Z''_F = 0.0091\{\dot{m}\}^2_w$$

The total head lost to friction in the pipe is

$$Z_F = (Z'_F + Z''_F) = (0.028 + 0.0091)\{\dot{m}\}^2_w = 0.0371\{\dot{m}\}^2_w$$

The minor losses Z_{FM} now include the entrance to the pipe, the gate valve, the sudden enlargement and the exit loss. Using Table 11.1 with the appropriate upstream and downstream water velocities, gives

$$\text{(entrance)} = \frac{0.5\{C'^2\}_w}{2g} = 4.14 \times 10^{-4}\{\dot{m}\}^2_w$$

$$\text{(gate valve)} = \frac{0.2\{C'^2\}_w}{2g} = 1.65 \times 10^{-4}\{\dot{m}\}^2_w$$

$$\text{(sudden expansion)} = \frac{((0.125^2/0.1^2) - 1)^2\{C''^2\}_w}{2g} = 1.07 \times 10^{-4}\{\dot{m}\}^2_w$$

$$\text{(exit)} = \frac{\{C''^2\}_w}{2g} = 3.39 \times 10^{-4}\{\dot{m}\}^2_w$$

The total minor losses are

$$Z_{FM} = 10.25 \times 10^{-4}\{\dot{m}\}^2_w$$

THE STEADY FLOW ENERGY EQUATION APPLIED TO PIPE FLOW

Apply the SFEE between the water surface levels in each reservoir as before. Considering Equation (11.11):

$\{p_1\}_w = \{p_2\}_w =$ atmospheric pressure

$\{C_1\}_w = \{C_2\}_w = 0$ (because the level of the water in the reservoirs will only slowly fall or rise)

$z_1 - z_2 = 30$ m

$Z_F + Z_{FM} = (0.0371 + 10.25 \times 10^{-4})\{\dot{m}\}_w^2 = 0.038\,125\{\dot{m}\}_w^2$

$\dot{W}_{12} = 0$ (because there is no pump)

Substituting in Equation (11.11) gives

$$30 = 0.038125\{\dot{m}\}_w^2$$

$$\therefore \{\dot{m}\}_w = 28.1 \text{ kg/s}$$

Increasing the pipe diameter by 25% improves the flow rate of the water by 75%, so the extra capital cost of the larger pipe may well be a good investment. Increasing the diameter of only half the pipe length by 25% increases the flow by 23%, but such a halfway measure is unlikely to be justified. In all cases, the minor losses are less than 3% of the energy required to overcome friction effects.

11.9.2 Pipes in parallel

If the mass flow of fluid is divided between two or more pipes in parallel, but downstream all the pipes join together into one again, the continuity equation must still apply such that the total mass flow rate of fluid upstream and downstream of the pipes in parallel is equal to the sum of the mass flow rates of fluid passing through each pipe. But between the upstream junction where the flow is divided and the downstream junction where the pipes all join together again, the pressure drop must be equal in each of the pipes. This is because there cannot be a discontinuity of pressure at either junction. Therefore, the sum of the head lost to friction and minor losses must be the same in each of the pipes in parallel, and the mass flow rate and velocity of fluid in each pipe will adjust themselves to ensure that this happens.

EXAMPLE 11.9

A horizontal pipe carries water of density 1000 kg/m³ at the rate of 25 kg/s. At a junction in the pipe, the flow is divided into three other pipes running in parallel. Pipe 1 is of diameter 25 mm, length 15 m and has a friction factor of 0.005; pipe 2 is of diameter 50 mm, length 12 m and has a friction factor of 0.0075; pipe 3 is of diameter 100 mm, length 20 m and has a friction factor of 0.01. The three pipes join together at a downstream junction on the same level as the upstream junction, and the water flow continues in one pipe at the same rate as before (Figure 11.12). Determine the head lost to friction and the flow rate of water through each pipe in parallel, neglecting minor losses.

Solution

The head lost to friction between the upstream and downstream junctions in each of the three pipes in parallel must be the same. Thus, from Equation (11.8), Z_F is

$$Z_F = \frac{4f_1 L_1 \{C_1^2\}_w}{2g D_1} = \frac{4 \times 0.005 \times 15 \{C_1\}_w^2}{2g \times 0.025}$$

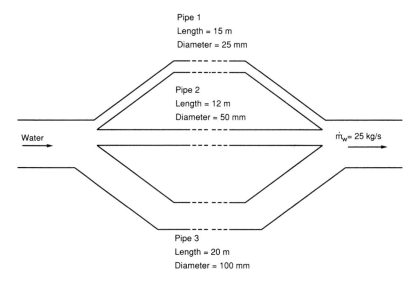

Fig. 11.12 Example 11.9—water flowing through three pipes in parallel.

and

$$Z_F = \frac{4f_2 L_2 \{C_2^2\}_w}{2g D_2} = \frac{4 \times 0.0075 \times 12 \{C_2\}_w^2}{2g \times 0.05}$$

and

$$Z_F = \frac{4f_3 L_3 \{C_3^2\}_w}{2g D_3} = \frac{4 \times 0.01 \times 20 \{C_3\}_w^2}{2g \times 0.1}$$

$$\therefore \quad Z_F = 0.61 \{C_1^2\}_w = 0.37 \{C_2^2\}_w = 0.41 \{C_3^2\}_w \qquad (11.13)$$

From the continuity equation, the total flow rate of water $\{\dot{m}\}_w$ is the sum of the flow rates of water in each pipe. Equation (8.3) gives

$$\{\dot{m}\}_w = \{\dot{m}_1 + \dot{m}_2 + \dot{m}_3\}_w$$

$$\therefore \quad \{\dot{m}\}_w = \{\rho\}_w \pi \times 0.25 (D_1^2 \{C_1\}_w + D_2^2 \{C_2\}_w + D_3^2 \{C_3\}_w)$$

$$\therefore \quad \frac{25 \times 4}{1000\pi} = 0.025^2 \{C_1\}_w + 0.05^2 \{C_2\}_w + 0.1^2 \{C_3\}_w$$

$$\therefore \quad 0.032 = 6.25 \times 10^{-4} \{C_1\}_w + 2.5 \times 10^{-3} \{C_2\}_w + 0.01 \{C_3\}_w \qquad (11.14)$$

Solving Equations (11.13) and (11.14) gives

$\{C_1\}_w = 2.0$ m/s $\{\dot{m}_1\}_w = 1.0$ kg/s

$\{C_2\}_w = 2.6$ m/s $\{\dot{m}_2\}_w = 5.1$ kg/s

$\{C_3\}_w = 2.4$ m/s $\{\dot{m}_3\}_w = 18.8$ kg/s

$Z_F = 2.4$ m

More flow goes through pipe 3 even though it is the longest, because it has the largest diameter.

328 THE STEADY FLOW ENERGY EQUATION APPLIED TO PIPE FLOW

11.9.3 Branched pipes

Pipe systems in which a number of pipes branch off from one pipe at a junction can be difficult to analyse if the fluid flow directions are not known. The principle of the equation of continuity must be followed, as must the fact that the pressure at a junction where a number of pipes meet can only have one value, but beyond that each application must be considered in its own right.

EXAMPLE 11.10

A farmer collects water of density 1000 kg/m³ in a large storage tank situated at the top of a tower. The water level is 15 m above the ground. A 100 mm diameter pipe of length 250 m and with a friction factor of 0.01, leads from the bottom of the tank to a field where the water is used for irrigation purposes. There is a gate valve in the pipeline situated near the exit (Figure 11.13). Accounting for minor losses, what is the flow rate of the water when the gate valve is fully open?

At a later stage, the farmer decides to irrigate two fields at the same time. At a point 150 m along the pipe, the farmer installs a T-junction. On one side of the T-junction is connected the remaining 100 m length of 100 mm diameter pipe leading to the first field. On the other side of the T-junction is connected a 170 m length of 100 mm diameter pipe leading to the second field. However, this second bit of pipe is of different material to the original pipe and has a friction factor of 0.008. It also has a gate valve fitted near its exit (Figure 11.14). What is the flow rate of water to each of the fields, accounting for minor losses, when both gate valves are fully open?

Solution

The head lost to friction in the pipe Z_F is given by Equation (11.8) as

$$Z_F = \frac{4fL\{C^2\}_w}{2gD} = \frac{4 \times 0.01 \times 250\{C\}_w^2}{2g \times 0.1} = 5.1\{C\}_w^2$$

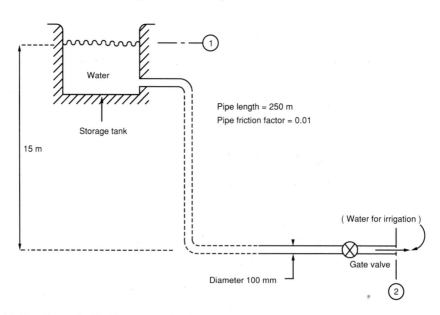

Fig. 11.13 *Example 11.10—water flowing from a storage tank through a pipeline for irrigation in a field.*

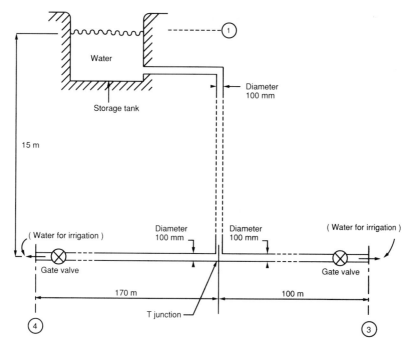

Fig. 11.14 *Example 11.10—water flowing from a storage tank through pipelines used to irrigate two fields.*

The minor losses Z_{FM} include the entrance to the pipe (assumed to be sharp) and the gate valve. Using Table 11.1 with an upstream velocity of $\{C\}_w$, the minor losses are

$$Z_{FM} = \frac{(0.5 + 0.2)\{C\}_w^2}{2g} = 0.036\{C\}_w^2$$

Note that the exit loss is not included because it appears in the SFEE.

Apply the SFEE between the surface level of the water in the tank (state 1) and the outlet of the pipe (state 2). Considering Equation (11.11):

$\{p_1\}_w = \{p_2\}_w =$ atmospheric

$\{C_1\}_w = 0$ (because the water level in the tank will drop only slowly)

$z_1 - z_2 = 15$ m

$\{C_2\}_w = \{C\}_w$ (because the water leaves the pipe at the same velocity with which it travels through the pipe)

$(Z_F + Z_{FM}) = (5.1 + 0.036)\{C\}_w^2 = 5.136\{C\}_w^2$

$\dot{W}_{12} = 0$ (because there is no pump)

Substituting in Equation (11.11) gives

$$15 = \frac{\{C\}_w^2}{2g} + 5.136\{C\}_w^2$$

$$\therefore \{C\}_w = 1.7 \text{ m/s}$$

The mass flow rate of water $\{\dot{m}\}_w$ can be determined from the continuity equation (Equation (8.3)) as follows:

$$\{\dot{m}\}_w = A_{xs}\{\rho C\}_w = 0.25\pi \times 0.1^2 \times 1000 \times 1.7 = 13.4 \text{ kg/s}$$

When the pipe is divided into two at the T-junction, let the 100 m length pipe conditions have a superscript ′ and the 170 m length pipe conditions a superscript ″. The head lost to friction in the 150 m length of 100 mm diameter pipe before the T-junction Z_F is given by Equation (11.8) as

$$Z_F = \frac{4fL\{C^2\}_w}{2gD} = \frac{4 \times 0.01 \times 150\{C\}_w^2}{2g \times 0.1} = 3.06\{C\}_w^2$$

The head lost to friction in the first branch Z'_F is given by Equation (11.8) as

$$Z'_F = \frac{4f'L'\{C'^2\}_w}{2gD'} = \frac{4 \times 0.01 \times 100\{C'\}_w^2}{2g \times 0.1} = 2.04\{C'\}_w^2$$

The head lost to friction in the second branch Z''_F is given by Equation (11.8) as

$$Z''_F = \frac{4f''L''\{C''^2\}_w}{2gD''} = \frac{4 \times 0.008 \times 170 \times \{C''\}_w^2}{2g \times 0.1} = 2.77\{C''\}_w^2$$

The minor losses in the pipe before the T-junction Z_{FM} are due to the entrance to the pipe and entrance to the T-junction. From Table 11.1 with an upstream velocity of $\{C\}_w$, the minor losses are

$$Z_{FM} = \frac{(0.5 + 0.4)\{C\}_w^2}{2g} = 0.046\{C\}_w^2$$

The minor losses in the first branch Z'_{FM} are due to the T-junction exit and the gate valve. From Table 11.1 with an upstream velocity of $\{C'\}_w$, the minor losses are

$$Z'_{FM} = \frac{(1.8 + 0.2)\{C'\}_w^2}{2g} = 0.102\{C'\}_w^2$$

The minor losses in the second branch Z''_{FM} are due to the T-junction exit and the gate valve. From Table 11.1 with an upstream velocity of $\{C''\}_w$, the minor losses are

$$Z''_{FM} = \frac{(1.8 + 0.2)\{C''\}_w^2}{2g} = 0.102\{C''\}_w^2$$

Apply the SFEE from the surface level of the water in the tank to the outlet of the first branch (state 3). Considering Equation (11.11):

$$\{p_1\}_w = \{p_3\}_w = \text{atmospheric pressure}$$

$$\{C_1\}_w = 0 \text{ (because the water level in the tank will drop only slowly)}$$

$$z_1 - z_3 = 15 \text{ m}$$

$$\{C_3\}_w = \{C'\}_w$$

$$(Z_F + Z'_F + Z_{FM} + Z'_{FM}) = \{C\}_w^2(3.06 + 0.046) + \{C'\}_w^2(2.04 + 0.102)$$

$$\therefore (Z_F + Z'_F + Z_{FM} + Z'_{FM}) = 3.106\{C\}_w^2 + 2.142\{C'\}_w^2$$

$$\dot{W}_{13} = 0 \text{ (because there is no pump)}$$

Substituting in Equation (11.11) gives

$$15 = \frac{\{C'\}_w^2}{2g} + 3.106\{C\}_w^2 + 2.142\{C'\}_w^2$$

$$\therefore \quad 15 = 3.106\{C\}_w^2 + 2.193\{C'\}_w^2 \tag{11.15}$$

Apply the SFEE from the surface level of the water in the tank to the outlet of the second branch (state 4). Considering Equation (11.11):

$$\{p_1\}_w = \{p_4\}_w = \text{atmospheric pressure}$$

$$\{C_1\}_w = 0 \text{ (because the water level in the tank will drop only slowly)}$$

$$z_1 - z_4 = 15 \text{ m}$$

$$\{C_4\}_w = \{C''\}_w$$

$$(Z_F + Z_F'' + Z_{FM} + Z_{FM}'') = \{C\}_w^2(3.06 + 0.046) + \{C''\}_w^2(2.77 + 0.102)$$

$$\therefore \quad (Z_F + Z_F'' + Z_{FM} + Z_{FM}'') = 3.106\{C\}_w^2 + 2.872\{C''\}_w^2$$

$$\dot{W}_{14} = 0 \text{ (because there is no pump)}$$

Substituting in Equation (11.11) gives

$$15 = \frac{\{C''\}_w^2}{2g} + 3.106\{C\}_w^2 + 2.872\{C''\}_w^2$$

$$\therefore \quad 15 = 3.106\{C\}_w^2 + 2.923\{C''\}_w^2 \tag{11.16}$$

The continuity equation shows that the mass flow rate of water in the pipe before the T-junction is equal to the mass flow rate of water in the first branch plus the mass flow rate of water in the second branch of the pipe. Thus, Equation (8.3) gives

$$\{\dot{m}\}_w = \{\dot{m}'\}_w + \{\dot{m}''\}_w$$

$$\therefore \quad D^2\{C\}_w = D'^2\{C'\}_w + D''^2\{C''\}_w$$

$$\therefore \quad \{C\}_w = \{C'\}_w + \{C''\}_w \tag{11.17}$$

Solving Equations (11.15), (11.16) and (11.17) gives

$$\{C\}_w = 1.99 \text{ m/s}$$

$$\{C'\}_w = 1.07 \text{ m/s}$$

$$\{C''\}_w = 0.92 \text{ m/s}$$

$$\{\dot{m}\}_w = 15.6 \text{ kg/s}$$

$$\{\dot{m}'\}_w = 8.4 \text{ kg/s}$$

$$\{\dot{m}''\}_w = 7.2 \text{ kg/s}$$

The total flow rate of water leaving the tank is increased by 16% but, of course, the flow rate to the first field is reduced by 37%.

EXAMPLE 11.11

An upper reservoir of water of density 1000 kg/m³, in which the level of water is maintained at 100 m above sea-level, is used to feed two other reservoirs, a middle one in which the water level is maintained at 60 m above sea-level, and a lower one in which the water level is maintained at 40 m above sea-level. One pipe of diameter 150 mm and length 1.8 km leads from the upper reservoir to a junction, where it divides into two other pipes of diameter 100 mm, one of which is of length 1.2 km and leads to the middle reservoir, and the other, of length 2.2 km, leads to the lower reservoir (Figure 11.15). If the friction factor of all the pipes is 0.01, determine the mass flow rate of water in each pipe, ignoring minor losses.

Solution

The head lost to friction in the initial pipe before the junction Z_F, from Equation (11.8), is

$$Z_F = \frac{4fL\{C^2\}_w}{2gD} = \frac{4 \times 0.01 \times 1.8 \times 10^3 \{C\}_w^2}{2g \times 0.15} = 24.47\{C\}_w^2$$

Let the conditions in the pipe from the junction to the middle reservoir have a superscript'. The head lost to friction, from Equation (11.8), is

$$Z_F' = \frac{4f'L'\{C'^2\}_w}{2gD'} = \frac{4 \times 0.01 \times 1.2 \times 10^3 \{C'\}_w^2}{2g \times 0.1} = 24.47\{C'\}_w^2$$

Let the conditions in the pipe from the junction to the lower reservoir have a superscript". The head lost to friction, from Equation (11.8), is

$$Z_F'' = \frac{4f''L''\{C''^2\}_w}{2gD''} = \frac{4 \times 0.01 \times 2.2 \times 10^3 \{C''\}_w^2}{2g \times 0.1} = 44.85\{C''\}_w^2$$

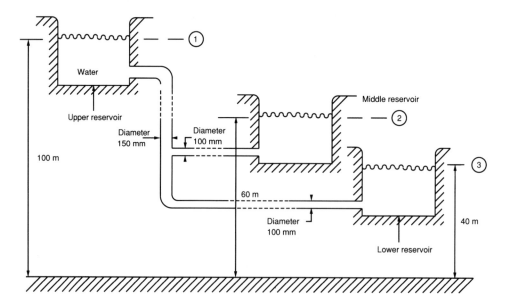

Fig. 11.15 *Example 11.11—water flowing from an upper reservoir into a middle and lower reservoir.*

Apply the SFEE from the surface level of the water in the upper reservoir (state 1) to the surface level of the water in the middle reservoir (state 2). Considering Equation (11.2):

$\{p_1\}_w = \{p_2\}_w =$ atmospheric pressure

$\{C_1\}_w = \{C_2\}_w = 0$ (because the water levels in the reservoirs will only rise or fall slowly)

$z_1 - z_2 = 100 - 60 = 40$ m

$(Z_F + Z'_F) = 24.47\{C\}_w^2 + 24.47\{C'\}_w^2$

$\dot{W}_{12} = 0$ (because there is no pump)

Substituting in Equation (11.2) gives

$$40 = 24.47\{C\}_w^2 + 24.47\{C'\}_w^2 \tag{11.18}$$

Apply the SFEE from the surface level of the water in the upper reservoir to the surface level of the water in the lower reservoir (state 3). Considering Equation (11.2):

$\{p_1\}_w = \{p_3\}_w =$ atmospheric pressure

$\{C_1\}_w = \{C_3\}_w = 0$ (because the water levels in the reservoirs will only rise or fall slowly)

$z_1 - z_3 = 100 - 40 = 60$ m

$(Z_F + Z''_F) = 24.47\{C\}_w^2 + 44.85\{C''\}_w^2$

$\dot{W}_{13} = 0$ (because there is no pump)

Substituting in Equation (11.2) gives

$$60 = 24.47\{C\}_w^2 + 44.85\{C''\}_w^2 \tag{11.19}$$

The continuity equation shows that the mass flow rate of water in the pipe before the junction equals the mass flow rate in the pipe after the junction leading to the middle reservoir plus the mass flow rate in the pipe after the junction leading to the lower reservoir. Thus, Equation (8.3) gives

$$\{\dot{m}\}_w = \{\dot{m}' + \dot{m}''\}_w$$

$$\therefore D^2\{C\}_w = D'^2\{C'\}_w + D''^2\{C''\}_w$$

$$\therefore 0.15^2\{C\}_w = 0.1^2\{C'\}_w + 0.1^2\{C''\}_w \tag{11.20}$$

Solving Equations (11.18), (11.19) and (11.20) by trial and error gives the approximate solution:

$\{C\}_w = 0.85$ m/s $\{\dot{m}\}_w = 15.0$ kg/s

$\{C'\}_w = 0.96$ m/s $\{\dot{m}'\}_w = 7.5$ kg/s

$\{C''\}_w = 0.96$ m/s $\{\dot{m}''\}_w = 7.5$ kg/s

In problems such as this, if the direction of flow is not known, for example, it could be that water from the middle reservoir flows into the lower reservoir, the solution is made more difficult. Generally, if the flow direction is assumed wrong, the answer will turn out to be a nonsense.

11.10 COMPUTER PROGRAMS

In complex systems of pipes and ducts, the calculations to determine the head lost to friction in the pipework or ductwork, and that due to minor losses, can become quite tedious. In this case it is much better to use a commercially available computer program to calculate the total head loss, and there are a number on the market written for this purpose. Similarly, analysis of situations which involve combinations of pipes can be eased with the use of a suitable computer program, particularly where a trial and error solution is required.

11.11 APPLICATION IN THERMAL SYSTEMS

Chapters 9, 10 and 11 have discussed what happens to a fluid as it flows from one component to another in a thermal system. It is now possible to measure its flow rate, determine the forces that arise as a consequence of its motion, around a bend, or in a straight duct, or over a body, and also to evaluate the changes that occur in the fluid property values as a consequence of those forces. In other words, the complete thermal system has been subject to examination, at least to a first approximation.

12 The Second Law of Thermodynamics

It is not necessary to know the second law of thermodynamics in order to predict the transfer of work and heat in a thermal system, but it is certainly important to understand the consequences of the law when a system is being assessed for its feasibility. The study of thermofluids rests upon the first and second laws, and they both say something about the relationship between the work and the heat transfer that exist when a fluid is changing its state. The first law equates the net work transfer with the net heat transfer when the fluid completes a cycle in a closed system, but the second law is more specific about how much work can be converted to heat and how much heat can be converted to work.

Further, just as the fluid property internal energy was shown to exist from a consideration of the first law, so the fluid property entropy can be shown to exist from a consideration of the second law.

The second law of thermodynamics may be stated as follows: 'it is impossible for an engine cycle to transform all of the heat supplied to it into work'. As with the first law, there are a number of conditions expressed in the second law which need clarifying. Firstly, the fluid must complete a cycle, in other words the fluid properties must return to their initial state, however loosely this is interpreted. Secondly, the second law refers to an engine which is defined as a thermal system which has a net output of work as a consequence of a given heat supply. And thirdly, the second law places some restriction upon the transfer of the heat supplied to the engine; specifically it says that not all the heat supplied can be converted into work output. Some of the heat supplied must be rejected. It does not, though, place any restriction upon the transfer of work into heat. There is nothing in the second law to prevent all the work supply to a thermal system being converted into heat, as happens in a refrigerator, heat pump or brake shoe. In that sense, work transfer is a more valuable form of energy than heat transfer in a thermal system.

If the second law of thermodynamics is accepted, and it is based upon observation of what happens in all sorts of possible thermal systems, there are a number of implications, or corollaries, which must follow as a consequence. These can be deduced from a consideration of both laws, utilising some of the theoretical concepts developed in earlier chapters.

12.1 IMPLICATIONS OF THE SECOND LAW OF THERMODYNAMICS

12.1.1 The efficiency of an engine

Engines can be made to work with an energy supply other than a heat supply, for example a turbine in a hydroelectric scheme is driven by the pressure of the water in a dam. The

336 THE SECOND LAW OF THERMODYNAMICS

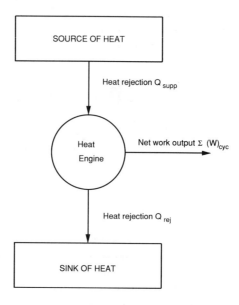

Fig. 12.1 *A heat engine produces a net work output from a heat supply and must reject some heat according to the second law of thermodynamics.*

second law of thermodynamics, however, is specifically concerned with a heat engine in which a certain heat supply is converted into a work output, as shown in Figure 12.1.

From the definition of overall thermal efficiency in Section 4.2.1, for the case of a heat engine, the useful net energy output is the net work output from the engine $-\sum(W)_{cyc}$ (the negative sign is included to comply with the sign convention and make the efficiency positive because the net work output will be negative), and the energy input which must be paid for is the heat supply Q_{supp} which, in most cases, arises from the combustion of a fuel. Hence, the overall thermal efficiency η is

$$\eta = \frac{-\sum(W)_{cyc}}{Q_{supp}} \tag{12.1}$$

Given that the fluid in the heat engine completes a cycle in producing the work output as a consequence of the heat supply, from the first law of thermodynamics, the net work output is equal to the sum of the heat supplied Q_{supp} and the heat rejected Q_{rej} in the cycle, but of opposite sign. Thus:

$$\eta = \frac{Q_{supp} + Q_{rej}}{Q_{supp}} = 1 + \frac{Q_{rej}}{Q_{supp}} \tag{12.2}$$

According to the sign convention adopted, the heat rejected will end up with a negative sign as it is a heat output from the system, hence the efficiency will not be greater than 100%! However, the second law of thermodynamics specifically says that not all the heat supplied to the engine can be converted to work output, some of it must be rejected. Therefore, the implication is that the efficiency of any heat engine must be less than 100%.

Further, heat flows naturally from a 'source' at a given temperature to a 'sink' at a lower temperature, where the source may be considered as an inexhaustible reservoir which

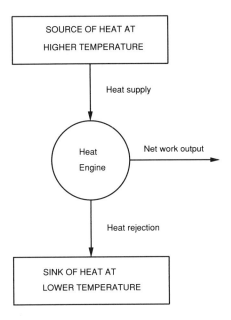

Fig. 12.2 *A heat engine requires a temperature difference between the source and sink of heat in order to produce a net work output.*

can supply any amount of heat at a constant temperature and the sink as an inexhaustible reservoir which can absorb any amount of heat at a constant temperature. In practice, they could be assumed to be a fuel burning continually and a river in full flow respectively. If there is to be a heat supply and a heat rejection in accordance with the second law, the heat supply must be from a source at a higher temperature than the sink to which the heat is rejected. Thus, for a heat engine to work successfully, it must operate between a source and sink of heat at different temperatures. Or, to put it another way, a temperature difference is required before an engine can produce a net output of work (Figure 12.2). Hence, it is not possible for an engine to operate from a source of heat such as a river or the atmosphere alone, because there must be a sink at a lower temperature to which some of the heat has to be rejected.

12.1.2 The heat pump

If the second law of thermodynamics is accepted, it must also follow that: 'it is impossible to construct a system which will operate in a cycle and transfer heat from a cooler to a hotter body without work being done on the system by the surroundings'. This is known as the Clausius statement of the second law after Clausius who carried out many of the early investigations which helped to clarify the issues surrounding the conversion of work and heat. The statement can be shown to be true by proving that the opposite is not possible. In the first instance, suppose that the opposite is correct, as in Figure 12.3.

System A takes heat Q from the cold reservoir. There is no work input from the surroundings. From the first law, it must deliver the same heat Q to the hot reservoir. A heat engine, system B, may also be made to operate between the same reservoirs but in

338 THE SECOND LAW OF THERMODYNAMICS

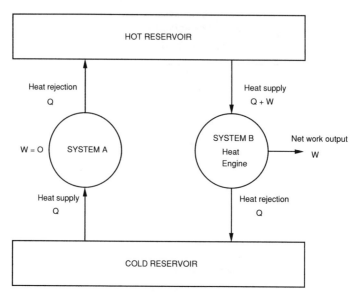

Fig. 12.3 *The Clausius statement of the second law of thermodynamics says that it is impossible to transfer heat from a low-temperature source to a higher-temperature sink without a work input.*

the opposite direction. It is of such a size that it rejects the same heat Q while providing a net work output W. From the first law, the heat supplied to the heat engine B must be $(Q + W)$.

For the combined system, the cold reservoir is balanced because heat Q is delivered to it by system B, and simultaneously taken from it by system A. In the hot reservoir, system A delivers heat Q while system B takes heat $(W + Q)$. There is a net supply of heat of $(W + Q) - Q = W$ from the source. But as system A requires no work transfer, there is a net output of work W for the combined system, in fact the work output from heat engine B. In other words, the combined system succeeds in having a net supply of heat W and completely converting it into a net output of work W, which is contrary to the second law of thermodynamics. Therefore, system A must have a net work input from the surroundings and the Clausius statement of the second law is verified.

A thermal system which transfers heat from a cold reservoir to a hot reservoir while having a work input is called a Heat Pump, as described in Section 4.3.3. Its overall thermal efficiency was shown to be greater than 100% in Example 4.8 and was called the coefficient of performance. This can be confirmed by reference to Equation (12.2), because the efficiency of a heat pump must be the inverse of the efficiency of a heat engine, and the latter must have a value less than 100% according to the second law (Figure 12.4)

Consideration of a heat pump and the Clausius statement of the second law of thermodynamics, enables the concept of reversibility to be explained in terms of heat transfer. In Section 3.3.2, it was stated that a reversible non-flow process was an ideal which could not be met because of the effects of friction. Similarly, in Chapter 8, a reversible flow process was not attainable for the same reason, but that this problem could be overcome by the use of an isentropic efficiency. As far as heat transfer is concerned, for a process to be reversible, the same heat must be transferred into the fluid as it goes from state 2 back to

IMPLICATIONS OF THE SECOND LAW OF THERMODYNAMICS   339

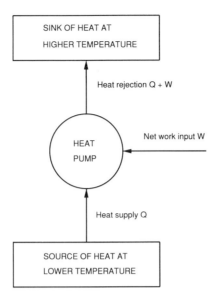

Fig. 12.4 *A heat pump transfers heat from a low-temperature source to a higher-temperature sink as a consequence of a work input.*

state 1 as is transferred out of the fluid as it goes from state 1 to state 2. But heat transfer naturally occurs from a source at a given temperature to a sink at a lower temperature. It can only be transferred from a sink at a given temperature to a source at a higher temperature through the additional input of a work transfer, according to the Clausius statement.

12.1.3 The efficiency of a reversible engine

The concept of a reversible process, and reversible cycle which is made up of reversible processes, represents an ideal state of affairs. In a reversible engine, the fluid must pass through a reversible cycle in producing a net work output from a supply of heat. But from a consideration of the second law of thermodynamics, it can now be stated that: 'it is impossible to construct an engine operating between only two reservoirs which will have a higher efficiency than a reversible engine operating between the same two reservoirs'. As before, this can be shown to be so by imagining the opposite to be true, as in Figure 12.5, and proving that it is impossible.

Figure 12.5 shows two heat engines connected between the same heat reservoirs. Heat engine I is an irreversible engine. It has a heat supply Q, produces a net work output W and by the first law rejects heat $(Q - W)$. Heat engine R is a reversible engine. It too produces a net work output W. But, if it has an efficiency less than heat engine I, it must require a greater heat supply $(Q + Q')$. From the first law, it will reject heat $(Q + Q' - W)$.

Now reverse heat engine R and make it operate as a reversible heat pump, as in Figure 12.6. Heat engine I cannot be reversed because it is irreversible. The work output from I becomes the work input for R so that there is no net work transfer in the composite system. From the cold reservoir, there is a net heat transfer out of $Q + Q' - W - (Q - W) = Q'$, and in the hot reservoir, there is a net heat transfer in of $Q + Q' - Q = Q'$. In other words,

340 THE SECOND LAW OF THERMODYNAMICS

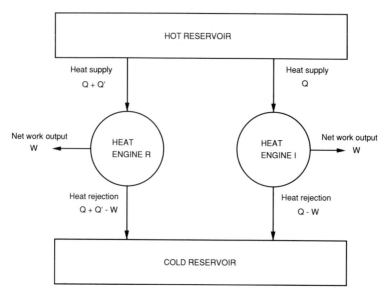

Fig. 12.5 *Two heat engines, one reversible and the other irreversible, operating between the same heat reservoirs.*

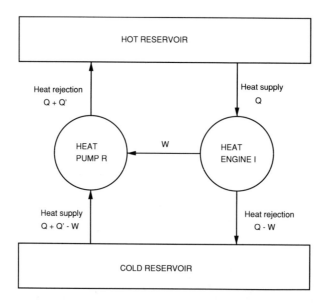

Fig. 12.6 *An irreversible heat engine and a reversible heat pump operating between the same heat reservoirs.*

the composite system succeeds in transferring heat Q' from a cold reservoir to a hot reservoir without the input of any work. This is impossible according to the Clausius statement of the second law and the conclusion must be that the reversible engine in Figure 12.4 has the higher efficiency, as stated.

IMPLICATIONS OF THE SECOND LAW OF THERMODYNAMICS

It must also follow that all reversible heat engines operating between the same source and sink have the same efficiency and that this efficiency must depend upon the temperature of the hot and cold reservoirs, T_{source} and T_{sink} respectively, as this is the only feature common to them all. By implication, the temperature of the working fluid in the reversible engine must also be T_{source} at the source and T_{sink} at the sink. Therefore, the overall thermal efficiency of a reversible heat engine η_r can be written:

$$\eta_r = \Phi(T_{sink}, T_{source}) \quad (12.3)$$

and from Equation (12.2), the overall thermal efficiency of any engine η is

$$\eta = 1 + \frac{Q_{rej}}{Q_{supp}}$$

The second law imposes a restriction upon the function chosen in Equation (12.3). Consider the three reversible heat engines A, B and AB in Figure 12.7.

Reversible heat engine AB has a heat supply Q_{supp}, has a net work output W_{AB} and rejects heat Q_{rej}. Reversible heat engine B has the same heat supply but has a net work output W_B and rejects heat Q_{int} to an intermediate sink at a temperature T_{int} somewhere between the temperature of the source and sink of reversible heat engine AB. Reversible heat engine A takes the heat rejected by reversible heat engine B, has a net work output W_A and rejects the same amount of heat as reversible heat engine AB to the sink.

For each engine, Equations (12.2) and (12.3) give

$$\frac{(Q_{rej})_{AB}}{(Q_{supp})_{AB}} = \Phi(T_{sink}, T_{source})$$

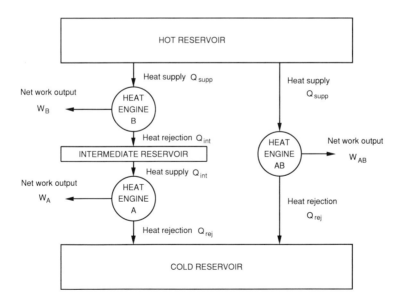

Fig. 12.7 *Three reversible heat engines.*

$$\frac{(Q_{int})_B}{(Q_{supp})_B} = \Phi(T_{int}, T_{source})$$

$$\frac{(Q_{rej})_A}{(Q_{int})_A} = \Phi(T_{sink}, T_{int})$$

As Q_{supp}, Q_{int} and Q_{rej} have the same value for each reversible heat engine:

$$\frac{Q_{rej}}{Q_{supp}} = \frac{Q_{rej}}{Q_{int}} \times \frac{Q_{int}}{Q_{supp}}$$

$$\therefore \quad \Phi(T_{sink}, T_{source}) = \Phi(T_{sink}, T_{int})(T_{int}, T_{source})$$

As the left-hand side of the equation is a function of T_{sink} and T_{source} only, so must be the right-hand side of the equation. The only way this is possible is if T_{int} cancels out and this can only be achieved if the function is of the ratio of the temperatures of the source and sink between which the reversible heat engine is operating. Therefore, for any reversible heat engine operating between a source of heat at a given temperature and a sink of heat at a lower temperature, it can be written:

$$\frac{Q_{rej}}{Q_{supp}} = \Phi\left(\frac{T_{sink}}{T_{source}}\right) \tag{12.4}$$

The simplest function to define in Equation (12.4) is

$$\frac{Q_{rej}}{Q_{supp}} = -\frac{T_{sink}}{T_{source}}$$

where the minus sign is used in accordance with the sign convention adopted and sign rule B. Any function is appropriate, it will only affect the magnitude of the scale of temperature which is used. The overall thermal efficiency of a reversible heat engine becomes

$$\eta_r = \frac{T_{source} - T_{sink}}{T_{source}} \tag{12.5}$$

All reversible heat engines operating between the same source and sink must have the same efficiency as given by Equation (12.5), and this must be the maximum efficiency possible. It is known as the Carnot efficiency after Carnot, who proposed the first and only truly reversible cycle for a heat engine.

Such an efficiency sets a value against which a real engine can be judged. For example, in a steam plant as shown in Figure 4.15, if the superheated steam temperature (T_{source}) at exit from the boiler is 900 K, which is the metallurgical limit for the materials of which the turbine is made, and the heat is rejected in the condenser to the atmosphere at a temperature (T_{sink}) of 300 K, the Carnot efficiency is 66.7%. An actual operating efficiency of about 40%, which is achievable in modern steam plant, compares well with this maximum ideal limit.

The Carnot efficiency may also be applied to a reversible heat pump except that it is written the other way round and called the coefficient of performance. Thus, the Carnot COP_{hp} is

$$\text{Carnot } COP_{hp} = \frac{T_{source}}{T_{source} - T_{sink}}$$

The refrigerant in a typical domestic heat pump works between temperatures of 70 and −10 °C which gives a Carnot COP_{HP} of 4.3.

From the definition of the overall thermal efficiency of a reversible heat engine in terms of the source and sink temperatures, it must follow that the efficiency of any reversible heat engine operating from or to any intermediate temperature must be less than that of a reversible heat engine operating between the source and sink alone. Also, as the efficiency of an irreversible heat engine is less than the efficiency of a reversible heat engine, the temperature of the working fluid in the irreversible engine is less than T_{source} at the source and greater than T_{sink} at the sink. It only approaches the temperatures T_{source} and T_{sink} in the limit when the heat engine becomes reversible.

12.1.4 Absolute scale of temperature

In Section 2.4, it was noted that all practical thermometers only give the same reading at the reference temperatures and that, in between these fixed points, the reading depends upon the way the medium used in the thermometer varies in some way with temperature. The common factor of all reversible heat engines is the temperature of the source and sink. The efficiency of the engine depends upon these temperatures alone and is independent of the nature of the engine or the fluid that may be used in it. Therefore, the use of reversible heat engines gives rise to the idea that an absolute scale of temperature can be defined which is independent of any particular thermometric substance, and which will provide an absolute zero of temperature.

A series of reversible heat engines can be imagined such that each successive engine operates from the sink of the preceding engine. If each engine has a net work output W, the increment of temperature between the source and the sink is the same for all the engines.

The system proposed implies an absolute zero of temperature but, in doing so, disobeys the second law of thermodynamics. The last engine must reject some heat according to the second law, but this would mean that the final temperature must be greater than zero. Therefore, an absolute zero of temperature can only be a conceptual limit.

Although it is not possible to build reversible heat engines, meaning that an absolute scale of temperature is hypothetical, by choosing the temperature of the triple point of water as 273.16 K, it is found that when the ice and steam points of water at NTP are measured on the 'most accurate' thermometer available (a constant volume gas thermometer), the difference is found to be 100 K within 0.005 K. In other words, the absolute scale of temperature coincides almost exactly with the empirical Celsius scale, which is useful. This is, in fact, what has been agreed internationally, at a conference on weights and measures held in 1954.

12.1.5 The fluid property entropy

In order to show the existence of the fluid property entropy, it is necessary to consider what is called the Clausius inequality statement, which is that: 'whenever a system undergoes a cycle, $\sum (dQ/T)_{cyc}$ is zero if the cycle is reversible and negative if the cycle is irreversible'.

Consider two heat engines operating in a cycle between the same two fixed temperature reservoirs, the source and the sink. One engine R is reversible and the other I irreversible, as shown in Figure 12.8. The efficiency of the irreversible engine η_{irr} is less than the efficiency of the reversible engine η_r. Also, the efficiency of the irreversible engine can only be

344 THE SECOND LAW OF THERMODYNAMICS

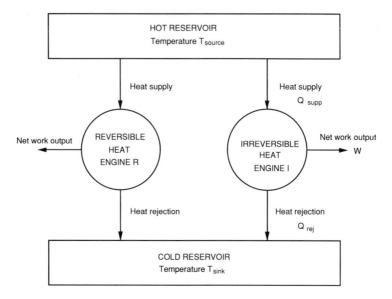

Fig. 12.8 *A reversible and irreversible heat engine operating between the same source and sink of heat.*

expressed in terms of the heat supply and heat rejection, in accordance with Equation (12.2), whereas the efficiency of the reversible engine can be expressed in terms of the temperature of the source and sink, in accordance with Equation (12.5). Thus:

$$\eta_{irr} < \eta_r \quad \therefore \quad \frac{Q_{supp} + Q_{rej}}{Q_{supp}} < \frac{T_{source} - T_{sink}}{T_{source}}$$

$$\therefore \quad \frac{Q_{rej}}{Q_{supp}} < -\frac{T_{sink}}{T_{source}} \quad \therefore \quad \frac{Q_{rej}}{T_{sink}} + \frac{Q_{supp}}{T_{source}} < 0$$

$$\therefore \quad \sum (Q/T) < 0$$

Mathematically, for a cycle, this can be written:

$$\sum (dQ/T)_{cyc} \leqslant 0 \qquad (12.6)$$

The limit is when the irreversible heat engine becomes reversible and $\sum (dQ/T)_{cyc}$ becomes zero. The Clausius inequality is shown to be true.

Now consider two reversible cycles, one in which the fluid goes from state 1 to state 2 by reversible non-flow process X and returns to state 2 by reversible non-flow process Z, and the other in which the fluid goes from state 1 to state 2 by reversible non-flow process Y but returns to state 2 again by reversible non-flow process Z. The processes can be drawn on a graph of two properties, call them property A and property B, as shown in Figure 12.9.

For reversible cycle XZ:

$$\sum (dQ/T)_{cycXZ} = \sum (dQ/T)_X + \sum (dQ/T)_Z$$

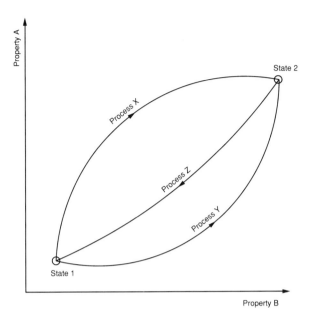

Fig. 12.9 *Graph of fluid property A against fluid property B for a fluid undergoing either reversible cycle XZ or reversible cycle YZ.*

For reversible cycle YZ:

$$\sum (dQ/T)_{\text{cycYZ}} = \sum (dQ/T)_{\text{Y}} + \sum (dQ/T)_{\text{Z}}$$

Obviously, $\sum(dQ/T)_{\text{Z}}$ is the same in both equations. And, by the Clausius inequality, as both cycles are reversible:

$$\sum (dQ/T)_{\text{cycXZ}} = \sum (dQ/T)_{\text{cycYZ}} = 0$$

Therefore, a comparison of the two equations shows that

$$\sum (dQ/T)_{\text{X}} = \sum (dQ/T)_{\text{Y}}$$

In other words, $\sum(dQ/T)$ is independent of the route taken by the fluid, or process, and must be a property of the fluid. It is called the entropy S and is an extensive property. Thus:

Entropy: S (units J/K)

Specific entropy: s (units J/kgK)

where $S = ms$.

When a fluid is going from state 1 to state 2, the entropy change mathematically is given by

$$S_2 - S_1 = \int_1^2 \frac{dQ}{T} \qquad (12.7)$$

THE SECOND LAW OF THERMODYNAMICS

This equation is, of course, the same as Equation (3.2) where the existence of the property entropy was guessed at after consideration of the equivalence of the formulae for work and heat transfer in a non-flow process. Now it can be seen that there must be a fluid property entropy as a consequence of the second law of thermodynamics. In Section 3.3.2, it was stated that the formulae for work and heat transfer in a non-flow process applied to reversible processes only. This was demonstrated by examining the effects of friction, which is mainly applicable to the work transfer. But it can now be seen that this restriction is equally applicable to the heat transfer because Equation (12.7) has been derived on the basis of reversible processes only. This does not mean, as discussed in Chapter 3, that the property entropy is limited to reversible processes. It is an extensive property like volume, and exists whatever the circumstances. It is only that Equations (3.2) and (12.7) are applicable to reversible processes only.

In this way, entropy is very like the property internal energy. One arises as a consequence of the first law of thermodynamics, and the other as a consequence of the second law. Neither can be measured directly and only changes in their value can be obtained as a fluid undergoes a process. Entropy, though, has considerably more importance in the thermodynamic scheme of things, as it can be used indirectly as a measure of the efficiency of a process. Also, the NFEE which includes the internal energy, Equation (5.1), applies to any non-flow process, so that the internal energy change of a fluid may be determined from a knowledge of the work and heat transfer during a change of state. Equations (12.7) and (3.2) which define the entropy only apply to reversible non-flow processes. Therefore, the entropy change of a fluid has to be determined through a knowledge of the other property values which can be measured, as explained in Section 6.2 for a perfect gas, and in Section 7.3, for water/steam.

12.1.6 Reversible and irreversible processes

It has already been shown in Chapter 8 that a process efficiency can be defined by making a comparison of what actually happens in a process with what would happen if the process was isentropic. In fact, as a consequence of the second law of thermodynamics, the implications for the entropy of a fluid are significant because it can proved that: 'the entropy of any closed system which is thermally isolated from the surroundings, either increases or, if the process is reversible, remains constant during a change of state'.

Consider a fluid undergoing a cycle, from state 1 to state 2 by an irreversible adiabatic non-flow process, and back from state 2 to state 1 by a reversible but not adiabatic non-flow process. As before, the cycle can be plotted on a graph of two of the fluid properties, property A and property B, as in Figure 12.10.

When going from state 1 to state 2, as the process is adiabatic, there is no heat transfer. Thus

$$Q_{12} = 0 \quad \therefore \quad \int_1^2 \frac{dQ_{12}}{T} = 0$$

Note that this is not the property entropy because the formula for entropy only applies to reversible processes and this is an irreversible adiabatic.

When returning from state 2 back to state 1, which is a reversible process, Equation (12.7) gives

$$\int_2^1 \frac{dQ_{21}}{T} = S_1 - S_2$$

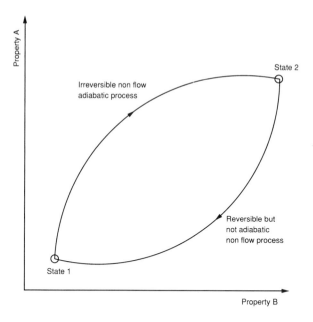

Fig. 12.10 *Graph of fluid property A against fluid property B for a fluid going from state 1 to state 2 by an irreversible adiabatic non-flow process, and from state 2 back to state 1 by a reversible but not adiabatic non-flow process.*

But from the Clausius inequality, as the complete cycle is irreversible, Equation (12.6) gives

$$\sum \left(\frac{dQ}{T}\right)_{cyc} < 0 \quad \therefore \quad \int_1^2 \frac{dQ_{12}}{T} + \int_2^1 \frac{dQ_{21}}{T} < 0$$

$$\therefore \quad 0 + (S_1 - S_2) < 0 \quad \therefore \quad S_2 > S_1$$

Hence, the entropy of a fluid as it follows an irreversible adiabatic non-flow process, which is what happens as it goes from state 1 to state 2, always increases. If the process were a reversible adiabatic, in other words an isentropic process, the entropy would remain constant. Therefore, the conclusion is as stated above, namely that the entropy of any closed system which is thermally isolated from the surroundings, either increases or, if the process is reversible, remains constant during a change of state.

The significance of this statement may not be immediately apparent. Initially it implies that a knowledge of the entropy at states 1 and 2 will reveal the type of adiabatic process followed in that, if the entropy remains constant, the process is reversible, and if it increases, it is irreversible. The implications are important because it depends upon where the boundary is drawn for the definition of adiabatic.

Consider a given mass of saturated steam in a container at a temperature of 100 °C and at a pressure of 101.325 kPa, changing phase to become saturated water. The latent heat of vaporisation is transferred out of the fluid accompanied by a local decrease in entropy which can be determined from a measurement of the other fluid properties. If the process is assumed to be reversible, from Equation (12.7) there is a decrease in entropy because the heat transfer Q is negative while the temperature T remains constant. But if the system boundary is taken as the room in which the steam container is placed rather than the walls of the container itself, it can be assumed that the room, due to its size, acts as a heat reservoir so that an addition or subtraction of a small amount of heat is insufficient to change its temperature. The room can be taken as an adiabatic system. If the process is reversible, the entropy will remain constant. In real life, all processes are condemned to be irreversible. If the system boundary is taken as the room, the entropy of the room must be increasing. If this argument is extended to the universe, the entropy of the universe must also be increasing.

This might seem unimportant until it is recalled that the property entropy is used as a measure of the efficiency of a process, in particular steady flow processes. For nozzles, diffusers, turbines and compressors, the steady flow process was assumed to be adiabatic initially and the efficiency was defined in terms of what actually happened to the fluid in some way compared to what would happen if the process was isentropic. The efficiency was called the isentropic efficiency. In real adiabatic systems, the entropy will increase, and the more it increases the lower will be the efficiency. The fact that the entropy of the universe is increasing implies that the efficiency of the universe is decreasing, in other words that the universe is steadily becoming more chaotic. An engineer who improves the efficiency of a turbine, for example, only succeeds in slowing the rate of increase of entropy in the universe and so the rate of descent into chaos!

The formulae for W and Q in non-flow processes apply only when the processes are reversible. But it is also possible to postulate what is called internally reversible processes. When a system undergoes a process which can be reversed, but in doing so the surroundings undergo an irreversible change, the process is said to be internally reversible. In this case the formulae for W and Q are applicable, as are all the other equations which can be derived from them. Most processes occurring in a piston cylinder reciprocating mechanism can be assumed to be internally reversible to a close approximation.

However, in the free expansion of a perfect gas (Joule's law), Section 6.1.2 and Example 6.2, it was stated that the internal energy was a function of the temperature of the gas. The free expansion was assumed to be an adiabatic process, but it is very much an irreversible process with a consequent increase in entropy. The formula for Q in a non-flow process, Equation (3.2) or (12.7), does not apply.

In steady flow processes, generally the assumption was made that the process was adiabatic and the isentropic efficiency term introduced to correct the answer. This is not possible, though, in two special cases, throttling and the adiabatic mixing of fluids.

Across a throttling device, Section 8.7, the enthalpy of the fluid was shown to remain constant, assuming the process to be adiabatic. But it is also highly irreversible due to the considerable turbulence of the fluid as it is throttled, and the entropy will increase.

Mixing processes, as in Section 8.8, were also assumed to be adiabatic but again are irreversible due to the fluid turbulence, resulting in an entropy increase.

For neither of these processes was an isentropic efficiency defined and it is not possible to do so because both are highly irreversible.

13

Problems

13.2.1 The air pressure at the bottom of a mountain of height 3000 m is 100 kPa. What is the pressure at the top of the mountain when:
(a) The density of the air is constant at 1.2 kg/m³?   Ans: 64.7 kPa
(b) The air temperature may be assumed constant at 280 K and the density is related to the air temperature by the equation $p = \rho \times 287T$?   Ans: 69.3 kPa

13.2.2 A static pressure tapping is fitted to a horizontal pipeline at a particular upstream cross-section and another static pressure tapping at a downstream cross-section.
(a) The static pressure tappings are connected to a simple U-tube mercury manometer. If the difference in the levels of the mercury either side of the U-tube is 40 mm and the relative density of mercury is 13.6, what is the static pressure drop of the fluid flowing in the pipe when it is
  (i) Water of density 1000 kg/m³?   Ans: 4944.2 Pa
  (ii) Oil of relative density 0.8?   Ans: 5022.7 Pa
  (iii) Alcohol of relative density 0.7?   Ans: 5602 Pa
  (iv) Acid of relative density 1.4?   Ans: 4787.3 Pa
  (v) Carbon tetrachloride of relative density 1.6?   Ans: 4708.8 Pa
  (vi) Atmospheric air of relative density 0.0012?   Ans: 5336.2 Pa
  (vii) Compressed air of relative density 0.006?   Ans: 5334.3 Pa
(b) The static pressure tappings are connected to an inverted U-tube air manometer. If the difference in the levels of the air either side of the U-tube is 40 mm and the density of the air is assumed constant at 1.2 kg/m³, what is the static pressure drop of the fluid flowing in the pipe when it is
  (i) Water of density 1000 kg/m³?   Ans: 391.9 Pa
  (ii) Oil of relative density 0.8?   Ans: 313.5 Pa
  (iii) Alcohol of relative density 0.7?   Ans: 274.2 Pa
  (iv) Acid of relative density 1.4?   Ans: 548.9 Pa
  (v) Carbon tetrachloride of relative density 1.6?   Ans: 627.4 Pa

13.2.3 The left-hand limb of a simple U-tube mercury manometer is connected to a static pressure tapping in the side of a water pipe, and the right-hand limb is open to the atmosphere. The water flowing in the pipe has a density of 1000 kg/m³ and the mercury has a density of 13 600 kg/m³.
(a) If the height from the static pressure tapping to the level of mercury in the left-hand limb is 200 mm and the mercury in the right-hand limb is at a height of 180 mm above that in the left-hand limb, what is the gauge static pressure of the water in the pipe?   Ans: 22.1 kPa
(b) To protect the mercury, a height of 150 mm of alcohol of relative density 0.7 is poured on top of the mercury in the right-hand limb. If the position of the mercury does not change in the U-tube, what is now the gauge static pressure of the water in the pipe?   Ans: 23.1 kPa

13.2.4 A pipe is inclined at some angle to the horizontal such that the fluid, which is water of density 1000 kg/m³, will flow upwards through it. Two static pressure tappings in the side of the pipe, 500 mm apart, are connected to a simple U-tube mercury manometer. The relative density of mercury is 13.6.
(a) If the water in the pipe is stationary, what is the static pressure difference of the water if the difference in levels of the mercury either side of the U-tube is 50 mm and the pipe is inclined at the following:
  (i) 30 degrees to the horizontal?   Ans: 9.5 kPa
  (ii) 45 degrees to the horizontal?   Ans: 10.6 kPa
  (iii) 60 degrees to the horizontal?   Ans: 11.5 kPa
(b) When the water is flowing upwards through the pipe, determine the percentage increase in the height difference of the mercury in the manometer if the static pressure difference increases by 10% at each angle of inclination. Ans: 14%, 15.6%, 16.9%

13.2.5 A U-tube mercury manometer is made of glass of internal diameter 10 mm. The right-hand limb is open to the atmosphere where the pressure is 100 kPa. A 50 mm internal diameter chamber is fitted on to the left-hand limb in order to contain a differential transformer which will provide an electrical output. The mercury partially fills the chamber and the transformer floats on top of the mercury. If the density of the mercury is 13 600 kg/m³, what static pressure is exerted upon the transformer such that the level of mercury in the chamber moves downwards by a distance of 12 mm?   Ans: 141.6 kPa

13.2.6 A U-tube mercury manometer of internal diameter 10 mm has both ends enlarged to form chambers of 50 mm internal diameter. An equal volume of water, of density 1000 kg/m³, sits on top of the mercury, of relative density 13.6, in each limb and partially fills each chamber. The right-hand limb of the manometer is open to the atmosphere. What gauge pressure is exerted on the water in the left-hand limb such that the difference in height of the mercury either side of the U-tube is 75 mm?   Ans: 9.3 kPa

13.2.7 Calculate the Reynolds number and state whether the flow would normally be laminar or turbulent in each of the following steady flows in a pipeline:
(a) Fluid; water, temperature 10°C, density 999.7 kg/m³, dynamic viscosity 0.0013 kg/m s, velocity 1.15 m/s. Pipe; commercial steel pipe, inside diameter 25.2 mm.   Ans: 22286, turbulent
(b) Fluid; water, temperature 20°C, density 998.7 kg/m³, dynamic viscosity 0.001 002 kg/m s, velocity 14.1 mm/s. Pipe; drawn copper tube, inside diameter 1.51 mm.   Ans: 21.2, laminar
(c) Fluid; oil, temperature 80°C, density 850 kg/m³, dynamic viscosity 27×10^{-3} kg/m s, velocity 0.36 m/s. Pipe; galvanised iron pipe, internal diameter 0.72 m.   Ans: 8160, turbulent
(d) Fluid; air, temperature 52°C, density 1.086 kg/m³, dynamic viscosity 1.807×10^{-5} kg/m s, velocity 10.8 m/s. Pipe; smooth plastic pipe, internal diameter 455 mm.   Ans: 295330, turbulent;

13.2.8 In a journal bearing, oil of density 800 kg/m³ and dynamic viscosity 0.25 kg/m s is fed into the gap between a shaft of diameter 150 mm, rotating at 15 rev/s, and a stationary bush housing of diameter 155 mm and length 100 mm. In practice, the shaft adopts an eccentric position within the bush but, for the purposes of determining the friction force on the bush, assume that the shaft lies concentric within it. The velocity gradient in the oil has a profile given by the equation

$$-C = Xr + Y$$

where C is the velocity of the oil in metres per second at any radius r and X and Y are constants. Determine the value of X and Y and the power required to overcome the friction forces generated in the oil film if laminar conditions apply.   Ans: 2827.4, 219.1, 18.9 W

13.2.9 In a thrust bearing, the lower end of a rotating vertical shaft of diameter 125 mm is separated from a stationary surface by an oil film, which supports the thrust exerted by the shaft. The oil film is actually of variable thickness but, in a first analysis, it may be assumed to be of constant thickness of 0.3 mm. The oil has a dynamic viscosity of 0.2 kg/m s. If the shaft is rotating at 2000 rev/min, determine the power absorbed in overcoming the fluid friction.   *Ans*: 701 W

13.2.10 A hydraulic dynamometer for measuring power consists of a thin disc 250 mm in diameter rotating in a casing which is full of water. The disc rotates at 3000 rev/min and the frictional resistance on the surface of the disc per unit area F is given by the formula:
$$F = \text{const}. \, C^2$$
where C is the linear velocity at any point in metres per second and F is in newtons per square metre. If the power absorbed is 18 kW, determine the value of the constant in the above formula.   *Ans*: 7.57

13.3.1 A fluid in a piston cylinder mechanism undergoes a constant pressure expansion while producing a work output of 10 kJ. The pressure is 2.5 MPa and the fluid expands through a volume ratio of 4. Find the initial and final volume of the fluid.   *Ans*: 1.33×10^{-3} m^3, 5.33×10^{-3} m^3

13.3.2 A fluid in a piston cylinder mechanism is compressed from a pressure of 100 kPa and a volume of 0.002 m^3 to a pressure of 900 kPa. If the fluid follows the relationship:
$$pV^{1.2} = \text{const}.$$
during the compression, what is the magnitude and direction of the work transfer?   *Ans*: + 442.2 J

13.3.3 Ammonia refrigerant condenses from the vapour to the liquid phase in the condenser of a refrigeration plant at the constant temperature of 50 °C. The decrease in specific entropy of the ammonia is determined as being 3.258 kJ/kg K. If the mass of refrigerant is 0.06 kg, what is the magnitude and direction of the heat transfer?   *Ans*: −63.1 kJ

13.3.4 It is proposed to build an engine, based upon a reciprocating piston cylinder mechanism, in which the fluid completes a cycle by undergoing the following four non-flow reversible processes:

1 State 1 to state 2—constant pressure expansion which produces a work output of 55 kJ;
2 State 2 to state 3—constant volume drop in pressure to a pressure of 100 kPa;
3 State 3 to state 4—constant pressure compression at a pressure of 100 kPa;
4 State 4 to state 1—constant volume rise in pressure to the initial state.

If the net work output from the cycle is 23 kJ and the cycle is completed 10 times per second, determine:

(a) The swept volume of the cylinder.   *Ans*: 0.32 m^3
(b) The pressure of the fluid as it goes from state 1 to state 2.   *Ans*: 171.8 kPa
(c) The power output of the engine.   *Ans*: 230 kW

13.3.5 It is proposed to build an engine in which 0.075 kg of a fluid undergoes the following four non-flow reversible processes in completing a cycle in a piston cylinder mechanism:

1 State 1 to state 2—heat addition at constant temperature during which the specific entropy of the fluid increases by 4.448 kJ/kg K;
2 State 2 to state 3—constant entropy decrease in temperature;
3 State 3 to state 4—heat rejection of 96.7 kJ at constant temperature during which the specific entropy of the fluid decreases through the same amount as it increased when going from state 1 to state 2;
4 State 4 to state 1—constant entropy rise in temperature back to state 1.

If the net heat transfer during the cycle is 54.4 kJ, determine the maximum and minimum fluid temperatures.
Ans: 452.9 K, 289.9 K

352 PROBLEMS

13.3.6 A reciprocating piston cylinder engine is to be built in which the fluid undergoes the following four reversible non-flow processes in completing a cycle:

1. State 1 to state 2 — constant temperature compression at a temperature of 300 K during which the heat rejected is 6.5 kJ;
2. State 2 to state 3 — constant volume heat addition of 8.8 kJ until the temperature reaches 865 K;
3. State 3 to state 4 — constant temperature expansion to the volume at state 1 during which there is a further heat supply of 18.7 kJ;
4. State 4 to state 1 — constant volume heat rejection of 8.8 kJ back to the initial state.

If the mass of fluid is 0.005 kg, determine:
(a) The change in the specific entropy of the fluid as it goes from state 1 to state 2.   Ans: −4.3 kJ/kg K
(b) The direction and magnitude of the work transfer as the fluid goes from state 2 to state 3.   Ans: 0 kJ
(c) The change in the specific entropy of the fluid as it goes from state 3 to state 4.   Ans: + 4.3 kJ/kg K
(d) The net heat transfer when the fluid completes one cycle.   Ans: + 12.2 kJ

13.3.7 A reciprocating piston cylinder engine is to be built which uses 0.02 kg of a certain fluid. The fluid undergoes the following four non-flow reversible processes in order to complete a cycle:

1. State 1 to state 2 — constant temperature compression from a pressure of 100 kPa and a temperature of 290 K to a pressure of 2.4 MPa during which the specific entropy decreases by 6.607 kJ/kg K;
2. State 2 to state 3 — constant pressure heat addition of 177.5 kJ until the temperature is 2000 K and the volume has increased by 0.03 m^3;
3. State 3 to state 4 — constant temperature expansion back to the initial pressure during which there is a heat input of 264.3 kJ;
4. State 4 to state 1 — constant pressure heat rejection back to the conditions at state 1 during which there is a work input of 72 kJ.

Determine:
(a) The magnitude and direction of the heat transfer as the fluid goes from state 1 to state 2.   Ans: −38.3 kJ
(b) The magnitude and direction of the work transfer as the fluid goes from state 2 to state 3.   Ans: + 72 kJ
(c) The change in specific entropy of the fluid as it goes from state 3 to state 4.   Ans: 6.607 kJ/kg K increase
(d) The change in volume of the fluid as it goes from state 4 back to state 1.   Ans: 0.72 m^3
(e) The magnitude of the heat transfer as the fluid goes from state 4 back to state 1 if the net heat transfer in the cycle is 226 kJ.   Ans: −177.5 kJ

13.3.8 A reciprocating piston cylinder engine is to be built in which the fluid undergoes the following four reversible non-flow processes:

1. State 1 to state 2 — constant entropy compression which requires a work input of 260 kJ/kg;
2. State 2 to state 3 — constant volume heat addition of 976 kJ/kg;
3. State 3 to state 4 — constant entropy expansion until the volume of the fluid is the same as at state 1, which produces a work output of 815 kJ/kg;
4. State 4 to state 1 — constant volume drop in pressure back to the initial state which produces a heat output of 420 kJ/kg.

Determine:
(a) The net heat transfer per kilogram of fluid when the fluid completes a cycle.   Ans: 555 kJ/kg
(b) The net work transfer per kilogram of fluid when the fluid completes a cycle.   Ans: 555 kJ/kg

13.3.9 A reciprocating piston cylinder engine is to be built in which 0.005 kg of fluid undergoes the following four reversible non-flow processes in completing the cycle 30 times per second:

1. State 1 to state 2 — constant entropy compression which requires a work input of 2.2 kJ;
2. State 2 to state 3 — constant pressure expansion at a pressure of 5 MPa until the specific volume of the fluid has increased by 0.115 m^3/kg, necessitating a heat input of 10 kJ;

3   State 3 to state 4—constant entropy expansion until the volume is equal to its initial value at state 1, during which there is a work output of 4.9 kJ;
4   State 4 to state 1—constant volume drop in pressure back to the initial state.

(a) Determine the magnitude and direction of the work transfer as the fluid goes from state 2 to state 3.   *Ans:* −2.9 kJ
(b) Determine the power output of the engine.   *Ans:* 168 kW
(c) If the net heat transfer when the fluid completes one cycle is 5.6 kJ, determine the magnitude and direction of the heat transfer when the fluid goes from state 4 back to state 1.   *Ans:* + 4.4 kJ

13.3.10 A reciprocating piston cylinder engine is to be built in which the fluid undergoes the following four non-flow reversible processes operating in a cycle:

1   State 1 to state 2—constant entropy compression of the fluid from a pressure of 100 kPa and a specific volume of 0.85 m³/kg to a specific volume of 0.085 m³/kg, during which the fluid obeys the law $pv^{1.4}$ = const.;
2   State 2 to state 3—constant pressure heat addition until the specific volume of the fluid is 0.23 m³/kg;
3   State 3 to state 4—constant entropy expansion of the fluid back to the initial pressure of 100 kPa during which the fluid obeys the law $pv^{1.4}$ = const.;
4   State 4 to state 1—constant pressure heat rejection back to the conditions at state 1.

Determine the magnitude and direction of the work transfer per kilogram of fluid in each process.
Ans: + 321.3 kJ/kg, −364 kJ/kg, −868.7 kJ/kg, + 145 kJ/kg

13.4.1 A four-stroke single cylinder compression ignition engine is run in the laboratory and the following results obtained:

Net brake load	= 240 N
Effective brake wheel radius	= 0.55 m
Engine speed	= 10 rev/s
Effective average fluid cylinder pressure	= 510 kPa
Cylinder bore	= 15 cm
Stroke	= 20 cm
Rate of consumption of fuel oil	= 5×10^{-4} kg/s
Calorific value of fuel	= 44000 kJ/kg

Calculate the indicated power, the brake power, the friction power and the overall thermal efficiency of the engine.   *Ans:* 9 kW, 8.3 kW, 0.7 kW, 37.7%

13.4.2 A bomb calorimeter is to be used to determine the calorific value of a particular brown coal; 6 g of the coal is enclosed in a small chamber which is filled with oxygen. The chamber is surrounded by a water bath, insulated from the atmosphere. The coal is set fire to by an electrical fuse and the heat released from the combustion reaction heats the 10 kg of water from a temperature of 20.8–24.2 °C. If the specific heat of the water at constant pressure at its average temperature is 4.182 kJ/kg K, determine the calorific value of the brown coal.   *Ans:* 23698 kJ/kg

13.4.3 A performance test on a four-cylinder, four-stroke petrol engine at a speed of 3000 rev/min revealed the following results:

Brake load	= 200 N
Friction power	= 13.2 kW
Dynamometer torque arm radius	= 0.5 m
Fuel consumption	= 280 cc/min
Calorific value of fuel	= 40 000 kJ/kg
Relative density of fuel	= 0.8

(a) Calculate the brake power, indicated power and the overall thermal efficiency of the engine at this speed.   *Ans:* 31.4 kW, 44.6 kW, 21%

(b) It is decided to recondition the engine such that it will produce the same brake power output at the same speed but with a reduced energy input. A total of $5000 is available for the reconditioning and a payback period of 1 year is specified. If the engine runs for 5000 hours per year, determine the reduced fuel flow rate given that the petrol costs $0.07/kWh.   *Ans*: 253.2 cc/min

13.4.4 A domestic house has a requirement for 8 kW of heat. At present this is supplied by electric under floor heating which is turned on for 3.5 hours every morning and 5.5 hours every evening for 212 days a year. As the electricity costs $0.12/kWh, this is turning out to be rather expensive on running costs for the householder and it is decided to consider installing one of the following alternative systems:

1. An oil-fired central heating system with a capital cost of $1800. The oil boiler will provide the required 8 kW of heat, but it is only 70% efficient and the oil costs $0.06/kWh.

2. A natural gas-fired central heating system with a capital cost of $1600. The natural gas boiler will provide the required 8 kW of heat, but it is only 80% efficient and the natural gas costs $0.04/kWh.

3. An electric-driven heat pump system of capital cost $2500 using electricity at $0.12/kWh. However, in providing the required 8 kW of heat, the heat pump has a coefficient of performance of 3, equivalent to an efficiency of 300%.

4. An electric night storage heating system of capital cost $600. These will provide the required 8 kW at an efficiency of 100%, but they will only operate for 7 hours during the night when they make use of cheaper electricity costing $0.05/kWh. The other 2 hours of heating must be provided by the existing under floor system utilising electricity at the daytime rate of $0.12/kWh.

(a) Determine the simple payback period of each system compared to the existing under floor heaters.
Ans: 3.44 years, 1.5 years, 2.05 years, 0.72 year

(b) If the householder stayed for 5 years after making the change, what would be the cash savings generated by each system which the householder could put in the bank?   *Ans*: $816.7, $3742.4, $3605.6, $3555.2

13.4.5 A domestic central heating system consists of a natural gas boiler, an electric-driven water pump and five radiators. The pump is driven by a 500 W electric motor and the energy transfer efficiency to the pump is 60%. Each radiator dissipates 1 kW of heat and the interconnecting pipework a further 1 kW.

(a) Assuming that the central heating system obeys the first law of thermodynamics, what is the heat transfer to the water in the boiler?   *Ans*: 5.7 kW

(b) If the water in the boiler is heated from 50–70°C and the specific heat of water at constant pressure is constant and equal to 4.2 kJ/kg K, what is the mass flow rate of water flowing through the system?   *Ans*: 0.068 kg/s

(c) If the boiler has an energy transfer efficiency of 70%, the boiler is operated for 3500 hours per year and the natural gas fuel costs $0.04/kWh, what is the annual running cost of the boiler?   *Ans*: $1140 per year

13.4.6 A nuclear reactor generates 3000 MW of heat. The heat is transferred in a heat exchanger of energy transfer efficiency 75% into steam which is expanded in a turbine in order to produce a power output. The steam is condensed in a condenser, releasing 1800 MW of heat, and pumped back through the heat exchanger by a feed pump which requires 3% of the power output from the turbine. Determine:

(a) The net power output from the plant.   *Ans*: 450 MW
(b) The power output from the turbine.   *Ans*: 464 MW
(c) The overall thermal efficiency of the plant.   *Ans*: 15%

13.4.7 A combined domestic refrigerator/heat pump serves the dual purpose of cooling the kitchen larder and providing hot water. The electric motor driving the compressor operates for approximately one-third of the day, and has an electrical power input of 0.225 kW. The energy transfer efficiency to the compressor is 85%. Heat leakage from the kitchen to the larder, where the evaporator is situated, is 0.3 kW. All the heat rejected at the condenser is taken by water in a domestic hot-water tank which is heated from 10–60°C.

(a) How long will it take to heat up a tank of water containing 25 kg of water if the specific heat of water at constant pressure is constant and equal to 4.2 kJ/kg K?   *Ans*: 14873 s
(b) Once the tank is heated up, what mass flow rate of hot water can be supplied?   *Ans*: 1.68×10^{-3} kg/s
(c) What is the daily running cost of the unit if electricity is priced at $0.12/kWh?   *Ans*: $0.216 per day

13.4.8 The manufacturers of an air-to-air heat pump proudly announce in their sales brochure that, at a given atmospheric air temperature, their unit has a COP of 4.5. In fact, they have forgotten to mention a few crucial details and this is the COP based upon the refrigerant. The electric motor driving the compressor has an output of 3.2 kW, but there is an energy transfer efficiency between the two components of 70%. There is an energy transfer efficiency at the condenser of 85%. In addition there is a fan fitted to the condenser which absorbs a further 0.5 kW.

(a) What is a more realistic COP for the unit based upon the actual energy input and output?   *Ans*: 2.32
(b) What is the actual annual running cost of the unit if it operates for 2000 hours per year and the electricity costs $0.12/kWh?   *Ans*: $888 per year

13.4.9 A closed cycle gas turbine system with two turbines is to be used in a nuclear power plant to produce electricity. The fluid in the closed cycle gas turbine system is compressed in a rotary compressor from a low to a high-pressure and is then heated at the constant high-pressure in a heater by the 450 MW of heat available from the nuclear reaction. However, the heater only has an energy transfer efficiency of 80%. The high-pressure fluid is firstly expanded in turbine 1 to an intermediate pressure, which produces a power output of 56 MW. All this power is used to drive the compressor. The fluid is further expanded in turbine 2 to the low pressure required at entry to the compressor, producing a work output at an alternator of 72 MW. Before entering the compressor, the fluid is cooled in a cooler by water which circulates through cooling towers. The cooler has an energy transfer efficiency of 82% and requires 3000 kg/s of water of specific heat 4.2 kJ/kg K. Determine:

(a) The temperature rise of the cooling tower water in the cooler.   *Ans*: 18.7 °C
(b) The overall thermal efficiency of the closed cycle gas turbine system.   *Ans*: 16%

13.4.10 A stationary open cycle gas turbine plant consists of a compressor, a combustion chamber, and a turbine coupled to an alternator. Air from the atmosphere is compressed in the compressor to a higher pressure and temperature, requiring a work input. In the combustion chamber, there is a heat transfer to the air from the combustion of natural gas, causing its temperature to rise. The products of combustion are expanded in the turbine, which produces an electrical work output at the alternator and exhaust to the atmosphere. Some of the work output produced in the turbine is used to drive the compressor through an interconnecting shaft. The following data applies to the open cycle gas turbine

Temperature of the air in the atmosphere	= 290 K
Temperature of the exhaust from the turbine	= 750 K
Specific heat of the exhaust from the turbine	= 1.1 kJ/kg K
Mass flow rate of exhaust from the turbine	= 120 kg/s
Energy transfer efficiency between turbine and alternator	= 90%
Mass flow rate of natural gas into combustion chamber	= 2 kg/s
Calorific value of natural gas	= 50 000 kJ/kg;
Energy transfer efficiency in combustion chamber (i.e. the efficiency at which the heat is transferred to the air)	= 90%
Cost of natural gas	= $0.02/kWh
Power required to drive compressor	= 10% of power output at turbine

Use the first law of thermodynamics to determine:
(a) The power output at the alternator.   *Ans*: 16352 kW
(b) The annual running cost of the plant if it operates for 5000 hours per year.   *Ans*: $10 million
(c) The efficiency of the plant.   *Ans*: 26.3%
(d) The annual profit if the electricity produced at the alternator is sold to consumers at the rate of $0.12/kWh.   *Ans*: $5.8 million

13.5.1 It is required to boil 1.8 kg of ammonia refrigerant under constant pressure conditions and the possibility exists to achieve this at five different pressures. The increase of specific internal energy and specific volume of the ammonia as it changes from the liquid to the vapour phase at each pressure is given in Table 13.1. How much heat must be supplied to the ammonia at each pressure in order to boil it?   *Ans*: As in Table 13.2

Table 13.1   Problem 13.5.1.

Pressure of the ammonia at which it will boil (kPa)	Change of specific volume of the ammonia during boiling (m³/kg)	Change of specific internal energy of the ammonia during boiling (kJ/kg)
158.8	0.74	1 225.1
268	0.45	1 183.7
429.5	0.29	1 138.6
658.5	0.19	1 093.8
972.2	0.13	1 044.7

Table 13.2   Problem 13.5.1—results.

Pressure of the ammonia at which it will boil (kPa)	Heat supply required (kJ)
158.8	2 416.7
268	2 347.7
429.5	2 273.7
658.5	2 194
972.2	2 108

13.5.2 A mass of 1.35 kg of a fluid in a container of fixed volume is being cooled under constant volume conditions. The decrease in specific internal energy of the fluid is known to be 14.2 kJ/kg. The container is surrounded by a water jacket in which 1.6 kg of water of specific heat 4.2 kJ/kg K are heated up from a temperature of 20 °C. If the energy transfer efficiency from the fluid to the water is 85%, what is the final temperature of the water?
Ans: 22.4 °C

13.5.3 A fluid in a piston cylinder mechanism is being compressed at the constant temperature of 150 °C. The decrease in specific entropy and specific internal energy of the fluid is 0.685 kJ/kg K and 18 kJ/kg respectively. If the work input required for the compression is 81.5 kJ, and the process is assumed reversible, determine the mass of fluid.   *Ans*: 0.3 kg

13.5.4 A fluid is made to undergo a reversible isentropic non-flow compression process from state 1 to state 2 during which the pressure increases from 100 kPa to 1.2 MPa. This requires a work input per kilogram of fluid of 219.5 kJ/kg. The fluid is made to return from state 2 to state 1 by means of a reversible non-flow polytropic process for which the index of expansion is 1.4. If the specific volume of the fluid at state 2 is 0.144 m³/kg, determine:
(a) The change in the specific internal energy of the fluid as it goes from state 1 to state 2.   *Ans*: + 219.5 kJ/kg
(b) The specific volume of the fluid at state 1.   *Ans*: 0.85 m³/kg
(c) The magnitude and direction of the heat transfer per kilogram of fluid as it goes from state 2 to state 1.
   Ans: 0 kJ/kg

13.5.5 A heat transfer of 15 kJ is supplied to a fluid undergoing a reversible polytropic expansion process in a piston cylinder mechanism, during which the internal energy of the fluid decreases by 27.5 kJ. The fluid expands through a volume ratio of 8 from an initial pressure of 2.4 MPa, and the index of expansion is 1.4. What is the initial and final volume of the fluid?   *Ans*: 0.0125 m^3, 0.1 m^3

13.5.6   A reciprocating piston cylinder engine is to be built in which the fluid undergoes the following four reversible non-flow processes:

1   State 1 to state 2—constant entropy compression during which the specific internal energy of the fluid increases by 259.7 kJ/kg;
2   State 2 to state 3—constant volume heat addition during which the specific internal energy of the fluid increases by 976.2 kJ/kg;
3   State 3 to state 4—constant entropy expansion until the volume of the fluid is the same as at state 1, during which the specific internal energy of the fluid decreases by 815.6 kJ/kg;
4   State 4 to state 1—constant volume drop in pressure back to the initial state during which the specific internal energy of the fluid decreases by 420.3 kJ/kg.

(a)   Determine the net heat transfer per kilogram of fluid when the fluid completes a cycle.   *Ans*: 555.9 kJ/kg
(b)   Show that the first law of thermodynamics is valid by finding that the net work transfer per kilogram of fluid when the fluid completes a cycle plus the net heat transfer equals zero.

13.5.7   A reciprocating piston cylinder engine is to be built in which the fluid undergoes the following four reversible non-flow processes in completing a cycle:

1   State 1 to state 2—isothermal compression at a temperature of 300 K during which the specific entropy of the fluid decreases by 0.597 kJ/kg K and the specific internal energy of the fluid remains constant;
2   State 2 to state 3—isochoric heat addition to a temperature of 900 K during which the specific internal energy of the fluid increases by 430.8 kJ/kg;
3   State 3 to state 4—isothermal expansion to the volume at state 1 during which the specific internal energy of the fluid again remains constant;
4   State 4 to state 1—isochoric heat rejection back to the initial state.

If the net work transfer per kilogram of fluid from the engine when the fluid completes one cycle is 295.5 kJ/kg, determine:

(a)   The magnitude and direction of the work transfer per kilogram of fluid as it goes from state 3 to state 4.   *Ans*: +474.6 kJ/kg
(b)   The magnitude and direction of the heat transfer per kilogram of fluid as it goes from state 4 to state 1.   *Ans*: −430.8 kJ/kg

13.5.8   A reciprocating piston cylinder engine is to be built in which 0.005 kg of fluid undergoes the following four reversible non-flow processes in completing the cycle 30 times per second:

1   State 1 to state 2—polytropic compression from a fluid pressure of 100 kPa and specific volume of 0.8 m^3/kg to a fluid pressure of 5 MPa, the index of compression being 1.2;
2   State 2 to state 3—constant pressure expansion until the specific volume of the fluid doubles, the fluid temperature increasing from 534.8 to 1069.7 K;
3   State 3 to state 4—polytropic expansion until the fluid specific volume is equal to its initial value at state 1 and the internal energy of the fluid decreases by 1538.7 J, the index of expansion being 1.2;
4   State 4 to state 1—constant volume drop in pressure back to the initial state.

If the specific heat of the fluid at constant pressure remains constant at 1.005 kJ/kg K as the fluid goes from state 2 to state 3, determine:

(a)   The net work transfer as the fluid completes a cycle.   *Ans*: 2005 J
(b)   The power output of the engine.   *Ans*: 60.15 kW
(c)   The overall thermal efficiency of the engine.   *Ans*: 47.5%

13.5.9 A reciprocating piston cylinder engine is to be built in which 0.015 kg of fluid undergoes the following four non-flow reversible processes operating in a cycle:

1. State 1 to state 2—constant entropy compression of the fluid during which there is a work input of 323.1 kJ/kg;
2. State 2 to state 3—constant pressure heat addition to the fluid of 753.75 kJ/kg during which there is a work output of 215.25 kJ/kg;
3. State 3 to state 4—constant entropy expansion of the fluid back to the initial pressure during which there is a work output of 646.2 kJ/kg;
4. State 4 to state 1—constant pressure heat rejection from the fluid of 301.5 kJ/kg back to the conditions at state 1 during which there is a work input of 86.1 kJ/kg.

If the specific heat of the fluid at constant pressure is constant and equal to 1.005 kJ/kg K during both constant pressure processes, determine:

(a) The temperature rise of the fluid as it goes from state 2 to state 3.   *Ans*: 749.5 K
(b) The temperature decrease of the fluid as it goes from state 4 to state 1.   *Ans*: 300 K
(c) The magnitude and direction of the change of internal energy of the fluid in each process.   *Ans*: + 4.8 kJ, + 8.1 kJ, −9.7 kJ, −3.2 kJ
(d) Confirm that the net change in internal energy of the fluid when it completes one cycle is zero.
(e) The net work transfer from the fluid when it completes one cycle.   *Ans*: 6.78 kJ
(f) Confirm that the net heat transfer to the fluid plus the net work transfer from the fluid when it completes one cycle equals zero.
(g) The power output of the engine if it completes 20 cycles per second.   *Ans*: 135.7 kW
(h) The overall thermal efficiency of the engine.   *Ans*: 60%

13.5.10 A reciprocating piston cylinder engine is to be built which uses 0.02 kg of a certain fluid. The fluid undergoes the following four non-flow reversible processes in order to complete a cycle:

1. State 1 to state 2—isothermal compression from a pressure of 100 kPa to a pressure of 1.5 MPa during which there is a heat rejection from the fluid of 691.4 kJ/kg, a work input of 654.4 kJ/kg and the specific volume of the fluid decreases from 2.406 m^3/kg at state 1 to 0.152 m^3/kg at state 2;
2. State 2 to state 3—isobaric heat addition of 548 kJ/kg during which the temperature of the fluid is raised to 500°C and the specific volume of the fluid increases by 0.0831 m^3/kg;
3. State 3 to state 4—isothermal expansion back to the initial pressure during which the specific entropy of the fluid increases by 1.265 kJ/kg K, the specific internal energy of the fluid increases by 11.0 kJ/kg and the specific volume increases from 0.2351 to 3.565 m^3/kg;
4. State 4 to state 1—constant pressure heat rejection back to the conditions at state 1.

Determine:

(a) The change in enthalpy of the fluid as it goes from state 1 to state 2.   *Ans*: −1.0 kJ
(b) The change in enthalpy of the fluid as it goes from state 2 to state 3.   *Ans*: 11.0 kJ
(c) The magnitude and direction of the work transfer as the fluid goes from state 2 to state 3.   *Ans*: −2.5 kJ
(d) The change in enthalpy of the fluid as it goes from state 3 to state 4.   *Ans*: + 0.3 kJ
(e) The magnitude and direction of the work transfer as the fluid goes from state 3 to state 4.   *Ans*: -19.3 kJ
(f) The net work transfer when the fluid completes a cycle.   *Ans*: + 6.4 kJ
(g) The change in enthalpy of the fluid as it goes from state 4 back to state 1.   *Ans*: −10.3 kJ

13.6.1 Hydrogen, assumed to be a perfect gas with $R = 4.157$ kJ/kg K, is expanded in a piston cylinder mechanism from a temperature of 650 K, a pressure of 800 kPa and a volume of 0.12 m^3, producing a work output of 1560 kJ/kg. Assuming a non-flow reversible isothermal process, determine:

(a) The direction and magnitude of the heat transfer per kilogram of hydrogen.   *Ans*: + 1560 kJ/kg

(b) The change in internal energy of the hydrogen. *Ans*: 0 kJ
(c) The final volume of the hydrogen. *Ans*: 0.21 m^3
(d) The final pressure of the hydrogen. *Ans*: 457 kPa
(e) The mass of hydrogen required to produce a work output of 20 kJ. *Ans*: 0.0128 kg
(f) The change of entropy of the hydrogen. *Ans*: 0.031 kJ/K

13.6.2 Oxygen, assumed to be a perfect gas with R = 260 J/kg K and c_v = 658 J/kg K, undergoes a reversible non-flow constant pressure expansion process in a piston cylinder mechanism. The initial pressure of the oxygen is 500 kPa, the initial temperature 450 K and the final volume 0.15 m^3. If the work output due to the expansion is 22 kJ, determine:

(a) The initial volume of oxygen. *Ans*: 0.106 m^3
(b) The mass of oxygen in the cylinder. *Ans*: 0.453 kg
(c) The final temperature of the oxygen. *Ans*: 636.8 K
(d) The direction and magnitude of the heat transfer. *Ans*: + 77.7 kJ
(e) The change in entropy of the oxygen. *Ans*: 0.144 kJ/K

13.6.3 A perfect gas, contained in vessel of fixed volume of 0.2 m^3, is cooled from a pressure of 700 kPa to a pressure of 100 kPa. The gas is air for which c_v = 0.718 kJ/kg K and R = 0.287 kJ/kg K, and the mass of air is 0.25 kg. If the cooling may be considered to be a reversible constant volume non-flow process, determine:

(a) The initial temperature of the air. *Ans*: 1951 K
(b) The final temperature of the air. *Ans*: 278.7 K
(c) The change in specific internal energy of the air. *Ans*: −300.2 kJ
(d) The heat transfer from the air. *Ans*: −300.2 kJ
(e) The change in specific entropy of the air. *Ans*: −1.4 kJ/kg K

13.6.4 Helium, considered to be a perfect gas with R = 2.079 kJ/kg K and γ = 1.67, undergoes an isentropic compression in a piston cylinder mechanism. The initial temperature of the helium is 0°C, initial pressure 100 kPa, and final pressure 800 kPa. If the mass of helium is 0.1 kg, determine:

(a) The final temperature of the helium. *Ans*: 628.75 K
(b) The initial specific volume of the helium. *Ans*: 5.68 m^3/kg
(c) The final specific volume of the helium. *Ans*: 1.63 m^3/kg
(d) The direction and magnitude of the work transfer. *Ans*: + 110 kJ

13.6.5 A piston cylinder mechanism contains air, assumed to be a perfect gas, for which R = 287 J/kg K and c_p = 1005 J/kg K. The initial pressure of the air is 1 MPa and the initial temperature 550 K. It is intended to expand the air to a temperature of 320 K and in so doing produce a work output of 330 kJ/kg. If the expansion may be considered to be a polytropic non-flow reversible process, determine:

(a) The index of expansion. *Ans*: 1.2
(b) The final pressure of the air. *Ans*: 38.8 kPa
(c) The initial volume of the air. *Ans*: 0.158 m^3/kg
(d) The final volume of the air. *Ans*: 2.37 m^3/kg
(e) The direction and magnitude of the heat transfer if the mass of air is 0.05 kg. *Ans*: + 8.24 kJ
(f) The change in specific entropy of the air. *Ans*: 0.388 kJ/kg K

13.6.6 0.4 m^3 of a fluid, initially at a pressure of 100 kPa and temperature 125°C, undergoes a reversible non-flow isothermal compression through a volume ratio of 4. Determine the changes in the internal energy, the entropy, the heat transferred and the work done in the process if the fluid is air, considered a perfect gas.
Ans: 0, −139.3 kJ/K, −55.4 kJ, +55.4 kJ

13.6.7 A reciprocating piston cylinder engine is to be built based upon the ideal Joule cycle. The fluid, assumed to be a perfect gas, undergoes the following four non-flow reversible processes operating in a cycle:

1. State 1 to state 2—isentropic compression of the fluid from a pressure of 100 kPa and a temperature of 290 K through a volume ratio of 10 to 1;
2. State 2 to state 3—constant pressure heat addition until the temperature of the fluid is 2000 K;
3. State 3 to state 4—isentropic expansion of the fluid back to the initial pressure of 100 kPa;
4. State 4 to state 1—constant pressure heat rejection back to the conditions at state 1.

(a) Determine the net heat transfer per kg of fluid per cycle and the efficiency of the cycle if the fluid is air for which $c_p = 1.01$ kJ/kg K and $\gamma = 1.4$.   Ans: 772.6 kJ/kg, 60.2%

(b) Determine the net heat transfer per kg of fluid per cycle and the overall thermal efficiency of the cycle if the fluid is helium for which $c_p = 5.19$ kJ/kg K and $\gamma = 1.67$.   Ans: 2626.4 kJ/kg, 78.6%

13.6.8 A reciprocating piston cylinder engine is to be built which uses hydrogen as the fluid. The hydrogen, assumed to be a perfect gas, undergoes the following four non-flow reversible processes operating in a cycle:

1. State 1 to state 2—isothermal compression of the fluid from a pressure of 100 kPa and a temperature of 300 K to a pressure of 600 kPa;
2. State 2 to state 3—constant volume heat addition until the pressure of the fluid is 2000 kPa;
3. State 3 to state 4—isentropic expansion of the fluid to a pressure of 160 kPa;
4. State 4 to state 1—constant volume heat rejection back to the conditions at state 1.

For the hydrogen, assume constant values of $R = 4.157$ kJ/kg K and $\gamma = 1.41$. Determine:

(a) The work transfer and the heat transfer per kilogram of hydrogen in each process.
Ans: $W_{12} = 2235$ kJ/kg, $W_{23} = 0$ kJ/kg, $W_{34} = -5272$ kJ/kg, $W_{41} = 0$ kJ/kg, $Q_{12} = -2235$ kJ/kg, $Q_{23} = 7097$ kJ/kg, $Q_{34} = 0$ kJ/kg, $Q_{41} = -1744$ kJ/kg

(b) The overall thermal efficiency of the cycle.   Ans: 42.8%

13.6.9 A reciprocating piston cylinder engine is to be built based upon the ideal Carnot cycle. The fluid is helium, assumed a perfect gas, for which $R = 2.079$ kJ/kg K and $\gamma = 1.67$. It is intended to use 0.006 kg of helium and for it to remain entirely within the engine as it undergoes the following four reversible non-flow processes in completing a cycle, which it does 20 times per second:

1. State 1 to state 2—isentropic compression from a pressure of 250 kPa and temperature of 300 K to a pressure of 5 MPa;
2. State 2 to state 3—isothermal heat addition of 11.4 kJ;
3. State 3 to state 4—isentropic expansion to a temperature of 300 K;
4. State 4 to state 1—isothermal heat rejection back to the initial state.

Determine:

(a) The maximum cycle temperature.   Ans: 997.9 K
(b) The pressure of the helium after the heat has been added.   Ans: 200 kPa
(c) The minimum cycle pressure.   Ans: 100 kPa
(d) The change of specific entropy during the heat addition.   Ans: 1.904 kJ/kg K
(e) The power produced by the engine.   Ans: 159.4 kW
(f) The overall thermal efficiency of the cycle.   Ans: 69.9%

13.6.10 A reciprocating piston cylinder engine is to be built based upon the ideal dual cycle, which is a combination of the ideal Otto cycle and ideal Diesel cycle. The fluid is air, assumed a perfect gas, for which the gas constant is 287 J/kg K, and the specific heat at constant volume 718 J/kg K. The mass flow rate of the air is 0.05 kg/s as it undergoes the following five reversible non-flow processes:

1. State 1 to state 2—adiabatic compression from an initial pressure of 100 kPa and specific volume 0.8 m³/kg through a volume ratio of 7;

2   State 2 to state 3 – constant volume heat addition until the temperature of the air is 1500 K;
3   State 3 to state 4 – constant pressure heat addition until the temperature is 2500 K;
4   State 4 to state 5 – adiabatic expansion until the volume is equal to its initial value at state 1;
5   State 5 to state 1 – constant volume drop in pressure back to the initial state.

(a) Evaluate the power output of the engine and the cycle efficiency. Ans: 41.8 kW, 50.7%

(b) If the fuel to be used has a calorific value 43 000 kJ/kg and costs $0.05/kWh, determine the mass flow rate of fuel and the annual running cost of the engine if it is to be run 7500 hours per year and there are no losses due to the combustion. Ans: 1.92×10^{-3} kg/s, $30 900 per year

13.6.11 The performance of a reciprocating air compressor is under examination. The air, considered a perfect gas, enters the cylinder at the rate of 1×10^{-3} kg/s with a pressure of 100 kPa and a temperature of 300 K. The discharge valve opens when the air, for which $c_v = 0.718$ kJ/kg K and $R = 0.287$ kJ/kg K, has reached a pressure of 10 MPa. Assuming a reversible non-flow polytropic process for which the index of compression n has the values of 1.6, 1.4, 1.2, 1.0, and 0.8, calculate:

(a) The air volume V_2 and temperature T_2 after compression.
(b) The work input required for each compression.
(c) The heat transfer in each compression.

Ans: As in Table 13.3

Table 13.3 Problem 13.6.11 — results.

n	1.6	1.4	1.2	1.0	0.8
V_2 (m³)	0.24×10^{-3}	0.19×10^{-3}	0.15×10^{-3}	0.10×10^{-3}	0.06×10^{-3}
T_2 (K)	711.4	579.2	440.3	300	168.7
W (J)	228.3	232.5	235	230.3	220
Q (J)	114.3	0	−118.0	−230.3	−329.5

13.6.12 A reciprocating piston cylinder engine is to be built based upon the ideal Ericsson cycle. The fluid is helium, assumed to be a perfect gas, for which $R = 2.079$ kJ/kg K and $c_p = 5.19$ kJ/kg K. The helium, of mass 0.02 kg, undergoes the following four non-flow reversible processes operating in a cycle:

1   State 1 to state 2 – isothermal compression from a pressure of 100 kPa and a temperature of 290 K to a pressure of 2.4 MPa;
2   State 2 to state 3 – constant pressure heat addition until its temperature is 2000 K;
3   State 3 to state 4 – isothermal expansion back to the initial pressure;
4   State 4 to state 1 – constant pressure heat rejection back to the conditions at state 1.

(a) Determine the net work output of the engine when the helium is made to complete one cycle. Ans: 226 kJ

(b) Confirm that the heat output as the helium goes from state 2 to state 3 under constant pressure conditions is numerically the same as the heat input when the helium goes from state 4 to state 1, again under constant pressure conditions.

(c) Determine the overall thermal efficiency of the engine assuming that all the heat output in one constant pressure process is utilised as heat input in the other constant pressure process. Ans: 85.5%

(d) Determine the volume of the helium at each of the four states and decide if such a design of engine is a practical possibility. Ans: 0.12 m³, 0.005 m³, 0.035 m³, 0.83 m³

13.7.1 Using the steam tables, determine the following properties of water/steam:

(a) Saturated water, $p = 1$ MPa; $h = ?$
(b) Steam, $p = 600$ kPa; $X = 0.8$; $T = ?$
(c) Saturated steam, $p = 2$ MPa; $h = ?$
(d) Steam, $p = 2$ MPa; $X = 0.8$; $h = ?$

(e) Saturated steam, $p = 1.2$ MPa; $v = ?$
(f) Steam, $p = 1.2$ MPa; $X = 0.75$; $v = ?$
(g) Steam, $p = 800$ kPa; $h = 2500$ kJ/kg; $X = ?$
(h) Steam, $p = 1.2$ MPa; $v = 0.12$ m³/kg; $X = ?$
(i) Steam, $p = 1$ MPa; $T = 300°C$; $h = ?$
(j) Steam, $p = 1$ MPa; $T = 270°C$; $h = ?$
(k) Steam, $p = 1$ MPa; $s = 7.0$ kJ/kg K; $v = ?$
(l) Steam, $T = 300°C$; $h = 3000$ kJ/kg; $p = ?$
(m) Steam, $T = 311°C$; $s = 5.0$ kJ/kg K; $h = ?$

Ans: 763 kJ/kg, 158.8 °C, 2799 kJ/kg, 2421 kJ/kg, 0.1632 m³/kg, 0.1224 m³/kg, 0.87, 0.735, 3052 kJ/kg, 2987 kJ/kg, 0.2422 m³/kg, 2.83 MPa, 2366 kJ/kg

13.7.2 Use the steam tables to complete Table 13.4. There is enough information to complete each line either by reading the answer directly from the tables, or using the tables to calculate the answer. When the steam is superheated, enter S/H in the dryness fraction column. Ans: As in Table 13.5

Table 13.4 Problem 13.7.2.

T (°C)	v (m³/kg)	X	h (kJ/kg)	u (kJ/kg)	s (kJ/kg K)	p (MPa)
50		1				
240.9		0.9				
300						1
	0.35					0.5
	0.56					0.5
		0				0.25
			3156			4
100	0.6					
250					6.926	
250				2700		
			3100			1
410						0.5
		0.9				2
	0.15					2

13.7.3 1.2 kg of steam at a pressure of 1 MPa and a temperature of 250 °C is expanded polytropically and reversibly to a pressure of 100 kPa according to the law $pv^{1.3} = $ const.. Calculate the work done and the heat transferred during the change of state in the non-flow process. Ans: −382 kJ, −346 kJ

13.7.4 Steam is present at a temperature of 250 °C and a pressure of 1 MPa in a closed vessel of fixed volume 0.14 m³. If the vessel is cooled so that the pressure falls to 360 kPa, determine the mass of steam, the final temperature, the dryness fraction and the heat transferred, assuming a reversible process.
Ans: 0.6 kg, 139.9 °C, 0.456, −737 kJ

13.7.5 0.4 m³ of a fluid, initially at a pressure of 100 kPa and a temperature of 125 °C, undergoes a reversible non-flow compression through a volume ratio of 4. Determine the changes in fluid internal energy, temperature, entropy and the heat and work transfer, if the fluid is steam and the process is a reversible isothermal.
Ans: −174 kJ, 0, −0.56 kJ/K, −224 kJ, +49.8 kJ

Table 13.5 Problem 13.7.2—results.

T (°C)	v (m³/kg)	X	h (kJ/kg)	u (kJ/kg)	s (kJ/kg K)	p (MPa)
50	12.05	1	2592	2444	8.078	0.01234
240.9	0.0528	0.9	2626	2446	5.792	3.4
300	0.2577	S/H	3052	2794	7.125	1
151.8	0.35	0.93	2600	2426	6.472	0.5
339.8	0.56	S/H	3147	2867	7.6	0.5
127.4	0.00107	0	535.25	535	1.607	0.25
375	0.07	S/H	3156	2876	6.683	4
100	0.6	0.36	1232	1171	3.484	0.1013
250	0.233	S/H	2943	2710	6.926	1
250	0.172	S/H	2930	2700	6.81	1.065
322.25	0.27	S/H	3100	2830	7.206	1
410	0.626	S/H	3293	2980	7.825	0.5
212.4	0.09	0.9	2608	2429	5.948	2
395.5	0.15	S/H	3239	2938	7.115	2

13.7.6 0.01 kg of steam initially has a dryness fraction of 0.5 and is at a pressure of 100 kPa. Determine its final condition for each of the following reversible non-flow processes:

(a) A work output of 1150 J at constant pressure. What is the magnitude and direction of the heat transfer in this case?   *Ans*: Superheated, + 12.6 kJ

(b) A heat supply at constant volume until the temperature is 200 °C. What is the magnitude of the heat supply in this case?   *Ans*: Superheated, + 11.9 kJ

(c) A heat supply at constant temperature until the specific entropy of the steam is 7.359 kJ/kg K. What is the magnitude of the heat supply and the work output in this case?   *Ans*: Saturated steam, +11.3 kJ, −847 J

(d) A work input at constant entropy until the pressure of the steam is 500 kPa. What is the magnitude of the work input in this case?   *Ans*: Mixture, + 1.39 kJ

(e) A polytropic compression according to the law $pv^{1.3} = $ const. until the pressure has doubled. What is the magnitude and direction of the work and heat transfer in this case?   *Ans*: Mixture, +136.7 J, −487.7 J

13.7.7 A throttling calorimeter is to be used to find the dryness fraction of mixtures of saturated water and saturated steam. The steam is throttled from a higher to lower pressure, taking it from the mixture stage to the superheated stage where the pressure and temperature are independent. The throttling calorimeter is first tried out on a sample steam mixture for which it is known that the dryness fraction is 0.9, the mixture pressure at entry is 500 kPa, and the superheated steam pressure and temperature at exit are 100 kPa and 120.8 °C respectively.

(a) Confirm that the enthalpy of the steam remains constant as it passes through the throttling calorimeter.
  Ans: $h = 2538.1$ kJ/kg = const.

(b) Find the dryness fraction of a sample of a mixture at 1.5 MPa if the pressure and temperature of the superheated steam at exit from the throttling calorimeter are 1 MPa and 250 °C respectively.   *Ans*: 0.958

13.7.8 A reciprocating piston cylinder steam engine is to be built based upon the ideal Carnot cycle. The steam undergoes the following four non-flow reversible processes operating in a cycle:

1 State 1 to state 2—isentropic compression of the steam from a temperature of 72 °C until the steam is a saturated liquid at a temperature of 230 °C;

2 State 2 to state 3—constant temperature heat addition until the steam is a saturated vapour at the same pressure and temperature;

PROBLEMS

3   State 3 to state 4 — isentropic expansion of the steam back to the initial temperature of 72 °C;
4   State 4 to state 1 — constant temperature heat rejection back to the conditions at state 1.
(a) Determine the dryness fraction of the steam at state 1 and state 4.   *Ans*: 0.776, 0.242
(b) Determine the work transfer per kilogram of fluid in each process.   *Ans*: + 161 kJ/kg, −196.8 kJ/kg, −617.1 kJ/kg, + 83.9 kJ/kg
(c) Determine the efficiency of the cycle.   *Ans*: 31.4%

13.7.9   A reciprocating steam compressor takes saturated steam from a well at a pressure of 200 kPa and compresses it in a reversible non-flow polytropic process to a pressure of 2 MPa. Determine the compressor work input required per kilogram of steam and the heat transfer per kilogram of steam if the index of compression is
(a) $n = 1.4$.   *Ans*: +412.2 kJ/kg, +140.8 kJ/kg
(b) $n = 1.2$.   *Ans*: +414.4 kJ/kg, −140.4 kJ/kg
(c) $n = 1.0$.   *Ans*: +407.8 kJ/kg, −525.8 kJ/kg

13.7.10   It is decided to use the Carnot cycle with steam as the fluid as the basis for a heat pump. The steam undergoes the following four non-flow reversible processes operating in a cycle:

1   State 1 to state 2 — isentropic compression of the steam from a temperature of 7 °C until the steam is a saturated vapour at a temperature of 86 °C;
2   State 2 to state 3 — constant temperature heat rejection until the steam is a saturated liquid at the same pressure and temperature;
3   State 3 to state 4 — isentropic expansion of the steam back to the initial temperature of 7 °C;
4   State 4 to state 1 — constant temperature heat supply back to the conditions at state 1.
(a) Determine the dryness fraction of the steam at state 1 and state 4.   *Ans*: 0.837, 0.117
(b) Determine the net heat transfer in the cycle if there is 0.005 kg of steam.   *Ans*: −2.5 kJ
(c) Determine the coefficient of performance of the heat pump.   *Ans*: 4.54

13.8.1   For each of the following cases of steady flow in a pipeline, determine C_2, the average velocity downstream:
(a) Water, density 1000 kg/m³, $\dot{m} = 6.5$ kg/s, $A_2 = 0.00196$ m².   *Ans*: 3.32 m/s
(b) Oil, $D_1 = 50$ mm, $\dot{V}_1 = 384$ litres/min, $D_2 = 100$ mm.   *Ans*: 0.82 m/s
(c) Air, $p_1 = 200$ kPa, $T_1 = 290$ K, $D_1 = 50$ mm, $C_1 = 12$ m/s, $p_2 = 100$ kPa, $T_2 = 300$ K, $D_2 = 50$ mm.   *Ans*: 24.8 m/s

13.8.2   Air at a pressure of 200 kPa and a temperature of 290 °C enters a pipeline with a uniform velocity of 15 m/s. The pipe is of constant diameter $D = 300$ mm. Further downstream, the velocity distribution across the pipe becomes parabolic and is represented by the equation:

$$C_2 = C_{max}\left(\frac{2z}{D}\right)^{1/7}$$

where C_2 is the downstream air velocity at a height z from the wall of the pipe and C_{max} is the maximum velocity which occurs on the centre line of the pipe. Determine C_{max} at a section where the air pressure is 180 kPa and the temperature 300 °C.   *Ans*: 21.0 m/s

13.8.3   A domestic natural gas boiler of energy transfer efficiency 70% heats 0.063 kg/s of water from a temperature of 50 °C to a temperature of 70 °C in order to supply a hot water tank and a central heating system.
(a) If the natural gas has a calorific value of 47 000 kJ/kg and costs $0.04/kWh, and the boiler runs for 3500 hours per year, determine the mass flow rate of natural gas consumed and the annual running cost of the boiler. Take the specific heat of water at constant pressure as 4.19 kJ/kg K.   *Ans*: 1.6×10^{-4} kg/s, $900 per year

(b) A modification is made to the natural gas boiler which turns it into the condensing type. In these, the water vapour in the exhaust products of combustion is condensed, releasing its latent heat. If the extra heat produced improves the energy transfer efficiency of the boiler to 85%, what is the annual saving in the running cost of the boiler and how much could be spent on the modification if a 3-year payback period is required?
Ans: $158.4 per year, $475.2

13.8.4 A condenser in a steam plant has an energy transfer efficiency of 80%. The cooling is provided by water from a nearby river, which is at a temperature of 20 °C, flowing through tubes in the condenser at the rate of 45.7 kg/s. The condenser is designed such that 2 kg/s of steam enters the shell containing the tubes in the saturated steam condition, condenses at constant pressure while releasing its latent heat and leaves in the saturated water condition. What is the minimum pressure at which the condenser can operate if the cooling water temperature from the river is not to exceed 40 °C and the specific heat of the water at constant pressure is 4.186 kJ/kg K?
Ans: 10 kPa

13.8.5 Air enters a convergent nozzle with a pressure of 150 kPa and a temperature of 300 K, and leaves at a pressure of 100 kPa. The isentropic efficiency of the nozzle is 95%.
(a) Assuming that the air inlet velocity is negligible, determine the exit velocity of the air from the nozzle if the air may be treated as a perfect gas with $\gamma = 1.4$ and $c_p = 1.005$ kJ/kg K.   *Ans*: 251 m/s
(b) If, in fact, the nozzle outlet diameter is one quarter of the inlet diameter, determine the actual value of the air inlet velocity and confirm that it has a negligible effect upon the air exit velocity.   *Ans*: 11.75 m/s

13.8.6 An exhaust-driven turbocharger on a car produces a power output of 2.0 kW from an exhaust mass flow rate of 0.015 kg/s. The exhaust enters the turbocharger at a pressure of 800 kPa and a temperature of 225 °C, and leaves at a pressure of 100 kPa. If the exhaust may be treated as a perfect gas with $\gamma = 1.3$ and $c_p = 1.1$ kJ/kg K, determine the isentropic efficiency of the turbocharger.   *Ans*: 64.2%

13.8.7 A centrifugal refrigeration compressor handles 0.052 kg/s of refrigerant vapour, which is compressed from a pressure of 0.4 MPa and a temperature of 0 °C to a pressure of 1 MPa and a temperature of 90 °C. If the refrigerant vapour may be treated as a perfect gas with $\gamma = 1.3$ and $c_p = 0.55$ kJ/kg K, determine the isentropic efficiency of the compressor and the work input required to drive it.   *Ans*: 71.4%, 2.6 kW

13.8.8 A stage of a steam turbine is supplied with steam at a pressure of 5 MPa and a temperature of 350 °C, and exhausts at a pressure of 550 kPa. The isentropic efficiency of the stage is 82%, and the steam consumption is 2270 kg/min. Determine the power output of the stage.   *Ans*: 14.4 MW

13.8.9 An industrial plant is to be built with a requirement for 50 MW of electrical power. You are asked by the management to carry out a thermodynamic and economic analysis of a system consisting of a coal-fired boiler, which takes in high-pressure saturated water and converts it to superheated steam at the same pressure, and a steam turbine generator set, which takes the high-pressure superheated steam at exit from the boiler and expands it down to a lower pressure, thereby producing electrical power at the generator. You are to compare the system with the cost of buying the electricity directly from the national grid at a price of $0.1/kWh. It may be assumed that the electrical power is required 24 hours a day, 7 days a week, for 48 weeks a year. The steam boiler and turbine system are to operate under the following conditions:

Steam turbine inlet temperature	= 500 °C
Steam turbine inlet pressure	= 12 MPa
Isentropic efficiency of steam turbine	= 90%
Steam turbine exhaust pressure	= 10 kPa
Boiler water inlet condition	= saturated water at a pressure of 12 MPa
Boiler energy transfer efficiency	= 75%
Calorific value of coal	= 32 450 kJ/kg

Cost of coal = $0.02/kWh
Capital cost of plant = $850 per kW output from the turbine

(a) Determine the mass flow rate of coal required in the boiler. Ans: 3.29 kg/s
(b) Determine the annual running cost of the steam system. Ans: $17.2 million per year
(c) Determine the SPP compared to the cost of purchasing the electricity directly from the national grid.
Ans: 1.84 years

13.8.10 A steam turbine is to be installed at a geothermal site. The steam comes from the ground at a pressure of 2 MPa and a temperature of 573 K. Initially the steam simply passes through the turbine and is exhausted to the atmosphere at a pressure of 100 kPa, and in doing so it produces 7500 kW of power from a steam mass flow rate of 60 000 kg/h.

(a) Under these conditions calculate the isentropic efficiency of the turbine. Ans: 78%
(b) To increase the power output by approximately 50%, it is decided to introduce a condenser, which operates at a pressure of 9 kPa, after the turbine. In other words, the steam now leaves the turbine at a pressure of 9 kPa. The isentropic efficiency of the turbine is expected to decrease by 5% as a consequence. Determine whether the increase in power will be met, assuming that the initial steam conditions and flow rate do not change.
Ans: Almost

13.8.11 An open cycle gas turbine system consists of three main components, a rotary compressor, a combustion chamber and a rotary turbine driving an alternator.

(a) The compressor takes in a steady mass flow rate of air of 35 kg/s from the atmosphere at a pressure of 101.3 kPa and a temperature of 288 K, and compresses it through a pressure ratio of 5.7 to 1. The isentropic efficiency of the compressor is 84%. What is the compressor power input required if the specific heat of the air at constant pressure $c_p = 1.005$ kJ/kg K and the ratio of the specific heats $\gamma = 1.4$? Ans: 7767 kW
(b) The high-pressure air passes through the combustion chamber under constant pressure conditions; 0.3 kg/s of natural gas fuel is burned in the air which causes the temperature to rise to 873 K. Assuming that the properties of the products of combustion remain the same as for the air alone, determine the heat supplied in the combustion chamber and the calorific value of the natural gas. Ans: 12921 kW, 43 070 kJ/kg
(c) The high-pressure products of combustion are expanded in the turbine, of isentropic efficiency 88%, back to atmospheric pressure. Again assuming that the properties of the products of combustion are the same as for the air alone, determine the power output of the turbine. Ans: 10 678 kW
(d) If some of the power output of the turbine is used to drive the compressor through an interconnecting shaft, what is the overall thermal efficiency of the system? Ans: 22.5%
(e) If the natural gas costs $0.03/kWh and the gas turbine is to be run for 7500 hours per year, what is the annual running cost of the system? Ans: $2.91 million per year

13.8.12 In a steam plant, 1.2 kg/s of steam at a pressure of 200 kPa and a temperature of 200 °C is bled from the main circuit.

(a) It enters a heat exchanger, is condensed and leaves as saturated water, while its pressure remains constant. How much heat is released by the steam in the heat exchanger? Ans: 2839.2 kW
(b) The saturated water passes through an expansion valve in order to reduce its pressure to 10 kPa. What is the specific enthalpy of the fluid after the expansion valve? Ans: 505 kJ/kg
(c) It enters a mixing chamber in which it combines with 9.5 kg/s of other steam at a pressure of 10 kPa and temperature 50 °C. What is the condition, specific enthalpy and temperature of the final mixture given that its pressure is also 10 kPa? Ans: Mixture, 2584 kJ/kg, 45.8 °C
(d) The mixture is condensed in a condenser to become saturated water at the pressure of 10 kPa. How much heat is released by the mixture in the condenser? Ans: 25.6 MW

13.9.1 In a vertical axis windmill, there are two aerofoil-shaped blades which are always made to face into the wind. In order to calculate the drag force on the blades, in the first instance they may be considered as being two flat plates, each of width 1 m, length 600 mm and negligible thickness. The air has a density of 1.25 kg/m³ and a dynamic viscosity of 1.7×10^{-5} kg/m s. When the wind speed is only 10 m/s, the boundary layer on both sides of each blade is laminar throughout its entire length, whereas when the wind speed is 100 m/s, the boundary layer on both sides of each blade may be taken as being turbulent throughout its entire length. Calculate the total drag force on the windmill blades for the two wind speeds. *Ans*: 0.33 N, 50.6 N

13.9.2 Wind of density 1.028 kg/m³ and dynamic viscosity 1.8×10^{-5} kg/m s blows over the smooth surface of a lake for a distance of 3 km. At the upwind shore of the lake, the wind velocity is uniform at 15 m/s. As the wind passes over the lake, a boundary layer forms. Given that the lake is so deep that its bottom exerts no retarding force on the water, what will be the force on the water per unit width of lake at the end of the 3 km distance? *Ans*: 327.8 N/m

13.9.3 A metal honeycomb type of flow straightener in a square section wind tunnel consists of 2500 square passages, each 20 by 20 mm and 180 mm long. Air of dynamic viscosity 2×10^{-5} kg/m s and density 1.1 kg/m³, approaches the straightener at a velocity of 3 m/s. Taking each surface of the honeycomb square to be a flat plate and ignoring the thickness of the metal and any three-dimensional effects in the corners, determine the drag force exerted by the air on the straightener. Assume that the boundary layer that forms on each surface of the honeycomb square is laminar throughout its length. *Ans*: 0.15 N

13.9.4 For a sphere falling in a fluid, the terminal velocity is reached when the gravitational forces downwards are balanced by the buoyancy forces and drag forces upwards. Determine the terminal velocity for a small particle of diameter 2 mm and density 1200 kg/m³ falling through oil of density 800 kg/m³ and dynamic viscosity 0.025 kg/m s. Assume laminar motion for which the coefficient of drag $\mathbf{C}_D$ is given by

$$\mathbf{C}_D = \frac{24}{Re}$$

where Re is the Reynolds number. *Ans*: 0.029 m/s

13.9.5 A new wing shape is being developed by an aircraft manufacturer. The wing has a 10 m span and a 2 m mean chord. It is decided to find the lift force and the drag force on the new wing shape when it is travelling at a velocity of 100 m/s in air of density 1.2 kg/m³ by carrying out a test on a one-twentieth scale model in a compressed air wind tunnel in which the air velocity is 300 m/s and air density 5.9 kg/m³. The lift and drag forces on the model wing are measured to be 7 and 0.5 kN respectively. If the lift and drag coefficients on the model wing are the same as on the full-scale wing, determine the lift and drag force on the new wing shape. *Ans*: 63.6 kN, 4.56 kN

13.9.6 Water of density 1000 kg/m³ flows at the rate of 0.1 m³/s through a horizontal pipe which tapers from a diameter of 150 mm to a diameter of 100 mm. The total pressure of the water is measured by a total pressure probe on the centre line of the pipe where the diameter is 150 mm and the static pressure of the water is measured from a static pressure tapping situated in the pipe wall where the diameter is 100 mm. The total pressure probe and the static pressure tapping are connected to either side of a U-tube manometer which uses mercury of density 13 600 kg/m³ as the liquid.
(a) Assuming there are no losses between the measuring positions, find the reading on the mercury manometer. *Ans*: 136 mm
(b) If the static pressure tapping and total pressure probe positions are reversed, what is now the reading on the mercury manometer? *Ans*: 659 mm

13.9.7 A pitot static tube, connected to an inclined manometer containing alcohol of relative density 0.8, is to be used to determine the velocity of an airflow. The manometer is inclined at 25° to the horizontal. If the air has a constant density of 1.2 kg/m^3, what is its velocity when the pitot static tube is pointing directly into the air-flow and the reading on the manometer scale is

(a) 50 mm of alcohol.   *Ans*: 16.6 m/s
(b) 100 mm of alcohol. *Ans*: 23.5 m/s
(c) 150 mm of alcohol. *Ans*: 28.8 m/s

13.9.8 The mass flow rate of compressed air flowing through a horizontal pipeline is determined by an orifice plate connected to a simple U-tube alcohol manometer. The relative density of the alcohol is 0.8. The manometer gives a reading of 200 mm of alcohol when the airflow in the pipeline is 3 kg/s and the density of the compressed air is 1.8 kg/m^3. Due to changing requirements, it is decided to use the pipeline to transport water of density 1000 kg/m^3 instead of compressed air, but to keep the same orifice installation for determining the mass flow rate. Mercury of relative density 13.6 replaces the alcohol in the U-tube manometer. What mass flow rate of water will give a difference in the levels of the mercury either side of the U-tube of 200 mm?   *Ans*: 281.5 kg/s

13.9.9 A venturi meter is situated in a horizontal section of a pipe which is used to transport water of density 1000 kg/m^3. The pipe diameter is 250 mm and the throat diameter of the venturi is 180 mm. A simple mercury U-tube manometer is employed to measure the pressure difference of the water in order to determine its mass flow rate. The coefficient of discharge of the venturi is 0.95.

(a) Determine the pressure difference across the venturi if the water flow rate is 130 kg/s.   *Ans*: 10.6 kPa
(b) Later it is found that the manometer scale was originally calibrated in kilopascals to measure air pressures, where the air density is 1.2 kg/m^3. What is the approximate percentage error in the reading of the pressure difference across the venturi if the density of mercury is 13 600 kg/m^3?   *Ans*: 7.3%

13.9.10 A nozzle meter is situated in a horizontal section of a pipe which is used to transport oil of density 800 kg/m^3. The pipe diameter is 100 mm and the vena contracta diameter of the nozzle is 50 mm. The coefficient of discharge of the nozzle is 0.97. A simple U-tube mercury manometer is employed to measure the pressure difference of oil across the nozzle in order to calculate its mass flow rate.

(a) For a difference in height of mercury either side of the U-tube of 45 mm, determine the mass flow rate of oil assuming that the density of mercury is 13 600 kg/m^3.   *Ans*: 5.915 kg/s
(b) Later, 5 mm of water of density 1000 kg/m^3 is discovered between the oil and the mercury in the left-hand limb of the U-tube manometer. For the same height difference of mercury, what is the true mass flow rate of the oil?   *Ans*: 5.909 kg/s

13.10.1 A jet of water of density 1000 kg/m^3 emits from a nozzle of diameter 80 mm with a velocity of 40 m/s in the horizontal direction and into the atmosphere. The jet is deflected by a curved plate through an angle of 60° in the horizontal plane and leaves at the same velocity. Determine the magnitude of the resultant force on the plate when:

(a) The plate is stationary.   *Ans*: 8037 N
(b) The plate is moving in the same initial direction as the jet with a velocity of 15 m/s.   *Ans*: 3142 N

13.10.2 A ship is to be driven at a velocity of 5 m/s by the reaction of a jet. It is proposed that the water enter the ship with negligible velocity at the bow and be discharged through a nozzle of diameter 950 mm at the stern. If the velocity of the jet relative to the ship is 7.5 m/s, what is the power required to propel the ship if the density of sea-water is 1025 kg/m^3?   *Ans*: 42.5 kW

13.10.3 A fluid flows downwards at the rate of 1 m^3/s through a vertical nozzle which tapers from a diameter of 600 mm to a diameter of 400 mm in a length of 1.5 m. The gauge pressure of the fluid at the inlet section is 400 kPa. Assuming that there are no losses in the nozzle, calculate the force on the nozzle when the fluid is

(a) Water of density 1000 kg/m^3.   *Ans*: 62.8 kN
(b) Oil of density 800 kg/m^3.   *Ans*: 62.7 kN

13.10.4 Oil of density 850 kg/m³ flows through a horizontal circular duct at the rate of 0.03 m³/s. At a certain point, the duct enlarges from a diameter of 50 mm to a diameter of 100 mm. The gauge static pressure of the oil before the change in diameter is 40 kPa. Assuming that the enlargement is frictionless, determine the force exerted upon the enlargement by the oil.   *Ans*: 675.5 N

13.10.5 An overflowing dam has an upstream water level of elevation 15 m, and average upstream water velocity of 2 m/s. Downstream of the dam, where the water velocity again becomes uniform, the water level is of elevation 3 m. If the density of the water is 1000 kg/m³, determine the force on the dam per unit width in the horizontal direction, neglecting friction effects.   *Ans*: 819.5 N/m

13.10.6 Air, assumed to have a constant density of 1.2 kg/m³, flows through a horizontal duct of cross-sectional area A_{xs} at 28 m/s. It is required to change the cross-sectional area of the duct from A_{xs} to $4A_{xs}$. Determine the rise in static and total pressure of the air from upstream to downstream of the change of area when this is achieved by the following.
(a) A smooth diffuser in which the flow may be assumed to be frictionless.   *Ans*: −12.6 Pa, 0
(b) A sudden enlargement in which the static pressure in the plane of change is equal to, and in the same direction as, that upstream.   *Ans*:− 176.4 Pa, 264.6 Pa

13.10.7 A flat perforated metal plate, placed perpendicular to an airflow in a duct, consists of a large number of holes, each of diameter 5 mm, drilled with their centres on a square-sided grid of side 40 mm. The air may be assumed to be of constant density of 1.2 kg/m³ and to be travelling upstream of the plate with a uniform velocity of 10 m/s. By treating each individual hole as an orifice plate, and assuming that the vena contracta is the same diameter as the hole, determine the pressure drop of the air across the plate.   *Ans*: 21.1 Pa

13.10.8 Turning vanes are fitted into a right-angle bend in a horizontal duct of uniform square cross-section, 0.35 by 0.35 m. The fluid is air, assumed to have a constant density of 1.2 kg/m³ and to be travelling at constant velocity of 6 m/s. The gauge static pressure of the air upstream of the right-angle bend is 125 kPa. If the flow through the bend can be assumed frictionless, determine the force exerted by the air on the turning vanes.   *Ans*: 21.7 kN

13.10.9 A pipe bend tapers from a diameter of 300 mm at inlet to a diameter of 150 mm at outlet. Water of density 1000 kg/m³ flows through the bend at the rate of 0.2 m³/s and is turned through an angle of 120 degrees in the vertical plane. The axis at inlet to the bend is horizontal and the centre of the outlet section of the bend is 1.5 m below the centre of the inlet section. The total volume of water contained in the bend is 0.1 m³. Neglecting the effects of friction, determine the magnitude and direction of the force exerted on the bend by the water given that the inlet static gauge pressure of the water is 130 kPa.   *Ans*: 11.9 kN, 11.1°

13.10.10 0.05 kg/s of air, assumed to be a perfect gas, receives a heat supply as it passes through a frictionless, horizontal diffuser of area ratio 5 : 1 so that the velocity of the air at entry to the diffuser is the same at exit. The air temperature and air pressure at entry are 300 K and 50 kPa respectively, and the specific heat of the air at constant pressure may be assumed constant and equal to 1.01 kJ/kg K. If the pressure ratio of the diffuser is 1 : 2:
(a) Determine the heat supply.   *Ans*: 151.5 kW
(b) Determine the force exerted by the air on the diffuser if the inlet area of the diffuser is 0.01 m².
   Ans: −5000 N

13.11.1 The energy required to overcome the force due to friction opposing the motion of a fluid flowing in a horizontal pipeline with a mass flow rate of 2.0 kg/s, is 120 W. Consider the fluid to be

1 Water of density 1000 kg/m³;
2 Oil of density 800 kg/m³;
3 Alcohol of density 700 kg/m³;
4 Acid of density 1400 kg/m³;
5 Carbon tetrachloride of density 1600 kg/m³.

For each fluid, determine:
(a) The pressure drop in kPa due to friction.   *Ans*: 60 kPa, 48 kPa, 42 kPa, 84 kPa, 96 kPa
(b) The pressure drop expressed as a head in metres of the flowing fluid.   *Ans*: 6.12 m
(c) The pressure drop expressed as a head in millimetres of mercury, which is of density 13 600 kg/m^3.
   Ans: 450 mm Hg, 360 mm Hg, 315 mm Hg, 630 mm Hg, 720 mm Hg

13.11.2   A dash pot is made up of a piston 100 mm long and of diameter 50 mm sliding in a cylinder containing oil of dynamic viscosity 0.09 kg/m s and density 800 kg/m^3. The piston has four holes drilled in it, each 2 mm in diameter and parallel to the axis, through which the oil can flow. Determine the force exerted on the piston when it is moving with a velocity of 0.25 m/s.   *Ans*: 2205 N

13.11.3   Oil of density 850 kg/m^3 and dynamic viscosity 0.9 kg/m s is pumped at the rate of 5 kg/s through a horizontal pipe, 75 mm in diameter. Determine the average shear stress on the wall of the pipe, and the maximum and average velocity of the oil in the pipe.   *Ans*: 255.6 N/m^2, 2.66 m/s, 1.33 m/s

13.11.4   Calculate the friction factor f for each of the following steady flows of fluid in a pipe using the equations:

$$f = \frac{16}{Re} \quad \text{for laminar flow}$$

$$4f = 0.0055 + 0.15\left(\frac{k}{D} + \frac{50}{Re}\right)^{0.33} \quad \text{for turbulent flow}$$

(a) Fluid: water, temperature 10 °C, density 999.7 kg/m^3, dynamic viscosity 0.0013 kg/m s, velocity 1.15 m/s. Pipe: commercial steel pipe, inside diameter 25.2 mm, $k = 0.046$ mm.   *Ans*: 0.00745

(b) Fluid: water, temperature 20 °C, density 998.7 kg/m^3, dynamic viscosity 0.001 002 kg/m s, velocity 14.1 mm/s. Pipe: drawn copper tube, inside diameter 1.51 mm, $k = 0$ mm.   *Ans*: 0.755

(c) Fluid: oil, temperature 80 °C, density 850 kg/m^3, dynamic viscosity 27×10^{-3} kg/m s, velocity 0.36 m/s. Pipe: galvanised iron pipe, internal diameter 0.72 m, $k = 0.15$ mm.   *Ans*: 0.008 43

(d) Fluid: air, temperature 52 °C, density 1.086 kg/m^3, dynamic viscosity 1.807×10^{-5} kg/m s, velocity 10.8 m/s. Pipe: smooth plastic pipe, internal diameter 455 mm, $k = 0$ mm.   *Ans*: 0.00351

13.11.5   In the basement of a building, there is a large tank full of old engine oil of density 880 kg/m^3 and dynamic viscosity 0.35 kg/m s. It is required to pump the oil through a vertical height of 20 m from the basement tank into drums fitted on to a truck in order to dispose of it. From the bottom of the tank, a 1 m long 75 mm diameter smooth suction pipe leads to an electric-driven pump. At the outlet of the pump, a 50 m long 75 mm diameter smooth delivery pipe takes the oil to the drums. With the pump switched on, the oil is delivered at the rate of 39 kg/s.

(a) What is the hourly running cost of the pump if it has an energy transfer efficiency of 70% and electricity costs $0.12/kWh. Ignore minor losses.   *Ans*: $9.05

(b) It is suggested that the flow rate of the oil could be doubled by incorporating a booster pump or by increasing the speed of the pump. Investigate the consequences of this upon the hourly running cost.   *Ans*: $76

13.11.6   In a given factory, rain-water is collected in a tank at a particularly suitable location. However, as the water is actually required elsewhere, the energy manager decides to install another tank at the same elevation 450 m away and to pump the rain-water between the two tanks. The water, of density 1000 kg/m^3 and dynamic viscosity 0.001 kg/m s, is to be pumped at the rate of 20 kg/s. A 100 mm diameter pipe is specified. The energy manager has to decide between installing a more expensive smooth pipe for which the friction factor is given by the Blasius equation, or a cheaper rough pipe for which the friction factor is 0.018. Given that the electric-driven pump, of energy transfer efficiency 75%, is estimated to run for 5000 hours per year and that electricity costs $0.1/kWh, how many extra dollars can be spent on the smooth pipe if a payback period of 2 years is required. Ignore minor losses.   *Ans*: $22 568

13.11.7 Estimate the pressure drop in a galvanised iron pipeline of internal diameter 35.7 mm in each of the following fittings for the two fluids:

1. Water of density 1000 kg/m³ flowing at the rate of 1.0 kg/s;
2. Air at a pressure of 101 kPa and a temperature of 20 °C flowing at a rate of 0.03 kg/s.

(a) A fully open globe valve. *Ans*: 5000 Pa, 3750 Pa
(b) A fully open gate valve. *Ans*: 100 Pa, 75 Pa
(c) A 90-degree elbow. *Ans*: 450 Pa, 337.5 Pa
(d) The sharp entry to the pipe from a reservoir. *Ans*: 250 Pa, 187.5 Pa
(e) The exit from the pipe into a reservoir. *Ans*: 500 Pa, 375 Pa
(f) A sudden enlargement to a 50.5 mm internal diameter pipeline. *Ans*: 125 Pa, 93.75 Pa
(g) A sudden contraction back to the 35.7 mm internal diameter pipeline from the 50.5 mm internal diameter pipeline. *Ans*: 105 Pa, 78.75 Pa

13.11.8 A tank contains water of density 1000 kg/m³ and dynamic viscosity 1.1×10^{-3} kg/m s. A 41.9 mm internal diameter horizontal straight pipe leads from a sharp entry at the bottom of the tank to a wide open gate valve, followed by an axial flow pump. The total input power to the pump is 220 W and it has an energy transfer efficiency of 65%. From the outlet of the pump, the same diameter pipe, initially horizontal, leads to a 90° elbow, a vertical section of height 17 m, another 90° elbow, and then a further horizontal section before the water discharges into the atmosphere. The total length of pipe is 140 m and the pipe friction factor is given by the empirical formula:

$$4f = 0.0055 + 0.15 \left(\frac{k}{D} + \frac{50}{Re} \right)^{0.33}$$

where $D = 41.9$ mm and k (the equivalent roughness) $= 0.3$ mm.

(a) Estimate the losses in the straight pipe, in the fittings and in the pump if the water flow rate is 0.86 kg/s. *Ans*: 2.5 m water, 0.05 m water, 9.1 m water
(b) Find the height of the water required in the tank to maintain this flow rate. *Ans*: 2.6 m

13.11.9 During a test made at the factory of a small water pump, the maximum mass flow rate that could be transmitted before the pressure at entry to the pump became too low for the pump to work properly was 5 kg/s, when the pump was connected to a suction pipe of diameter 0.1 m, length 3 m, friction factor 0.1, and vertical height from sump water level to pump inlet 2 m. It is proposed to use the pump in an installation with similar atmospheric conditions, but with a 50 m long suction pipe of the same diameter and friction factor, and with the pump inlet at the same level as the sump.

(a) Calculate the maximum mass flow rate of water that can be transmitted now within the pressure limitation imposed in the factory, ignoring minor losses. The density of water is 1000 kg/m³. *Ans*: 1.74 kg/s
(b) If the pump is to maintain the same flow rate in its new location as in the factory, where must the pump inlet be in relation to the sump water level? *Ans*: 1.9 m below

13.11.10 Water drains from a reservoir into the atmosphere through a pipe 60 m long. The first 15 m of pipe is 50 mm in diameter and has a friction factor of 0.005, and the rest of the pipe is 75 mm in diameter and has a friction factor of 0.01. Determine the difference in level between the water in the tank and the discharge at the pipe exit when a flow rate of 3 kg/s is maintained. Account for the minor losses encountered at the sharp entrance to the pipe and the sudden enlargement. *Ans*: 1.44 m

13.11.11 An upper reservoir contains water which is at a level of 60 m above the water in a lower reservoir. The two reservoirs are connected by a pipeline. The first 5000 m length of the pipeline is 500 mm in diameter. There is then a junction in the pipe from which two identical pipes, each 250 mm in diameter and 5000 m long, lead into the lower reservoir. If the friction factor of all the pipes is 0.01, determine the total flow rate of water possible from the upper to the lower reservoir, ignoring minor losses.   *Ans*: 0.11 m³/s

13.11.12   An upper reservoir contains water which is at a level of 10 m above the water in a lower reservoir. The two reservoirs are connected by a pipeline which is 750 m long and of diameter 500 mm. Due to the landscape to be negotiated between the two reservoirs, at a distance of 300 m from the upper reservoir the water in the pipeline is actually at a height of 4 m above that of the water in the upper reservoir itself. If the friction factor of the pipe is 0.01, determine the gauge pressure of the water at this highest point in the pipeline, ignoring minor losses.
Ans: −434 kPa

13.11.13   Water of density 1000 kg/m³ and dynamic viscosity 10^{-3} kg/m s flows through two pipes operating in parallel. Both have a diameter D of 1.0 m and are of the same length, but one pipe is smooth and the other pipe is rough of roughness size $k = 0.5$ mm. The friction factor in the smooth pipe is given by the Blasius equation, whereas the friction factor f in the rough pipe is given by

$$\frac{1}{\sqrt{f}} = 4 \log 10 \left(\frac{D}{2 \times k}\right) + 3.48$$

Determine the velocity of water in each pipe by trial and error if the total flow through the pipes is 37.9 kg/s.
Ans: 0.2 m/s, 0.19 m/s

13.11.14   A hydraulic power house supplies power to a factory, situated a distance of 6 km away, through three horizontal pipes, each of diameter 125 mm and friction factor 0.01. If the inlet pressure is maintained constant at 5 MPa and the density of water is 1000 kg/m³, determine:

(a) The delivery pressure if the head loss to friction is 10% of the head generated at the power house.
Ans: 4.5 MPa

(b) The increase of pressure that would be required at the power station to produce the same delivery pressure if one of the pipes is out of action and the total flow rate of water is maintained.   *Ans*: 5.62 MPa

Appendix 1 Formulae

CHAPTER 2 THERMOFLUID PROPERTIES

(2.1) The absolute pressure of a fluid is the sum of its gauge pressure and the local atmospheric pressure:

$$\text{Absolute pressure} = p_{ga} + p_{atm}$$

(2.3) The total pressure of a fluid is the sum of its static and dynamic pressures:

$$p_0 = p + 0.5\rho C^2$$

(2.8) The pressure variation with vertical height in a fluid is

$$p_2 - p_1 = \rho g(z_1 - z_2)$$

(2.13) The shear stress of a Newtonian fluid in flow is equal to the product of its dynamic viscosity and velocity gradient within certain limitations:

$$\tau = \mu \frac{dC}{dz}$$

(2.14) The kinematic viscosity of a fluid is the dynamic viscosity divided by the density:

$$\nu = \frac{\mu}{\rho}$$

(2.15) The Reynolds number of a fluid in a pipe is the product of its density, velocity and pipe diameter, divided by its dynamic viscosity:

$$Re = \frac{\rho C D}{\mu}$$

(2.16) The Reynolds number of a fluid flowing over a surface is the product of its density, velocity and length of surface that it is in contact with, divided by its dynamic viscosity:

$$Re = \frac{\rho C L}{\mu}$$

(2.17) The laminar shear stress of a Newtonian fluid in flow is equal to the product of its dynamic viscosity and velocity gradient:

$$\tau = \mu \frac{dC}{dz}$$

(2.18) The turbulent shear stress of a Newtonian fluid in flow is equal to the product of its eddy viscosity and velocity gradient:

$$\tau = \epsilon \frac{dC}{dz}$$

(2.19) The friction factor is the ratio of the shear stress to the dynamic pressure of the fluid:

$$f = \frac{\tau}{0.5\rho C^2}$$

CHAPTER 3 WORK AND HEAT IN A THERMAL SYSTEM

(3.1) The work transferred into or out of a fluid in a non-flow reversible process is the sum of the incremental product of its pressure and volume between two states:

$$W_{12} = mw_{12} = -\int_1^2 mp \, dv = -\int_1^2 p \, dV$$

(3.2) The heat transferred into or out of a fluid in a non-flow reversible process is the sum of the incremental product of its temperature and entropy between two states:

$$Q_{12} = mq_{12} = \int_1^2 mT \, ds = \int_1^2 T \, dS$$

CHAPTER 4 THE FIRST LAW OF THERMODYNAMICS

(4.1) The sum of the net work delivered to the surroundings and the net heat taken from the surroundings when any closed system goes through a cycle is zero:

$$\sum (Q)_{\text{cyc}} + \sum (W)_{\text{cyc}} = 0$$

(4.2) The brake power output of an engine, as recorded by a dynamometer, is the product of the torque exerted and the rotational speed:

$$\dot{W}_{\text{bp}} = -2\pi N_{\text{rev}} F L$$

(4.3) The indicated power output of a reciprocating engine is the product of the area of the engine fluid pressure against volume diagram and the number of cycles completed:

$$\dot{W}_{\text{ip}} = -A_{pV} N_{\text{cyc}}$$

The indicated power input of a reciprocating compressor is the product of the area of the compressor fluid pressure against volume diagram and the number of cycles completed:

$$\dot{W}_{\text{ip}} = A_{pV} N_{\text{cyc}}$$

(4.4) The friction power of an engine is the difference between the indicated power and the brake power, usually expressed as a heat output:

$$\dot{W}_F = \dot{W}_{ip} - \dot{W}_{bp} = \dot{Q}_F$$

(4.5) The specific heat of a fluid is the heat required to raise unit mass by one degree:

$$c = \frac{dQ}{m\,dT}$$

(4.6) For fluids with a constant specific heat, the heat transferred is the product of its mass, specific heat, and temperature difference:

$$Q_{12} = mc(T_2 - T_1)$$

(4.7) The heat released by the combustion of a fuel is the product of the mass and calorific value of the fuel:

$$Q_{comb} = \{m\}_{fuel} CV$$

(4.9) The annual running cost of a thermal system is the product of the energy input that has to be paid for, the number of hours per year that the system is run for, and the cost of the energy:

$$\$_{ARC} = E N_{hrs} \$_{kWh}$$

CHAPTER 5 THE NON-FLOW ENERGY EQUATION

(5.1) The NFEE for a fluid undergoing a non-flow process between two states shows that the sum of the net heat transfer and the net work transfer is equal to the change in its internal energy:

$$Q_{12} + W_{12} = U_2 - U_1 = m(u_2 - u_1)$$

(5.2) The work transfer in a reversible non-flow constant pressure process between two states is the product of the fluid pressure and its change of volume:

$$W_{12} = -p(V_2 - V_1) = -mp(v_2 - v_1)$$

(5.3) The heat transfer in a non-flow constant pressure process between two states is equal to the change in the fluid enthalpy:

$$Q_{12} = H_2 - H_1 = m(h_2 - h_1)$$

(5.4) The enthalpy of a fluid is the sum of its internal energy and the product of its pressure and volume:

$$H = mh = U + pV = m(u + pv)$$

(5.5) The work transfer in a reversible non-flow constant volume process between two states is zero:

$$W_{12} = 0$$

(5.6) The heat transfer in a reversible non-flow constant volume process between two states is equal to the change in the fluid internal energy:

$$Q_{12} = U_2 - U_1 = m(u_2 - u_1)$$

(5.7) The heat transfer in a reversible non-flow constant temperature process between two states is the product of the fluid temperature and its change in entropy:

$$Q_{12} = T(S_2 - S_1) = mT(s_2 - s_1)$$

(5.8) The heat transfer in a reversible non-flow constant entropy process between two states is zero:

$$Q_{12} = 0$$

(5.9) The work transfer in a reversible non-flow constant entropy process between two states is equal to the change in the fluid internal energy:

$$W_{12} = U_2 - U_1 = m(u_2 - u_1)$$

(5.10/11) In a polytropic non-flow process between two states, the relationship between the fluid pressure and volume is

$$pV^n = p_1 V_1^n = p_2 V_2^n = \text{const.}$$

$$pv^n = p_1 v_1^n = p_2 v_2^n = \frac{p_1}{\rho_1^n} = \frac{p_2}{\rho_2^n} = \text{const.}$$

(5.12) The work transfer in a reversible non-flow polytropic process between two states is

$$W_{12} = -\frac{p_2 V_2 - p_1 V_1}{1 - n} = -\frac{m(p_2 v_2 - p_1 v_1)}{1 - n}$$

(5.13) The work transfer in a reversible non-flow polytropic process between two states when the index of compression or expansion $n = 1$ is

$$W_{12} = -p_1 V_1 \ln \frac{V_2}{V_1} = -mp_1 v_1 \ln \frac{v_2}{v_1}$$

(5.14) The specific heat at constant pressure of a fluid undergoing a non-flow constant pressure process between two states is proportional to the rate of change of enthalpy of the fluid with temperature:

$$c_p = \frac{(dH)_p}{(m \, dT)_p}$$

(5.15) The specific heat at constant volume of a fluid undergoing a non-flow constant volume process between two states is proportional to the rate of change of internal energy of the fluid with temperature:

$$c_v = \frac{(dU)_v}{(m \, dT)_v}$$

(5.16) For a fluid with a constant specific heat at constant pressure undergoing a constant pressure process between two states, the change of enthalpy of the fluid is the product of its mass, its specific heat at constant pressure and its temperature difference:

$$(H_2 - H_1)_p = (mc_p(T_2 - T_1))_p$$

(5.17) For a fluid with a constant specific heat at constant volume undergoing a constant volume process between two states, the change of internal energy of the fluid is the product of its mass, its specific heat at constant volume and its temperature difference:

$$(U_2 - U_1)_v = (mc_v(T_2 - T_1))_v$$

CHAPTER 6 THE FLUID AS A PERFECT GAS

(6.1/2) The equation of state is a relationship between the pressure, volume and temperature of a perfect gas:

$$\frac{pV}{mT} = \frac{p\dot{V}}{\dot{m}T} = \frac{pv}{T} = \frac{p}{\rho T} = \text{const.} = R$$

$$\frac{p_1 V_1}{m T_1} = \frac{p_1 \dot{V}_1}{\dot{m} T_1} = \frac{p_1 v_1}{T_1} = \frac{p_1}{\rho_1 T_1} = \frac{p_2 V_2}{m T_2} = \frac{p_2 \dot{V}_2}{\dot{m} T_2} = \frac{p_2 v_2}{T_2} = \frac{p_2}{\rho_2 T_2}$$

(6.3) Joule's law states that the internal energy of a perfect gas is proportional to its temperature:

$$U = \Phi(T)$$

(6.4) For a perfect gas with a constant specific heat at constant volume undergoing a process between two states, the change in internal energy is the product of its mass, its specific heat at constant volume and its temperature difference:

$$U_2 - U_1 = mc_v(T_2 - T_1)$$

(6.5) The specific heat at constant volume of a perfect gas is proportional to the rate of change of its internal energy with temperature:

$$c_v = \frac{dU}{m\, dT}$$

(6.6) The enthalpy of a perfect gas is proportional to its temperature:

$$H = \Phi(T)$$

(6.7) For a perfect gas with a constant specific heat at constant pressure undergoing a process between two states, the change in enthalpy is the product of its mass, its specific heat at constant pressure and its temperature difference:

$$H_2 - H_1 = mc_p(T_2 - T_1)$$

FORMULAE

(6.8) The specific heat at constant pressure of a perfect gas is proportional to the rate of change of its enthalpy with temperature:

$$c_p = \frac{dH}{m\,dT}$$

(6.9) The specific heat at constant pressure of a perfect gas is equal to the sum of its specific heat at constant volume and its gas constant:

$$c_p = c_v + R$$

(6.10) The ratio of the specific heats of a perfect gas is

$$\gamma = \frac{c_p}{c_v}$$

(6.11/12/13) The change of specific entropy of a perfect gas undergoing a process between two states is given by

$$s_2 - s_1 = c_v \ln \frac{T_2}{T_1} + R \ln \frac{v_2}{v_1}$$

$$s_2 - s_1 = c_p \ln \frac{T_2}{T_1} - R \ln \frac{p_2}{p_1}$$

$$s_2 - s_1 = c_v \ln \frac{p_2}{p_1} + c_p \ln \frac{v_2}{v_1}$$

(6.14) For a perfect gas undergoing a reversible non-flow constant temperature process between two states, the amount of heat transfer is numerically the same as the amount of work transfer:

$$Q_{12} = -W_{12} = mRT \ln \frac{V_2}{V_1}$$

(6.15/16) For a perfect gas undergoing a reversible constant entropy process between two states, the relationship between the pressure and volume is

$$p_2 v_2^\gamma = p_1 v_1^\gamma = \text{const.}$$

$$p_2 V_2^\gamma = p_1 V_1^\gamma = \text{const.}$$

(6.17) For a perfect gas undergoing a non-flow process between two states, the relationship between its pressure, volume and temperature is

$$\frac{T_2}{T_1} = \frac{p_2^{(n-1)/n}}{p_1^{(n-1)/n}} = \frac{V_2^{1-n}}{V_1^{1-n}} = \frac{v_2^{1-n}}{v_1^{1-n}}$$

where $n = 0$ for a constant pressure process, $n = \infty$ for a constant volume process, $n = 1$ for a constant temperature process, $n = \gamma$ for a constant entropy process and n for a polytropic process.

(6.18) The NFEE for a perfect gas is

$$Q_{12} = mc_v(T_2 - T_1) - W_{12}$$

STEADY FLOW PROCESSES   379

CHAPTER 7   THE FLUID AS WATER/STEAM

(7.1) The dryness fraction of a mixture of saturated steam and saturated water is defined as

$$X = \frac{\text{Mass of saturated steam}}{\text{Mass of mixture}}$$

(7.3) The specific volume of a mixture of saturated steam and saturated water is

$$v = X v_g$$

(7.4) The specific entropy of a mixture of saturated steam and saturated water is

$$s = X s_g + (1 - X) s_f$$

(7.5) The specific internal energy of a mixture of saturated steam and saturated water is

$$u = X u_g + (1 - X) u_f$$

(7.6) The specific enthalpy of a mixture of saturated steam and saturated water is

$$h = X h_g + (1 - X) h_f$$

(7.7) The specific internal energy of unsaturated water is

$$u = u_f - c_v (T_{fg} - T)$$

(7.8) The specific enthalpy of unsaturated water is

$$h = h_f - c_p (T_{fg} - T)$$

CHAPTER 8   STEADY FLOW PROCESSES

(8.1/2/3/4) For a steady flow process, the conservation of mass flow, the continuity equation, is

$$\dot{m}_1 = \dot{m}_2 = \dot{m} = \text{const}.$$

$$\dot{m}_1 = \dot{V}_1 \rho_1 = \dot{m}_2 = \dot{V}_2 \rho_2 = \dot{m} = \text{const}.$$

$$\dot{m}_1 = \rho_1 C_1 (A_{xs})_1 = \dot{m}_2 = \rho_2 C_2 (A_{xs})_2 = \dot{m}$$

$$\dot{m}_1 = \frac{C_1 (A_{xs})_1}{v_1} = \dot{m}_2 = \frac{C_2 (A_{xs})_2}{v_2} = \dot{m}$$

(8.7/8) For a steady flow process, the conservation of energy, the SFEE is

$$\dot{Q}_{12} + \dot{W}_{12} = \dot{m}(h_2 - h_1) + 0.5\dot{m}(C_2^2 - C_1^2) + \dot{m}g(z_2 - z_1)$$

(8.9) For a boiler, combustion chamber, heat exchanger, condenser and evaporator, the SFEE shows that the heat transfer is equal to the product of the mass flow rate of fluid and the change in the specific enthalpy of the fluid:

$$\dot{Q}_{12} = \dot{m}(h_2 - h_1)$$

(8.10) When the fluid in a boiler, combustion chamber, heat exchanger, condenser or evaporator is a perfect gas, the SFEE becomes

$$\dot{Q}_{12} = \dot{m}c_p(T_2 - T_1)$$

(8.11) For a nozzle and diffuser, the SFEE shows that the sum of the specific enthalpy change and the kinetic energy change per mass flow rate of fluid are zero:

$$0 = (h_2 - h_1) + 0.5(C_2^2 - C_1^2)$$

(8.12) The efficiency of a nozzle is the ratio of the actual kinetic energy of the fluid at exit to the kinetic energy that it would have if it followed an isentropic process:

$$\eta_N = \frac{C_{act}^2}{C_{isen}^2}$$

(8.13) The efficiency of a diffuser is the ratio of the actual pressure rise of the fluid to the pressure rise that it could achieve if it followed an isentropic process:

$$\eta_{diff} = \frac{(p_2 - p_1)_{act}}{(p_2 - p_1)_{isen}}$$

(8.14) When the fluid in a nozzle or diffuser is a perfect gas, the SFEE becomes

$$0 = c_p(T_2 - T_1) + 0.5(C_2^2 - C_1^2)$$

(8.15) For a rotary turbine or compressor, the SFEE shows that the work transfer is equal to the product of the mass flow rate of the fluid and the change in the specific enthalpy of the fluid:

$$\dot{W}_{12} = \dot{m}(h_2 - h_1)$$

(8.16) The efficiency of a rotary turbine is the ratio of the actual work output to the work output that could be achieved if the fluid followed an isentropic process:

$$\eta_T = \frac{(\dot{W}_{12})_{act}}{(\dot{W}_{12})_{isen}} = \frac{(h_2 - h_1)_{act}}{(h_2 - h_1)_{isen}}$$

(8.17) The efficiency of a rotary compressor is the ratio of the work input required if the fluid followed an isentropic process to the actual work input required:

$$\eta_C = \frac{(\dot{W}_{12})_{isen}}{(\dot{W}_{12})_{act}} = \frac{(h_2 - h_1)_{isen}}{(h_2 - h_1)_{act}}$$

(8.18) When the fluid in a rotary turbine or compressor is a perfect gas, the SFEE becomes

$$\dot{W}_{12} = \dot{m}c_p(T_2 - T_1)$$

(8.19) For an expansion valve, the SFEE shows that the specific enthalpy of the fluid remains constant:
$$0 = h_2 - h_1$$

(8.20) When the fluid in the expansion valve is a perfect gas, the SFEE shows that its temperature remains constant:
$$0 = T_2 - T_1$$

(8.21) For a mixing chamber, the SFEE shows that the mass-dependent enthalpy of the mixture is the sum of the mass-dependent enthalpies of the constituents before mixing:
$$\dot{m}_2 h_2 = (\dot{m}_1 h_1)_A + (\dot{m}_1 h_1)_B$$

(8.22) When the fluids in a mixing process are perfect gases, the SFEE becomes
$$\dot{m}_2 (c_p)_2 T_2 = (\dot{m}_1 (c_p)_1 T_1)_A + (\dot{m}_1 (c_p)_1 T_1)_B$$

(8.23) The SFEE for an ideal fluid becomes the Bernoulli equation:
$$p_2 v_2 + 0.5 C_2^2 + g z_2 = p_1 v_1 + 0.5 C_1^2 + g z_1$$

(8.24) The relationship between the static and total temperature of a perfect gas is
$$T + \frac{0.5 C^2}{c_p} = \text{const.} = T_0$$

CHAPTER 9   EFFECTS OF A FLUID IN MOTION

(9.1) In laminar flow, the entry length in a pipe is proportional to the product of the fluid Reynolds number and the pipe diameter:
$$EL = 0.065 Re \dot{D}$$

(9.2) The thickness of a laminar boundary layer when a fluid is flowing over a flat plate is a function of the distance down the plate and the fluid Reynolds number at that distance. A typical relationship is
$$\delta = \frac{5.48 x}{Re^{0.5}}$$

(9.3) The thickness of a turbulent boundary layer when a fluid is flowing over a flat plate is a function of the distance down the plate and the fluid Reynolds number at that distance. A typical relationship is
$$\delta = \frac{0.37 x}{Re^{0.2}}$$

(9.4) The velocity profile in a laminar boundary layer when a fluid is flowing over a flat plate is a function of the distance above the plate and the boundary layer thickness. A typical relationship is
$$\frac{C}{C_{FS}} = \frac{2z}{\delta} - \frac{z^2}{\delta^2}$$

(9.5) The velocity profile in a turbulent boundary layer when a fluid is flowing over a flat plate is a function of the distance above the plate and the boundary layer thickness. A typical relationship is

$$\frac{C}{C_{FS}} = \frac{z^{1/7}}{\delta^{1/7}}$$

(9.6) The friction force generated on the surface of a plate as a fluid is flowing over it, is the sum of the product of the shear stress generated at the surface and the surface area of the plate that the fluid is in contact with:

$$F_F = b \int_0^L \tau_s \, dx$$

(9.7) The friction factor for turbulent flow over a flat plate is proportional to the fluid Reynolds number. A typical expression is

$$f = \frac{0.058}{Re^{0.2}}$$

(9.8) The drag coefficient for fluid flow over a body is the ratio of the force exerted by the fluid in the direction of motion of the fluid per unit area to the dynamic pressure:

$$C_D = \frac{F_D}{0.5 A \rho C^2}$$

(9.9) The lift coefficient for fluid flow over a body is the ratio of the force exerted by the fluid in the direction perpendicular to the fluid motion per unit area to the dynamic pressure:

$$C_L = \frac{F_L}{0.5 A \rho C^2}$$

(9.11) The theoretical mass flow rate of a fluid in a pipe, which passes through a change in area, thereby causing a change in the fluid pressure, is

$$\dot{m}_{theor} = \frac{\rho_1 (A_{xs})_1 (2g((p_1/\rho_1 g) - (p_2/\rho_2 g) + (z_1 - z_2)))^{0.5}}{(((\rho_1(A_{xs})_1)/(\rho_2(A_{xs})_2))^2 - 1)^{0.5}}$$

(9.13) The actual mass flow rate of a fluid in a pipe, which passes through a change in area, thereby causing a change in the fluid pressure, is equal to the product of the theoretical mass flow rate and the coefficient of discharge:

$$\dot{m}_{act} = C_d \dot{m}_{theor}$$

(9.14) The velocity of a fluid can be determined from a knowledge of its total and static pressures as in Bernoulli's equation:

$$C = \frac{(2(p_0 - p))^{0.5}}{\rho^{0.5}}$$

CHAPTER 10 THE STEADY FLOW MOMENTUM EQUATION

(10.1/2/3) The momentum force due to a fluid flowing in a given direction is equal to the product of the mass flow rate of fluid and its velocity change in that direction:

$$F_{M(x,y,z)} = (\dot{m}(C_1 - C_2))_{(x,y,z)}$$

(10.4/5/6) The pressure force due to a fluid flowing in a given direction is equal to the sum of the products of the fluid pressures in that direction and the cross-sectional areas on which they act:

$$F_{P(x,y,z)} = \sum (pA_{xs})_{(x,y,z)}$$

(10.7/8/9) The body force due to a fluid flowing in a given direction is equal to the product of the mass of fluid and the acceleration due to gravity in that direction:

$$F_{B(x,y,z)} = (mg)_{(x,y,z)}$$

(10.10/11/12) The total force in a given direction due to a flowing fluid is the sum of the momentum, pressure, and body forces acting in that direction:

$$F_{(x,y,z)} = (F_M + F_P + F_B)_{(x,y,z)}$$

(10.13) The total force caused by a flowing fluid is the square root of the sum of the squares of the forces in the x-, y- and z-directions:

$$F = (F_x^2 + F_y^2 + F_z^2)^{0.5}$$

CHAPTER 11 THE STEADY FLOW ENERGY EQUATION APPLIED TO PIPE FLOW

(11.2) The SFEE, modified for the flow of a fluid through a pipe, is

$$\frac{p_1}{\rho_1 g} + \frac{0.5C_1^2}{g} + z_1 = \frac{p_2}{\rho_2 g} + \frac{0.5C_2^2}{g} + z_2 + Z_F - \frac{\dot{W}_{12}}{\dot{m}g}$$

(11.5) The velocity profile of the laminar flow of an incompressible fluid in a pipe is

$$C = \frac{-0.25(0.25D^2 - r^2)dp}{\mu \, dx}$$

(11.6) The mean and maximum velocities of the laminar flow of an incompressible fluid in a pipe are a function of the pipe diameter, the pressure gradient in the direction of motion of the fluid, and the fluid dynamic viscosity:

$$C_{mean} = \frac{-D^2 dp}{32\mu \, dx} = 0.5 C_{max}$$

(11.8) The head lost to friction by an incompressible fluid flowing through a pipe is

$$Z_F = \frac{-dp}{\rho g} = \frac{4fLC^2}{D2g}$$

(11.9) The friction factor for an incompressible fluid in laminar flow through a pipe is proportional to the inverse of the fluid Reynolds number:

$$f = \frac{16}{Re}$$

(11.10) The friction factor for an incompressible fluid in turbulent flow through a pipe is a function of the fluid Reynolds number and the surface roughness of the pipe. A typical relationship for a smooth pipe is

$$f = \frac{0.079}{Re^{0.25}}$$

(11.12) The total head lost to friction by a fluid flowing through a pipe is the sum of the head lost due to pipe friction and that due to minor losses in bends, etc.:

$$Z_F + Z_{FM} = \frac{4fLC^2}{D2g} + \frac{k_{FM}C^2}{2g}$$

(11.13) The equivalent length of pipe which causes the same pressure drop in a fluid passing through it as a minor fitting is

$$L_{EQ} = \frac{k_{FM}D}{4f}$$

CHAPTER 12   THE SECOND LAW OF THERMODYNAMICS

(12.1/2) The overall thermal efficiency of a heat engine is the ratio of the net work output to the heat supplied

$$\eta = \frac{-\sum(W)_{cyc}}{Q_{supp}} = 1 + \frac{Q_{rej}}{Q_{supp}}$$

(12.5) The overall thermal efficiency of a reversible heat engine is given by the temperature ratio of the source and sink of heat:

$$\eta_{rev} = \frac{T_{source} - T_{sink}}{T_{source}}$$

(12.6) The Clausius inequality states that whenever a system undergoes a cycle, $\sum (dQ/T)_{cyc}$ is zero if the cycle is reversible and negative if the cycle is irreversible:

$$\sum (dQ/T)_{cyc} \leqslant 0$$

(12.7) The change of entropy of a fluid as it changes its state is the sum of the changes in the heat transfer during the process divided by the fluid temperature:

$$S_2 - S_1 = \int_1^2 \frac{dQ}{T}$$

Appendix 2 Nomenclature

ENGLISH SYMBOLS

Symbol	Meaning	Unit
A	Area	m²
A_{pV}	Area of a fluid pressure volume diagram	m²
A_s	Surface area	m²
A_{xs}	Cross-sectional Area	m²
BDC	Bottom dead centre position	
b	Breadth	m
C	Velocity	m/s
C	Coefficient	
c	Specific heat of fluid	J/kg K
C_D	Drag coefficient	
C_d	Coefficient of discharge	
C_{DN}	Downstream velocity of fluid	m/s
C_{FS}	Free stream velocity of fluid	m/s
C_L	Lift coefficient	
C_{max}	Maximum velocity of fluid	m/s
C_{mean}	Mean velocity of fluid	m/s
C_s	Velocity of surface	m/s
C_U	Upstream velocity of fluid	m/s
c_p	Specific heat at constant pressure of a fluid	J/kg K
c_v	Specific heat at constant volume of a fluid	J/kg K
COP	Coefficient of performance	
COP_{hp}	Coefficient of performance of a heat pump	
COP_{fridge}	Coefficient of performance of a refrigerator	
CV	Calorific value of fuel	J/kg
D	Diameter	m
d	Infinitesimal change — differential	
E	Energy transfer	kW
E_F	Energy transfer of fluid due to friction	W
EL	Entry length of fluid to pipe	m
F	Force	N

F_B	Body force of fluid	N
F_D	Drag force on body	N
F_F	Friction force	N
$F_{inertia}$	Inertia force of fluid	N
F_L	Lift force on body	N
F_M	Force due to rate of change of momentum of fluid	N
F_P	Force due to change in pressure of fluid	N
$F_{viscous}$	Viscous force of fluid	N
f	Friction factor	
g	Acceleration due to gravity	m/s^2
H	Enthalpy of fluid	J
h	Specific enthalpy of fluid	J/kg
KE	Kinetic energy	J or W
k	Height of pipe surface roughness	m
k_{FM}	Loss coefficient	
L	Length	m
LHS	Left-hand side	
L_{EQ}	Equivalent length of pipe	m
m	Mass of fluid	kg
$\dot{m}$	Mass flow rate of fluid	kg/s
MW	Molecular weight	
NFEE	Non-flow energy equation	
N	Number of	
NTP	Normal temperatures and pressures	
N_{cyc}	Number of cycles per second	cycles/s
N_{hrs}	Number of hours per year	hours/yr
N_{rev}	Number of revolutions per second	rev/s
n	Index of compression/expansion	
PE	Potential energy	J or W
p	Static pressure of fluid	Pa
p_{atm}	Atmospheric pressure	Pa
p_{dyn}	Dynamic pressure of fluid	Pa
p_{ga}	Gauge pressure of fluid	Pa
p_0	Total pressure of fluid	Pa
Q	Heat transfer	J
Q_{comb}	Heat released in the combustion of a fuel	J
$\dot{Q}$	Heat transfer per second	W
$\dot{Q}_c$	Heat transfer at condenser	W
$\dot{Q}_{comb}$	Heat released per second in the combustion of a fuel	W
$\dot{Q}_{cool}$	Heat transfer at cooler	W
$\dot{Q}_{ev}$	Heat transfer at evaporator	W

$\dot{Q}_F$	Friction power expressed as heat transfer	W
$\dot{Q}_i$	Heat input to thermal system	W
Q_{int}	Heat transfer to or from intermediate reservoir between source and sink of heat	J
Q_{supp}	Heat supplied to a heat engine	J
Q_{rej}	Heat rejected by a heat engine	J
q	Specific heat transfer	J/kg
R	Gas constant of a gas	J/kg K
RHS	Right-hand side	
r	Radial direction	
$\tilde{R}$	Universal gas constant	J/kmol K
Re	Reynolds number	
SFEE	Steady flow energy equation	
S	Entropy of fluid	J/K
s	Specific entropy of fluid	J/kg K
SPP	Simple payback period	years
T	Static temperature of a fluid	K
TDC	Top dead centre position	
T_{int}	Temperature of an intermediate reservoir of heat	K
T_0	Total temperature of a fluid	K
T_{sink}	Temperature of a heat sink	K
T_{source}	Temperature of a heat source	K
TQ	Torque	N m
U	Internal energy of fluid	J
u	Specific internal energy of fluid	J/kg
V	Volume of fluid	m^3
$\dot{V}$	Volumetric flow rate of fluid	m^3/s
v	Specific volume of fluid	m^3/kg
W	Work transfer	J
$\dot{W}$	Work transfer per second (power)	W
$\dot{W}_{alt}$	Work transfer to alternator	W
$\dot{W}_{bp}$	Brake power of engine	W
$\dot{W}_C$	Work transfer to compressor	W
$\dot{W}_{EM}$	Work transfer at electric motor	W
$\dot{W}_F$	Friction power of engine	W
$\dot{W}_{FP}$	Work transfer to feed pump	W
$\dot{W}_{ip}$	Indicated power of engine	W
$\dot{W}_T$	Work transfer from turbine	W
w	Specific work transfer	J/kg
X	Dryness fraction of a mixture of saturated vapour and saturated liquid	

x, y, z	Three-dimensional directions or distance in those directions	m
Z	Vertical height or head of fluid	m
Z_F	Head of fluid to overcome friction	m
Z_{FM}	Head of fluid to overcome minor losses	m

GREEK SYMBOLS

Symbol	Meaning	Unit
α	Kinetic energy correction factor	
β	Momentum correction factor	
γ	Ratio of the specific heat at constant pressure of a fluid to the specific heat at constant volume of the fluid	
Δ	Quantifiable change	
δ	Boundary layer thickness	m
ϵ	Eddy viscosity of fluid	kg/m s
μ	Dynamic viscosity of fluid	kg/m s
ν	Kinematic viscosity of fluid	m²/s
π		
ρ	Density of fluid	kg/m³
Φ	Function of	
θ	Angle	degrees
τ	Shear stress	Pa
τ_s	Shear stress at a surface	Pa
η	Overall thermal efficiency	%
η_C	Isentropic efficiency of compressor	%
η_{diff}	Isentropic efficiency of diffuser	%
η_e	Energy transfer efficiency	%
η_N	Isentropic efficiency of nozzle	%
η_{rev}	Efficiency of a reversible cycle	%
η_T	Isentropic efficiency of turbine	%

OTHER SYMBOLS

Symbol	Meaning	Unit
ln	Natural logarithm	
$	Cost	$
$_{ARC}$	Annual running cost	$/year
$_{kWh}$	Cost of a fuel or energy input	$/kWh

%	Percentage	%
$\sum$	Sum	
$\sum(\ )_{cyc}$	Sum in a cycle	
$\int$	Integral	
$\int_1^2$	Integral between state 1 and state 2	
$\int_0^L$	Integral between 0 and L	

SUPERSCRIPTS

Symbol	Meaning	Unit
$'$	Alternative value of something	
$''$	Alternative value of something	

SUBSCRIPTS

Symbol	Meaning	Unit
$\{\ \}_{air}$	Air as fluid	
$\{\ \}_{ex}$	Exhaust products of combustion as fluid	
$\{\ \}_{fuel}$	Fuel as fluid	
$\{\ \}_{He}$	Helium as fluid	
$\{\ \}_m$	Mercury as fluid	
$\{\ \}_{oil}$	Oil as fluid	
$\{\ \}_{st}$	Steam as fluid	
$\{\ \}_w$	Water as fluid	
act	Applies to actual conditions	
alt	Applies to alternator	
atm	Applies to the atmosphere	
B	Applies to a body force	
bp	Applies to brake power	
C	Applies to compressor	
c	Applies to a condenser	
comb	Applies to the combustion of a fuel	
cool	Applies to a cooler	
cyc	Applies to a cycle	
D	Applies to the drag force	
DN	Applies to downstream positions	
d	Applies to the coefficient of discharge	
diff	Applies to diffuser	
dyn	Applies to dynamic pressure	

EM	Applies to electric motor
EQ	Applies to an equivalent length
e	Applies to energy transfer
ev	Applies to evaporator
F	Applies to effects of friction
FM	Applies to the loss coefficient
FS	Applies to the free stream
FP	Applies to feed pump
f	Applies to saturated liquid
fg	Applies to mixture of saturated vapour and saturated liquid
fridge	Applies to refrigerator
g	Applies to saturated vapour
ga	Applies to a gauge
hp	Applies to heat pump
hrs	Applies to number of hours
inertia	Applies to the inertia force of a fluid
int	Applies to intermediate reservoir of heat
ip	Applies to indicated power
irr	Applies to irreversible process
isen	Applies to isentropic condition
L	Applies to a lift force
M	Applies to the momentum of the fluid
max	Applies to maximum conditions
mean	Applies to mean conditions
N	Applies to nozzle
o	Applies to total or stagnation conditions
P	Applies to the pressure of a fluid
p	Applies to constant fluid pressure
pV	Applies to the area of a fluid pressure versus volume diagram
r	Applies to reversible process
rej	Applies to heat rejected
s	Applies to a solid surface
sink	Applies to a sink of heat
source	Applies to a source of heat
supp	Applies to heat supplied
T	Applies to turbine
theor	Applies to theoretical conditions
U	Applies to upstream conditions
v	Applies to constant fluid volume

viscous	Applies to the viscous force of a fluid
X, Y, Z	Applies to process X, Y, Z
x, y, z	Applies to directions x, y, z
xs	Applies to a cross-sectional area
1, 2, 3, etc.	Applies to state 1, 2, 3 etc.
12, 34, etc.	Applies to work and heat transfer between states 1 and 2, 3 and 4, etc.
A, B	Applies to fluid A, B

CONVENTIONS

Integral	$Q_{12}, W_{12}, (V_2 - V_1)$
Differentials	dQ, dW, dV

Further Reading

Bayley, F. J. (1958) *An Introduction to Fluid Dynamics* George Allen and Unwin.
Boxer, G. (1970) *Engineering Thermodynamics, Theory, Worked Examples and Problems* Macmillan.
Burghardt, M. D. (1982) *Engineering Thermodynamics with Applications* 2nd edn, Harper and Row.
Carnot, S. N. L. (1824) *Reflexions sur la puissance motrice du feu et sur les machines propres a developper cette puissance* Bachelier (translation: Thurston, R. H., A. S. M. E., 1943).
Cengel, Y. A. and Boles, M. A. (1989) *Thermodynamics, An Engineering Approach* McGraw-Hill.
Clausius, R. (1879) *The Mechanical Theory of Heat* Macmillan.
Cravalho, E. G. and Smith, J. L. (1981) *Engineering Thermodynamics* Pitman.
Douglas, J. F. (1986) *Solution of Problems in Fluid Mechanics* Longman.
Douglas, J. F., Gasiorek, J. M. and Swaffield, J. A. (1985) *Fluid Mechanics*, 2nd edn, Pitman.
Duncan, W. J., Thom, A. S. and Young, A. D. (1975) *Mechanics of Fluids*, 2nd edn, Edward Arnold.
Eastop, T. D. and Croft, D. R. (1990) *Energy Efficiency for Engineers and Technologists* Longman.
Eastop, T. D. and McConkey, A. (1993) *Applied Thermodynamics for Engineering Technologists* 5th edn, Longman.
Fellinger, R. C. and Cook, W. J. (1985) *Introduction to Engineering Thermodynamics* Wm C. Brown.
Francis, J. R. D. (1975) *Fluid Mechanics for Engineering Students* 4th edn, Edward Arnold.
Holman, J. P. (1980) *Thermodynamics* 3rd edn, McGraw-Hill.
Joel, R. (1971) *Basic Engineering Thermodynamics* 3rd edn, Longman.
Keenan, J. H. (1956) *Thermodynamics* Wiley.
Kinsky, R. (1989) *Heat Engineering* 3rd edn, McGraw-Hill.
Kinsky, R. (1982) *Applied Fluid Mechanics* McGraw-Hill.
Massey, B. S. (1983) *Mechanics of Fluids* 5th edn, Van Nostrand Reinhold.
Moody, L. F. (1944) Friction factors for pipe flows. *Trans. Am. Soc. Mech. Engrs.* **66**, 671–84.
Moran, M. J. and Shapiro, H. N. (1993) *Fundamentals of Engineering Thermodynamics* 2nd edn, Wiley.
Nikuradse, J. (1933) Strömungsgesetze in rauhen Rohren. *VDI-Forschungsh.*, **361**, 1–22.
Pefley, R. K. and Murray, R. I. (1966) *Thermofluid Mechanics* McGraw-Hill.
Reynolds, A. J. (1971) *Thermofluid Dynamics* Wiley.
Reynolds, O. (1883) An experimental investigation of the circumstances which determine whether the motion of water shall be direct or sinuous, and the law of the resistance in

parallel channels. *Phil. Trans. R. Soc.*, **174**, 935–82.

Rogers, G. F. C. and Mayhew, Y. R. (1988) *Thermodynamic and Transport Properties of Fluids* 4th edn, Basil Blackwell.

Rogers, G. F. C. and Mayhew, Y. R. (1992) *Engineering Thermodynamics, Work and Heat Transfer* 3rd edn, Longman.

Schmidt, F. W., Henderson, R. E. and Wolgemuth, C. H. (1993) *Introduction to Thermal Sciences* 2nd edn, Wiley.

Spalding, D. B. and Cole, E. H. (1966) *Engineering Thermodynamics* 2nd edn, McGraw-Hill.

Wark, K. (1971) *Thermodynamics* 2nd edn, McGraw-Hill.

Van Wylen, G. A. and Sonntag, R. E. (1985) *Fundamentals of Classical Thermodynamics* 3rd edn, Wiley.

Index

Absolute pressure, 21, 28, 286, 291, 294, 295, 298
Absolute scale of temperature, 42, 343
Adiabatic, 123, 124, 150, 154, 203, 205, 212, 213, 221, 346, 347, 348
Adverse pressure gradient, 239, 240, 242
Aerofoil, 210, 242, 244, 245
Air, 133, 134, 138, 139, 150, 151, 152, 153, 154, 155, 156, 157, 158, 159, 160
Annual running cost, 95, 96, 97, 105, 106, 108, 109, 154, 157, 159, 198, 199, 200, 201, 312, 314, 315
Atmospheric pressure, 21, 27, 29, 30, 31, 32, 33, 34, 60, 67, 110, 164, 189, 261, 270, 280, 281, 291, 294, 295, 298, 303, 303, 313, 315, 317, 321, 322, 324, 326, 330, 331, 333
Axial flow turbine/compressor, 210, 211, 212

Barometer, 32, 33, 34
Bernoulli's equation, 227, 239, 240, 247, 248, 250, 258, 259, 290, 291, 293, 294, 295, 296, 298, 302
Body force, 271, 273, 274, 276, 277, 279, 281, 283, 285, 286, 287, 288, 289, 290, 292, 293, 298, 304
Boiler, 1, 2, 95, 96, 97, 101, 102, 103, 109, 110, 111, 118, 127, 192, 195, 196, 197, 198, 199, 200, 201, 222, 342
Boiling point, 41, 163, 164, 165, 173, 174, 175, 176, 177, 178, 179, 182
Bomb calorimeter, 89
Bore, 58
Bottom dead centre, 58
Boundary, 4, 5, 57, 58, 68, 69, 72, 78, 84, 97, 116, 137, 183, 184, 204, 229, 251, 267, 268, 269, 270, 271, 274, 286, 305, 347
Boundary layer, 229, 230, 231, 232, 233, 234, 235, 236, 237, 238, 239, 240, 242, 284

Boundary layer separation, 241, 242
Boundary layer thickness, 232, 238
Bourdon gauge, 39
Brake power, 85, 86, 87, 88, 97, 99
Branched pipes, 328

Calorific value, 89, 90, 97, 98, 102, 108, 120, 154, 196
Carnot, 117, 123, 342, 343
Carnot cycle, 121
Carnot COP, 342, 343
Carnot efficiency, 121, 342
Clausius, 337, 338, 339, 340
Clausius inequality, 343, 344, 345, 347
Clearance volume, 58, 62, 86, 118, 128, 130
Closed cycle gas turbine, 107, 108, 110, 111, 216, 222
Closed system, 5, 6, 8, 11, 58, 60, 83, 84, 97, 110, 114, 335, 346, 347
Coefficient of discharge, 249, 250, 251, 252, 253, 255, 256
Coefficient of performance, 107, 338, 342
Combined cycle plant, 111
Combined heat and power plant, 110, 111
Combustion, 2, 16, 58, 69, 89, 90, 95, 96, 97, 98, 99, 101, 107, 118, 120, 150, 153, 154, 157, 159, 163, 195, 196, 199, 281, 314, 315, 336
Combustion chamber, 9, 10, 58, 118, 196
Compressible fluid, 20, 189
Compressor, 2, 5, 6, 9, 10, 58, 60, 73, 76, 77, 78, 85, 86, 87, 88, 90, 91, 92, 93, 97, 104, 105, 106, 107, 108, 109, 128, 130, 146, 160, 211, 212, 215, 216, 217
Compressor efficiency, 216, 217
Condensation, 65, 118, 130, 192, 348
Condensation of water/steam, 110
Condenser, 101, 102, 103, 104, 105, 106, 107, 110, 197, 342

Condition, 15, 39, 41, 45, 53, 55
Connecting-rod, 58
Conservation of energy, 22, 84, 97, 114, 132, 189, 191
Constant entropy non flow process, 80, 116, 122, 127
Constant pressure non flow process, 116, 117, 127, 143, 144, 146
Constant temperature non-flow process, 80, 116, 121, 144
Constant volume non-flow process, 80, 116, 121, 144
Constant entropy non flow process with a perfect gas, 143
Constant volume gas thermometer, 343
Continuity equation, 185, 187, 188, 189, 202, 207, 208, 210, 226, 230, 247, 249, 250, 259, 260, 261, 263, 265, 273, 274, 275, 278, 279, 281, 283, 285, 287, 288, 289, 290, 293, 296, 297, 303, 304, 305, 309, 311, 313, 314, 316, 317, 319, 320, 321, 324, 325, 326, 327, 330, 331, 333
Continuum, 15
Control volume, 268, 269, 270, 271, 272, 273, 274, 275, 276, 277, 279, 281, 282, 283, 284, 285, 287, 288, 289, 291, 292, 294, 295, 297
Crankshaft, 57, 58, 60, 62, 69, 70, 75, 85, 86, 87, 90
Critical point, 164, 165, 168, 170, 173, 175, 180
Critical Reynolds number, 47, 48, 232
Cycle, 15
Cylinder volume, 58

Darcy, 307, 309
Degree Celsius, 39, 41, 42
Degree of hotness, 39
Degree Kelvin, 39, 41, 42, 66
Density, 18, 19, 20, 24, 25, 27, 28, 29, 30, 31, 32, 33, 34, 35, 36, 38, 44, 46, 47, 48
Diesel cycle, 117, 118, 120, 123, 132, 150, 151
Diffuser, 192, 201, 204, 205, 208, 209, 210, 212, 213, 239, 250, 251, 252, 289, 348
Diffuser efficiency, 204, 205
District heating scheme, 110, 111
Drag coefficient, 242, 243, 244, 245, 246
Drag force, 8, 241, 242, 243, 244, 246
Dryness fraction, 169, 170, 174, 175, 176, 178, 180, 181, 208, 214, 218, 221, 224
Dynamic pressure, 22, 26, 28, 30, 50, 202, 227, 243, 244

Dynamic viscosity, 15, 43, 44, 45, 47, 48, 49, 50, 51, 52, 53, 54, 55, 234, 236, 237, 309, 311, 312, 315
Dynamometer, 85, 86, 87, 97, 99

Eddies, 46, 240, 251
Eddy viscosity, 50
Efficiency, 4, 5, 9, 10, 11, 12, 13, 90, 92, 93, 100, 107, 111
Efficiency of an engine, 335
Electrical energy, 1, 2, 4
End effects, 51, 52, 53, 55
Energy, 1, 2, 3, 4, 5, 6, 7, 8, 9, 10, 11, 12, 13
Energy conversion processes, 2, 9, 10, 12, 13
Energy transfer efficiency, 77, 91, 92, 93, 94, 95, 96, 101, 102, 103, 105, 106, 107, 108, 109, 120, 121, 196, 197, 198, 199, 200, 201, 302, 303, 309, 310, 312, 314, 315, 316, 317, 319, 321
Energy triangle, 2, 6, 97, 222
Engine, 1, 2, 9, 57, 58, 60, 63, 65, 69, 71, 73, 79, 81
Enthalpy, 117, 118, 119, 127, 136, 137, 140, 141, 166, 167, 168, 177, 179, 193, 194, 195, 200, 203, 208, 220, 348
Entropy, 115, 117, 121, 122, 123, 124, 125
Entry length, 230, 231, 307
Entry nozzle, 261, 262, 263, 264
Entropy of a perfect gas, 139, 145
Equation of state, 42, 133, 134, 135, 136, 138, 140, 144, 146, 148, 149, 151, 152, 153, 155, 156, 157, 160, 188, 189, 205, 207, 226, 246, 247, 288
Equilibrium, 15
Equivalent diameter, 47, 231, 232
Equivalent length, 321, 323
Evaporator, 104, 105, 106, 107, 197
Expansion valve, 104, 217, 218, 219, 220, 222, 223
Extensive property, 25, 60, 66, 345, 346

Fan, 30, 31, 32, 192, 212, 213, 222, 262
First Law of Thermodynamics, 83, 84, 86, 87, 88, 90, 92, 94, 96, 97, 98, 99, 100, 102, 103, 104, 105, 106, 107, 108, 109, 110, 113, 114, 119, 130, 132, 139, 140, 149, 153, 157, 158, 336, 346
Fixed points, 41, 48, 343
Flat plate, 18, 232, 233, 234, 236, 237, 239, 272, 274, 276, 277, 279
Flow process, 58, 78, 197, 203, 204, 205

Flow measurement, 246, 247, 258, 264
Flow work, 22, 190, 191
Fluid, 4, 15, 16, 58, 65, 66, 83, 88, 113, 116, 117, 170, 183, 184, 192, 199, 266, 267, 269, 301, 335, 346, 347
Fluid mechanics, 4
Fluid properties, 15, 16, 58, 65, 66, 83, 88, 113, 116, 117, 170, 183, 184, 192, 199, 266, 267, 269, 301, 335, 346, 347
Force, 6, 7, 8, 267, 268, 269, 270, 271, 272, 273, 274, 275, 276, 277, 278, 279, 280, 281, 282, 283, 284, 285, 286, 287, 288, 289, 290, 291, 292, 293, 294, 295, 296, 297, 298, 299
Free stream, 229, 233, 238
Free stream velocity, 229, 230, 232, 234, 236, 237, 238, 243, 244, 284
Friction, 2, 6, 8, 9, 12, 70, 71, 86, 87, 92, 98, 99, 100, 124, 203, 204, 226, 227, 231, 237, 248, 249, 251, 262, 268, 285, 287, 295, 296, 297, 298, 301, 302, 306, 311, 312, 315, 318, 323, 326, 338, 346
Friction drag, 242, 243
Friction energy, 301, 302
Friction factor, 50, 231, 234, 237, 307, 308, 309, 312, 313, 314, 316, 317, 319, 320, 321, 323, 324, 326, 328, 332
Friction force, 44, 47, 50, 51, 52, 53, 54, 70, 71, 85, 86, 87, 116, 203, 226, 229, 231, 232, 233, 234, 235, 236, 237, 238, 239, 240, 242, 246, 251, 268, 273, 284, 285, 286, 287, 288, 301, 304, 306, 321
Friction power, 87, 88, 88, 98
Fully developed flow, 230, 231, 259

Gas, 15, 16, 17, 19, 20, 25, 36, 41, 42, 45
Gas calorimeter, 89
Gas constant, 133, 134, 135, 138, 139, 140, 141, 144, 147, 150, 154, 159
Gauge pressure, 21, 28, 30, 186, 290, 291, 292, 293, 294, 296, 297

Head, 282, 283, 284, 302, 306, 315, 317
Head lost to friction, 302, 303, 306, 307, 310, 312, 313, 314, 316, 317, 318, 320, 322, 324, 325, 326, 328, 330, 332, 334
Heat exchanger, 76, 77, 92, 93, 94, 96, 97, 100, 107, 108, 110, 118, 150, 186, 187, 192, 195, 196, 197, 199, 200, 201, 222, 225
Heat flow rate, 11

Heat engine, 336, 337, 338, 339, 340, 341, 342, 343, 344
Heat pump, 97, 104, 105, 106, 107, 111, 222, 335, 337, 338, 339, 340, 342, 343
Heat transfer, 81
Heat transfer in a flow process, 78
Heat transfer in a non flow process, 65, 66, 68, 72, 189
Hot wire anemometer, 265
Hydroelectric power station, 11

Ideal fluid, 44, 71, 227, 228, 239, 240, 290, 296, 302
Index of compression/expansion, 124, 125, 128, 159, 182, 222, 225
Inclined manometer, 35, 36
Incompressible fluid, 20, 24, 272, 285, 292, 295, 303, 304, 306, 307, 308
Indicated power, 86, 87, 88, 99
Inertia force, 46, 47, 48
Intensive property, 25
Internal energy, 46, 114, 115, 116, 117, 118, 119, 120, 121, 122, 123, 124, 125, 126, 127, 135, 136, 137, 140, 141, 145, 166, 167, 168, 177, 183, 189, 191, 203, 226, 227, 251, 301, 335, 346, 348
Irreversible process, 134, 136, 145, 223, 346, 348
Irreversibility,
Isentropic efficiency, 203, 204, 208, 212, 214, 215, 338, 348
Isentropic process, 203, 212, 347
Isobaric process, 117
Isochoric process, 119
Isothermal process, 122, 150

Joule, 7, 8, 11, 66, 135
Joule equivalent of heat, 83
Joule's Law, 122, 135, 136, 138, 145, 348

Kinematic viscosity, 43, 44, 49
Kinetic energy, 7, 22, 42, 46, 101, 189, 1291, 192, 193, 194, 198, 201, 202, 203, 204, 206, 209, 210, 211, 212, 213, 251, 271
Kinetic energy correction factor, 271

Laminar flow, 45, 46, 48, 49, 50, 231, 233, 234, 238, 302, 303, 304, 306, 307, 308
Laminar/viscous sub layer, 232
Latent heat, 2, 67, 110, 163
Latent heat of vaporisation, 163, 164, 347

INDEX

Lift coefficient, 242, 244, 245, 246
Lift force, 244, 245
Liquid, 15, 16, 18, 19, 24, 28, 29, 30, 31, 32, 33, 34, 35, 36, 45, 52, 53, 54, 55, 67, 101, 103, 104, 110, 125, 163, 165, 167, 175, 177, 186, 192, 196, 197, 199, 212, 246, 247, 317
Loss coefficient, 318, 319

Manometer, 24, 28, 29, 30, 31, 32, 34, 35, 36, 37, 38, 39, 250, 252, 255, 256, 259, 261, 262
Mass, 5, 6, 7, 8, 15, 16, 17, 18, 19, 22, 23, 24, 25, 46
Mass flow rate, 6, 7, 11, 12, 17, 19, 20, 22
Measurement of viscosity, 50
Mechanical energy, 2, 4, 6, 9, 12
Mercury in glass thermometer, 39, 40, 41
Minor losses, 318, 319, 320, 321, 322, 323, 324, 325, 326, 328, 329, 330, 332, 334
Mixing process, 219, 220, 348
Mixture, 15, 133, 159, 163, 164, 165, 166, 169, 170, 173, 174, 175, 177, 178, 180, 181, 182, 208, 210, 214, 215, 217, 218, 220, 221, 224
Molecular weight, 133, 134, 147
Molecules, 15, 16, 17, 18, 20, 21, 22, 26, 39, 42, 43, 45, 83, 114, 115, 116, 124, 134, 135, 137, 138, 233
Momentum correction factor, 271, 272
Momentum force, 268, 270, 272, 273, 276, 279, 280, 283, 284, 285, 287, 289, 292, 298, 304
Moody, 307
Moody diagram, 307, 308, 313

Newton, 7, 44, 86
Newton's Laws of Motion, 268
Nikuradse, 307
Non flow energy equation, 113, 114, 116, 118, 120, 122, 124, 126, 128, 130, 132
Non flow process, 58, 60, 61, 62, 63, 64, 65, 66, 67, 72, 73, 74, 75, 77, 78, 79, 80, 81, 113, 114, 115, 116, 117, 119, 120, 121, 122, 123, 124, 125, 127, 128, 131, 132, 183
Non flow process with perfect gas, 143
Non flow process with water/steam,
Nozzle, 192, 201, 202, 203, 204, 205, 206, 207, 208, 210, 212, 213, 225, 226, 249, 250, 251, 252, 261, 272, 273, 274, 276, 277, 278, 279, 282, 283, 284, 286, 289, 290, 291, 292, 293, 294, 298, 348
Nozzle efficiency, 203, 206, 207, 226
Nozzle meter, 249, 253, 254, 255, 256
NTP, 15, 17, 19, 133, 168, 343

Open cycle gas turbine, 9, 10, 110, 222
Open system, 5, 9, 10, 11, 17, 58, 78, 88
Orifice plate meter, 249, 256, 257, 258
Otto cycle, 117, 119, 120, 123, 154, 155, 159
Overall thermal efficiency, 90, 91, 95, 99, 100, 102, 103, 106, 107, 108, 109, 110, 111, 131, 132, 147, 150, 153, 157, 159, 193, 194, 336, 338, 341, 342, 343

Pascal, 7, 20, 25, 253, 263
Pelton wheel, 278
Perfect gas, 122, 125, 132, 133, 134, 135, 136, 137, 138, 139, 140, 141, 142, 143, 144, 145, 146, 147, 148, 150, 152, 154, 156, 158, 159, 160, 161, 163, 166, 167, 168, 188, 192, 193, 197, 205, 213, 215, 218, 220, 225, 226, 228, 229, 346, 348
Pipe flow, 230, 301, 302, 304, 306, 308, 310, 312, 314, 316, 318, 320, 322, 324, 326, 328, 330, 332, 334
Pipes in parallel, 326, 327
Pipes in series, 323
Piston, 1, 5, 6, 8, 9, 11, 57, 58, 59, 60, 62, 63, 69, 70, 71, 74, 75, 76, 77, 78, 85, 86, 87, 97, 116, 118, 121, 122, 123, 124, 125, 126, 128, 130, 141, 142, 146, 150, 154, 159, 160, 178, 203, 204, 222, 224, 225, 249, 250, 348,
Piston ring, 58
Pitot static tube, 258, 259, 260, 261, 262, 263, 264
Platinum resistance thermometer, 42
Polytropic non-flow process, 116, 124, 125, 126, 127, 143, 144, 145, 146, 222
Polytropic non flow process with perfect gas, 145
Potential energy, 7, 8, 11, 12, 189, 191, 192, 193, 195, 196, 198, 203, 212, 218, 220, 288, 303, 311, 312
Power, 1, 2, 4, 5, 11, 12, 57, 58, 79
Pressure, 15, 16, 18, 19, 20, 21, 22, 23, 24, 25, 26, 27, 28, 29, 30, 31, 32, 33, 34, 36, 37, 38, 39, 41, 42, 45
Pressure force, 270, 273, 278, 281, 283, 285, 286, 289, 292, 293, 298

Pressure gauge, 39, 248
Pressure transducer, 39, 86
Process, 15, 57, 58, 61, 66, 68, 69, 70, 71, 72, 73, 74, 78, 79
Property, 16, 17, 18, 20, 25, 39, 42, 44, 58, 60, 66, 68, 69, 72, 73, 78, 81
Pump, 60, 101, 102, 103, 110, 192, 195, 202, 212, 218, 219, 223, 227, 302, 303, 309, 310, 311, 312, 314, 315, 316, 317, 319, 321, 323, 324, 326, 329, 330, 331, 333

Rate of heat flow, 11
Ratio of specific heats of a perfect gas, 138
Reaction of a jet, 280, 281
Real fluid, 44, 70, 71, 227, 228, 240, 241, 249
Reciprocating piston cylinder mechanism, 9, 58, 60, 62, 63, 75, 113, 118, 121, 123, 125, 128, 229
Refrigerator, 84, 104, 105, 106, 107, 222, 229, 335
Relationship between specific heats of a perfect gas, 140
Relative density, 19, 30, 31
Reversibility, 71, 117, 338
Reversible cycle, 71, 339, 342, 344, 345
Reversible process, 71, 88, 117, 121, 125, 128, 129, 130, 131, 132, 140, 144, 148, 149, 150, 151, 152, 153, 155, 156, 157, 158, 159, 160, 180, 181, 182, 233, 346
Reynolds, 45, 46
Reynolds number, 46, 47, 48, 49, 50, 231, 232, 233, 234, 235, 236, 237, 238, 243, 244, 251, 307, 308, 309, 311, 313, 314, 316, 317
Rotameter, 264, 265
Rotary turbine/compressor, 101, 210, 213, 215, 278

Saturated, 45, 163, 165, 166, 167, 169, 170, 174, 175, 176, 177, 178, 180, 181, 182, 186, 187, 198, 199, 215, 216, 222, 223, 347
Saturated water and steam, 170, 173, 176, 178, 180, 181, 187, 198, 199, 201, 208, 214, 216, 218, 221, 223, 224
Saturation curve, 165, 168
Second Law of Thermodynamics, 39, 42, 65, 66, 83, 121, 335, 336, 337, 338, 339, 340, 342, 343, 344, 346, 348
Seebeck effect, 40
Sensible heat, 67, 89

Shear stress, 44, 45, 49, 50, 51, 54, 229, 233, 234, 235, 237, 274, 284, 304, 307
Sign convention, 60, 63, 64, 66, 74, 75, 76, 78, 81, 83, 85, 86, 98, 102, 105, 107, 108, 109, 190, 193, 195, 336, 342
Sign rule A, 74, 107
Sign rule B, 75, 76, 77, 78, 79, 92, 93, 99, 103, 105, 109, 119, 120, 121, 122, 124, 126, 179, 200, 301
Sign rule C, 76, 77, 87, 92, 93, 94, 96, 100, 103, 105, 106, 107, 109, 132, 150, 153, 157, 159, 194, 200, 201
Simple payback period, 95
Sink of heat, 337, 342, 344
Source of heat, 342, 344
Specific enthalpy, 117, 137, 138, 170, 171, 172, 173, 174, 175, 177, 180, 192, 193, 194, 195, 197, 198, 199, 201, 205, 207, 208, 209, 213, 214, 216, 218, 219, 220, 221, 223, 224
Specific entropy, 66, 68, 140, 142, 143, 170, 171, 172, 173, 174, 181, 205, 206, 207, 208, 209, 210, 213, 214, 215, 216, 217, 218, 219, 222, 223, 224, 345
Specific gravity, 19
Specific heat, 88, 89, 97, 98, 102, 106, 107, 108, 121, 127, 131, 132, 137, 138, 140, 154, 167, 177, 193, 197, 198, 200, 205, 206, 213, 218
Specific heat at constant pressure, 127, 136, 137, 138, 139, 146, 150, 159, 167, 192, 193, 197, 200, 205, 213, 215, 218, 220, 225
Specific heat at constant volume, 127, 136, 137, 138, 139, 140, 147, 150, 167
Specific heat transfer, 65
Specific heat of water steam,
Specific internal energy, 114, 137, 170, 171, 172, 174, 177, 180, 181, 182
Specific volume, 17, 18, 120, 121, 134, 135, 150, 151, 152, 154, 155, 156, 157, 158, 169, 170, 171, 172, 174, 175, 180, 182, 186, 187, 207, 208, 210, 222, 223, 224
Specific work transfer, 60, 224
Stagnation point, 27, 239, 240
Stalling, 244
State, 15, 16, 24
Static pressure, 21, 22, 23, 26, 27, 28, 29, 30, 31, 32, 34, 35, 36, 37, 38, 202, 227, 239, 240, 253, 254, 256, 258, 259, 260, 261, 262, 270, 271, 280, 286, 305
Static temperature, 42, 228

402 INDEX

Steady flow energy equation, 22, 42, 185, 191, 301, 302, 304, 306, 308, 310, 312, 314, 316, 318, 320, 322, 324, 326, 328, 330, 332, 334
Steady flow momentum equation, 267, 268, 270, 272, 274, 276, 278, 280, 282, 284, 286, 288, 290, 292, 294, 296, 298
Steady flow process, 183, 184, 185, 186, 188, 189, 190, 192, 194, 196, 198, 200, 202, 204, 206, 208, 210, 212, 214, 216, 218, 220, 222, 224, 226, 228, 348
Steam, 2, 6, 15, 16, 39, 41, 44, 67, 68, 95, 101, 102, 103, 110, 111, 125, 127, 132, 163, 164, 165, 167, 168, 169, 173, 174, 175, 178, 179, 180, 181, 182, 186, 187, 194, 195, 196, 198, 199, 200, 201, 201, 205, 207, 208, 209, 210, 213, 214, 215, 216, 216, 217, 218, 219, 221, 222, 223, 224, 225, 343, 347, 348
Steam plant, 97, 101, 103, 104, 110, 215, 229, 342
Steam tables, 170, 177
Stirling cycle, 121, 146, 148, 150
Stored energy, 2, 3, 4, 6
Streamline, 42, 43, 44, 45, 240, 241, 245, 267, 269, 270
Streamline body, 242, 244
Streamline flow, 42, 45, 272, 284
Stroke, 58, 63, 76, 86, 87, 98, 99, 124, 128
Supercritical steam, 114, 114, 137, 170, 171, 172, 174, 177, 180, 181, 182
Superheated steam, 164, 165, 166, 170, 171, 172, 177, 178, 180, 181, 182, 186, 187, 198, 213, 342
Surroundings, 4, 58, 71, 83, 87, 105, 107, 179, 184, 301, 302, 337, 338, 346, 347, 348
Swept volume, 58, 60, 119
System, 1, 4, 5, 6, 9, 10, 12
System boundary, 4, 5, 46, 84, 183, 184, 189, 192, 347, 348

Temperature, 15, 16, 19, 39, 40, 41, 42, 44, 45, 50
Thermal energy, 2, 8, 9, 10, 11
Thermal power station, 1, 2, 4, 5, 15, 58, 101
Thermal system, 6, 15, 16, 44, 50, 57, 58, 60, 62, 64, 65, 66, 68, 70, 72, 73, 74, 76, 78, 80, 83, 84, 89, 90, 91, 95, 97, 113, 117, 123, 127, 132, 133, 141, 143, 163, 166, 167, 168, 177, 179, 182, 185, 189, 190, 191, 192, 193, 195, 218, 192, 222, 228, 231, 232, 248, 266, 299, 301, 334, 335, 338
Thermistor, 42
Thermocouple thermometer, 40
Thermodynamics, 2, 4, 83, 170, 346
Thermofluids, 4, 5, 6, 7, 9, 10, 11, 12, 13, 15, 16, 18, 20, 21, 22, 24, 26, 28, 30, 32, 34, 36, 38, 40, 42, 44, 46, 48, 50, 52, 54, 60, 71, 88, 111, 117, 192, 269, 335
Thermometer, 39, 41, 42, 343
Top dead centre, 58
Torque, 50, 51, 52, 53, 54, 55, 85, 283, 284
Total drag, 242, 243
Total or stagnation pressure, 22
Total or stagnation temperature, 42, 228
Transition region, 48, 232
Triple point, 164, 165, 166, 167, 168, 170, 173, 343
Turbine efficiency, 213, 224
Turbine meter, 264, 265
Turbulent flow, 45, 46, 48, 49, 50, 232, 233, 234, 237, 238, 302, 307, 308
Two property rule, 16

U-tube manometer, 28, 29, 30, 31, 32, 34, 35, 36, 37, 38, 253, 254, 256, 257, 259, 261
Units, 1, 6, 7, 8, 9, 10, 11, 12
Universal gas constant, 133
Unsaturated water, 164, 165, 166, 168, 169, 170, 176, 177, 178

Valves, 5, 58, 60, 113, 192, 217, 218, 289, 321, 322, 328
Vapour, 15, 16, 17, 19, 60, 67, 101, 103, 104, 105, 106, 125, 133, 163, 165, 167, 168, 192, 197
Velocity, 17, 22, 26, 27, 28, 30, 42, 43, 44, 45, 46, 47, 48, 49, 50, 51, 54
Velocity profile, 26, 44, 51, 52, 53, 54, 55, 230, 233, 234, 235, 259, 260, 271, 284, 304, 307
Vena contracta, 253
Venturi meter, 204, 249, 250, 251, 252, 253
Viscometer, 50, 51, 52, 53, 54, 55
Viscous force, 44, 45, 46, 47, 51, 52, 233, 284, 304
Volume, 15, 16, 17, 18, 19, 20, 25, 28, 39, 41, 46
Volumetric flow rate, 19, 305
Vortices, 46, 251

INDEX **403**

Water/steam, 102, 163, 164, 165, 166, 167, 168, 169, 170, 172, 174, 176, 177, 178, 179, 180, 182, 229, 346
Water/steam in a flow process, 182
Water/steam in a non flow process, 178
Watt, 7, 11, 12
Weight, 16, 17, 18, 23, 110, 244, 271, 273, 293, 343
Work done, 7, 8, 9, 10, 11, 92, 126, 190
Work transfer, 7, 8, 10, 11, 57, 58, 59, 60, 61, 62, 63, 64, 65, 66, 68, 69, 70, 71, 72, 73. 74, 75, 76, 78, 79, 80, 189, 190, 193, 195, 202, 212, 218, 219, 222, 224, 225, 227, 228
Work transfer in a flow process, 78
Work transfer in a non flow process, 60, 61, 68, 71, 72, 74, 75

Zeroth Law of Thermodynamics, 39, 83